EUROPE'S NUCLEAR POWER EXPERIMENT

History of the OECD Dragon Project

EUROPE'S NUCLEAR POWER EXPERIMENT

History of the OECD Dragon Project

by

E. N. SHAW

PERGAMON PRESS

OXFORD · NEW YORK · TORONTO · SYDNEY · PARIS · FRANKFURT

U.K. Pergamon Press Ltd., Headington Hill Hall,
Oxford OX3 0BW, England

U.S.A. Pergamon Press Inc., Maxwell House, Fairview Park,
Elmsford, New York 10523, U.S.A.

CANADA Pergamon Press Canada Ltd., Suite 104,
150 Consumers Rd., Willowdale, Ontario M2J 1P9, Canada

AUSTRALIA Pergamon Press (Aust.) Pty. Ltd., P.O. Box 544,
Potts Point, N.S.W. 2011, Australia

FRANCE Pergamon Press SARL, 24 rue des Ecoles,
75240 Paris, Cedex 05, France

FEDERAL REPUBLIC OF GERMANY Pergamon Press GmbH, 6242 Kronberg-Taunus,
Hammerweg 6, Federal Republic of Germany

First edition 1983

British Library Cataloguing in Publication Data

Shaw, E.N.
Europe's nuclear power experiment
1. Organisation for Economic Co-operation and Development. *Dragon Project*—History
I. Title
621.48'3'072 TK9202

Library of Congress Cataloging in Publication Data

Shaw, E. N.
Europe's nuclear power experiment.
(An OECD/NEA report)
Bibliography: p.
1. Dragon Project. I. Title. II. Series.
TK9055.S5 1982 333.79'24'0604 82-3789

ISBN 0-08-029324-7

The opinions expressed and arguments employed in this publication are the responsibility of the author and do not necessarily represent those of the OECD.

Printed in Great Britain by A. Wheaton & Co. Ltd., Exeter

OECD/NEA

The Organisation for Economic Co-operation and Development (OECD) was set up under a Convention signed in Paris on 14th December 1960, which provides that the OECD shall promote policies designed:

— to achieve the highest sustainable economic growth and employment and a rising standard of living in Member countries, while maintaining financial stability, and thus to contribute to the development of the world economy;
— to contribute to sound economic expansion in Member as well as non Member countries in the process of economic development;
— to contribute to the expansion of world trade on a multilateral, non discriminatory basis in accordance with international obligations.

The Members of OECD are Australia, Austria, Belgium, Canada, Denmark, Finland, France, the Federal Republic of Germany, Greece, Iceland, Ireland, Italy, Japan, Luxembourg, the Netherlands, New Zealand, Norway, Portugal, Spain, Sweden, Switzerland, Turkey, the United Kingdom and the United States.

The OECD Nuclear Energy Agency (NEA) was established on 20th April 1972, replacing OECD's European Nuclear Energy Agency (ENEA) on the adhesion of Japan as a full Member.

NEA now groups all the European Member countries of OECD and Australia, Canada, Japan, and the United States. The Commission of the European Communities takes part in the work of the Agency.

The primary objectives of NEA are to promote co-operation between its Member governments on the safety and regulatory aspects of nuclear development, and on assessing the future role of nuclear energy as a contributor to economic progress.

This is achieved by:

— encouraging harmonisation of governments' regulatory policies and practices in the nuclear field, with particular reference to the safety of nuclear installations, protection of man against ionising radiation and preservation of the environment, radioactive waste management, and nuclear third party liability and insurance;
— keeping under review the technical and economic characteristics of nuclear power growth and of the nuclear fuel cycle, and assessing demand and supply for the different phases of the nuclear fuel cycle and the potential future contribution of nuclear power to overall energy demand;
— developing exchanges of scientific and technical information on nuclear energy, particularly through participation in common services;
— setting up international research and development programmes and undertakings jointly organised and operated by OECD countries.

In these and related tasks, NEA works in close collaboration with the International Atomic Energy Agency in Vienna, with which it has concluded a Co-operation Agreement, as well as with other international organisations in the nuclear field.

ACKNOWLEDGEMENTS

THE AUTHOR wishes to acknowledge the generous help he received from all the Dragon Signatories and Signatory countries in the form of both oral discussions and written comments. Those designated as "points of contact", in many cases went to great trouble to give guidance on relevant documentation and to explain personally the order and significance of events. In addition, many people in the Signatory countries contributed their personal recollections and suggestions.

All possible help was given by the OECD Nuclear Energy Agency from the Directors-General down, of particular assistance being the former Head of Information, Bruce Adkins, and W. T. (Gillie) Potter and his secretary Mme S. Quarmeau upon whom fell the task of co-ordinating the work and preparing the final text.

Compilation of the History, however, would not have been possible without the unstinting help of the first Chief Executive, Compton Rennie, and those who remained at Winfrith in the Dragon office, notably the second Chief Executive, Leslie Shepherd, his secretary Miss R. Howard and M. S. T. (Mike) Price, Especially the author would like to express his gratitude to Dr. Shepherd for his untiring patience and his readiness to analyse and discuss all aspects of the Project's development without seeking to impose a personal view-point or pass censorious judgement on the author's own interpretations.

All illustrations are reproduced by kind permission of the United Kingdom Atomic Energy Authority.

Finally, the author would like to record his appreciation of the sincere desire he found everywhere to give maximum exposure to the Dragon experience. This was to be seen in the general willingness to explore openly and in a critical manner, opinions and actions of past years. It was evident that all those who had come closely into contact with the Project regarded it as having real value and believed that this should not be forgotten. The author hopes these same people, on reading the History that follows, will feel that the trust they placed in him was not misplaced.

CONTENTS

INTRODUCTION

FOR MUCH of its 17 years of life, the Dragon Project was widely regarded as one of Europe's most successful collaborations in applied science and certainly the most important multinational technical collaboration in the field of nuclear energy. It was created in 1959 under the aegis of OEEC's European Nuclear Energy Agency (now the Nuclear Energy Agency of OECD) when the political desire for a co-ordinated (Western) Europe was running strongly and nuclear energy was seen as the technical elixir that would emancipate our age.

The Project was centred on the construction of a nuclear reactor that was to demonstrate the essential properties of a new system, designed to produce higher temperatures than could be achieved with the first generation of reactors. Dragon ceased to exist in March 1976, when only limited progress had been made towards a united Europe and nuclear energy was no longer regarded as a global panacea.

Altogether, 13 nations participated in Dragon, either individually or collectively through Euratom and the European Communities. Funding came from government sources and was channelled to the Project through the Signatories to the Dragon Agreements. The Signatories, seven in all, included governments as such, atomic energy commissions and Euratom.

As the Project evolved, the original, partially impracticable ideas, became translated into an operating system that offered a promise for application well beyond the early expectations. Similar research and development in the USA followed the European initiative. It led to the construction of a small experimental power station, and soon after this began operation, to the building of a demonstration plant. Orders for full scale stations followed and for a time, the American company that had been at the centre of these developments, seemed to have established a world market. Difficulties arose, and when commercial exploitation in the USA was effectively abandoned, the system had no other backers with either the strength or will to make the transition from the small scale experiment of Dragon to large scale construction. Sufficient funds could not be found amongst the Signatories to the Dragon Agreements to support the Project further and it was wound up.

For many, Dragon thereby demonstrated the three cardinal virtues desirable in an international venture: it was a successful political collaboration, it was a successful technical development and it was not immortal.

Towards the end, when it became clear that there would be no further prolongation, discussions were held in the managing board on information recovery at several levels, included in which was a proposal that a history of the Dragon Project should be written. Subsequently, while the former Chief Executive of the Project undertook the compilation of a technical report on developments during the Project's lifetime, the OECD Nuclear Energy Agency commissioned an independent History that would record and comment on the various phases of Dragon's creation, existence and final closure. The technical report has been produced under the title "A Summary and Evaluation of the Achievements of the Dragon Project and its Contribution to the Development of the High Temperature Reactor", Dragon Project Report 1000.

In this more general History, events are recorded with a view to examining the virtues with which the Project was reputed to be endowed, and the extent to which the Project's mission was accomplished. Draft chapters have been circulated to the Signatories for their correction and comment and their responses have been taken into account in preparing the final text. It should, nevertheless, be made clear that the manner in which the subject is approached, and the opinions expressed, are the author's own and do not necessarily reflect those of any Signatory. Much of the information presented has been extracted from Dragon's own documentation which has been carefully preserved. A great deal has also been gleaned from interviews and the personal files of people concerned with Dragon's formation and subsequent development. Extensive quotation has been avoided, partly because in many instances such a preference was expressed and partly because a continuous synthesis has been made, based on the recollections of many people.

POLITICAL COLLABORATION

Before considering Dragon's success as an international collaboration in any detail, some definition of the term is required. We are, it is to be hoped, past the stage in Europe when a multi-national activity is deemed to be a success simply because it is set up and major confrontations between nations are avoided. Something more positive is demanded both from the signatories and the project concerned. Among the Signatories, was there for example, evidence of the will to define an optimum objective that brought maximum benefit to the community as a whole, irrespective of whether in the shorter term this was less directly interesting to some of the partners than to others? It seems legitimate also to enquire whether once defined, a

Figure 1. Within the excavation for the DRAGON REACTOR EXPERIMENT (DRE) on 27 April 1960, on the occasion of the dedication by Sigvaard Eklund, Chairman of the Dragon Board of Management, in the presence of delegates from the Signatories, guests, and members of the Dragon Project and Winfrith staff.

Figure 2. The DRE compound when complete, with the reactor building to the left of the pylon and the fuel facilities to the right.

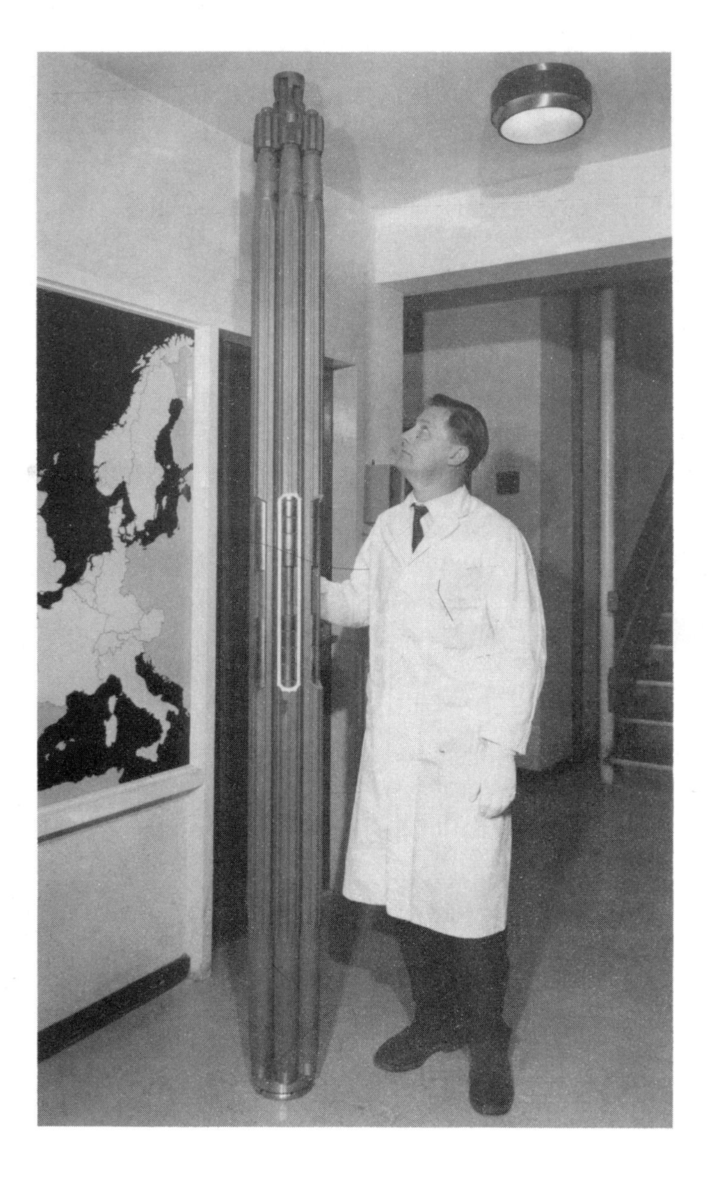

Figure 3. A seven-rod fuel element for the DRE, cut away to show the fuel inserts.

continuous effort was made to reach this objective, with the necessary flexibility and even generosity that put the Project before competing individual aims. At the same time, was a fair and reasonable hearing given to all the partners, and when differences arose, a solution sought which represented a technical optimum, and not one which merely provided an acceptable political compromise? Was the political will equal to the technical interest? When negotiating over contributions and budgets, however tough the bargaining, did the well-being of the Project take precedence over marginal national gains?

Political harmony amongst the Signatories had also to be reflected in the Project itself, by its clear identification with the objectives defined, and a programme of activity that was consonant with the means put at its disposal. One of the major criteria by which the Project will be judged, is how well the balance was struck between showing initiative, leadership and perseverance, and acting in a headstrong, arrogant and obstinate way. At the other end of the scale, the possibility of slackness or even incompetence cannot be excluded.

At the working level, political success implied a mobilizing of different national talents to produce a team that was stronger than a national team would be. A balance again was needed, this time between a distribution of responsibilities based entirely on merit (at the time of appointment) and one which ensured a just and harmonious distribution between nations, bearing in mind that the objective, however it was defined, was not only to accumulate information or experience, but diffuse it also. When in post, did the staff work together with verve and good humour, identifying their first loyalty with the Project even though they may have had 'representative' duties in addition?

These, then, are some of the questions we might pose in the 1980s, when we seek to evaluate the level of political success that was achieved in Dragon.

TECHNICAL DEVELOPMENT

Evaluation of the detailed technical results we shall leave to others and we shall make no attempt to summarise DPR/1000 – a document of some 660 pages. Here the approach is essentially non-technical but some consideration of the main technical aspects is required if we are to establish a basis on which performance can be assessed.

Seventeen years were spent by Dragon in the development of high temperature reactor technology. At the end of the seventies, no plans for the construction of a nuclear power station of the same type as Dragon's own reactor experiment remained in existence in Europe. This in itself is not a condemnation of Dragon's efforts. The Project must be judged against a changing background of aims and markets. It is too restrictive to require that every successful research and development project be followed by a production series. At the same time no project has the right to be called successful, just because it existed and produced a large number of technical papers.

Research and development is a lengthy procedure and it has to be recognised that changes to an established programme take a long time to work through and if they are made too abruptly, the value of work in hand may be lost. This is well known and it is, therefore, proper to enquire whether original objectives were appropriate to the time when they were formulated, and whether the means adopted were appropriate to the objectives. Subsequently, were both adjusted at a rate consistent with the extrinsic conditions and the intrinsic time factors? Was the knowledge that had been gained by both the Project itself and its contemporaries also taken into account? In other words, were the relevant technical questions being asked and were the best means being used to answer them?

One stage further was involved, namely the transmission of the information and experience back to the Signatories, each of which had its own view on their exploitation. Communication is part of the technical development process. However beautiful a piece of research may be, if it remains unknown or its significance is not appreciated, its merit is largely negated. Especially is this true in an international enterprise. Information is much more than written reports; it requires interaction. There had to be technical exchange between the Project and technical levels in the Signatories. However, the Project could only do so much; beyond that, an effort was required on the part of the Signatories themselves. Whereas a great responsibility rested on Dragon to make its work known and resist any tendency to regard itself as the 'owner' of the know-how it had acquired, it could go no further than it was allowed. Within these limitations, it can be asked how well did Dragon serve the Signatories as a source of both information and inspiration?

DURATION

All Signatories to the Dragon Agreement were, initially at least, keen that the collaboration should be of limited duration. To this extent Dragon's closure was foreordained. The first agreement was for a period of five years and subsequently six extensions were negotiated. It seems relevant, therefore, to ask how sensible this pattern was. Were so many extensions justified and was the final end the result of a mutual decision to cease activities?

Few projects will finish willingly of their own accord. Any organisation has built into it a will for continuation and Dragon was no exception to the general rule; nevertheless arguments in favour of continuation put forward by a project are not necessarily motivated only by self-interest. At best perhaps, one can look for a general consensus amongst the Signatories, and tacit acceptance by the Project that decisions were advised and not arbitrary. Did the technical judgement match the political realities? In addition, one can enquire whether in the end, the run-down was properly phased and whether the desire to extract the maximum from the joint programme was balanced by an awareness of the effort involved and whether due consideration was given to people who had devoted their professional careers to the Project.

These questions that have been singled out are not independent. Technical decisions are influenced by political factors and vice versa. Duration is not dissociated from political motivations nor evidently technical developments; they form an interlocking mosaic. However, they do provide us with a framework within which the achievements of Dragon can be set out and they allow us to establish a tentative scale against which these achievements can be measured.

CHAPTER 1

EARLY EFFORTS TOWARDS COLLABORATION IN EUROPE

OPPORTUNISM, apprehension, ambition, idealism, tradition, were all in evidence as the peoples of Europe sought to find their way through the turbulent days of the nineteen fifties. Erasing the misery of the forties, and creating conditions in which the nations of Western Europe could never again be at war with each other, were the primary objectives of those entrusted with reconstruction, as they contended with the tensions generated by the cold war and the threat of Soviet expansionism. Security and prosperity were the goals, and when principles rather than expediency were allowed to direct actions, two distinct – almost opposed – approaches could be discerned. As was apparent in the physical reconstruction, one side favoured restoration, rebuilding again what had existed before; the other preferred invention, creating from the ruins something revolutionary that might be better adapted to the new conditions and evolving social patterns.

Co-operation and collaboration **(1)*** were popular concepts, but their real significance and their implications as to means were personal and subject to change. Consistency was a rarity and, although a few staunch crusaders kept alive their image of a new united Europe that transcended sectarian interests, few leaders in any country were prepared to back a community policy when it ran contrary to one which preserved the illusion of unfettered national sovereignty. Winston Churchill lit the flame of European federalism by advocating, in his speech at Zürich University in September 1946, the creation of a sort of United States of Europe. Yet, just a few years later, when next he headed a British government, he showed himself no less insular than his predecessors of different political hue. Britain was not alone in resisting the pressure for a federal Europe. Other countries which had kept their borders intact during the war, were reluctant to see them breached in time of peace and, as economies were re-established elsewhere, national boundaries became steadily more rigid, and policies based on notions of national gain replaced those that had been designed to further the ideals of a unified sub-continent.

---

*Bold figures refer to References in Appendix 4.

Nevertheless, in the latter part of the forties and in the fifties, the will to co-operate was strong throughout Europe, and this was to manifest itself through agreements between nations on traditional lines, and the establishment of new institutions with federal-like powers. Both approaches were pursued side by side; new organisations were created in an ad hoc process, with little thought to overall strategy and little regard to their long-term function. It seemed that every sector, military, political, economic, social, scientific, industrial, etc., required its own special instruments of liaison.

One of the earliest European institutions to come into existence was the Organisation for European Economic Co-operation (OEEC) which was set up essentially to guide into the most effective channels, the massive economic aid that the United States was offering to Europe under the Marshall Plan. Inevitably, it was to the governments of the different European countries that the United States turned and OEEC was born on 16 April, 1948 as an inter-governmental body.

OEEC grouped together 18 European countries*, which pledged themselves "to combine their economic strength, to join together to make the fullest collective use of their individual capacities and potentialities, to increase their production, develop and modernise their industrial and agricultural equipment, expand their commerce, reduce progressively barriers to trade among themselves, promote full employment and restore or maintain the stability of their economies and general confidence in their national currencies". The United States and Canada participated as Associate Members. Great diversity existed amongst the members in regard to both their development status and needs in all the different branches of the economy. If any form of federalism had been suggested, it is clear that the new organisation could never have been founded. Instead, it was explicitly set down that each State could participate or not as it wished in the several programmes that would be instituted (Clause XIV).

OEEC's usefulness proved to extend beyond the scope of simply managing the American aid, and the organisation became established as the principal forum where the countries of Western Europe and North America (and later Japan) could consult each other on a wide range of economically significant problems. OEEC became thereby also a bulwark against federalism in Europe, providing an alternative to the closer unions that were explored.

In the defence field, the first move towards a joint approach came also in early 1948 with the signature of the Brussels Pact, whereby Belgium, France, the Netherlands, Luxembourg and the United Kingdom, agreed to

---

*Austria, Belgium, Denmark, France, Republic of Germany, Greece, Iceland, Ireland, Italy, Luxembourg, the Netherlands, Norway, Portugal, Spain, Sweden, Switzerland, Turkey, and the United Kingdom.

mutual military aid in the event of an attack. The following year, the more important North Atlantic Treaty Organisation (NATO) was formed, that included, in addition to the Pact countries, Canada and the USA, Denmark, Iceland, Italy, Norway and Portugal. NATO was enlarged in 1952 to include Greece and Turkey and again in 1955 to include Germany, soon after Germany had joined the Pact which, in October 1954 was re-named the Western European Union (WEU). The last event was most important for the recognition it gave to the new status of Germany, which was able from May 1955 to resume its place as an independent sovereign state, without external restrictions on its actions.

Although because of NATO, the Pact and the WEU provided little more than another meeting place for discussions of a largely military character, it was in talks between the initial members of the Pact that agreement was reached on the need for a closer co-operation in the economic and cultural fields. From this followed the Congress at The Hague in May 1948 where it was resolved to create a European parliament and establish a European charter of human rights. This led to the signature in London on 5 May, 1949 of a treaty setting up the Council of Europe, which initially included as members, the Brussels Pact countries plus Denmark, Ireland, Italy, Norway and Sweden. Later the majority of the countries of Western Europe were also to become members, but their addition did little to add weight to the Council's influence.

Significantly, the 'parliament' formed, was named the Consultative Assembly, and was comprised of delegates from national parliaments; it was given no powers and no explicit work load. Its Committee of Ministers was just what its title implied: a committee able to discuss and even make recommendations, but with no powers of execution. This was far from what the federalists in The Hague had hoped to see and they turned their attention towards more specific activities. Two names stand out at this time: those of Jean Monnet and Robert Schuman. Both men were imbued with a fervent belief in the need for a closer integration of the peoples of Europe. However, they had seen that unification on a broad vague front was unacceptable, so in May 1950, when Schuman was the Minister of Foreign Affairs in the French government, he put forward the proposal that just the production of coal and steel in France and Germany be placed under the control of a higher authority in an organisation open to other countries. In this way it was contended, order could be brought to industries that were in a chaotic state, and a recrudescence of German military activity made impossible. The proposal was warmly received in Germany, and the Benelux countries and Italy expressed a wish to be involved. The United Kingdom would take no part in the negotiations, and on 18 April, 1951, the Treaty instituting the European Coal and Steel Community (ECSC) was signed in Paris by six European countries only: Belgium, France, Germany, Italy,

Luxembourg and the Netherlands. Europe had its first international organisation with supranational powers.

These same six countries, responding to a British resolution adopted by the Consultative Assembly of the Council of Europe in August 1950, and contemplating the formation, once again, of a German army, signed a Treaty in May 1952, establishing the European Defence Community. This should have resulted in an even more determinedly federal action, namely the creation of a European force. Britain was ready to be associated with such a move but unwilling to be part of it. Majority opinion in France, also was against, and successive administrations found it prudent to avoid presenting the Treaty to parliament for ratification. When finally a vote was taken, in August 1954, France's participation was rejected. The Treaty became meaningless and Italy was able to avoid putting the issue to the test. During the period of indecision, a draft treaty that would have created a European Political Community of the Six was placed in limbo and the federalists were left with only the ECSC to mark their road to European integration. The non-federalists, many of whom it must be recognized, placed considerable emphasis on the need for co-operation in Europe, seemed to have gained the day.

CHAPTER 2

NUCLEAR ENERGY

STIMULATING the collaboration between European nations by the collective fear it engendered, and hampering it by the restrictions it imposed on joint military and political action, was the spectre of the atomic bomb. News of the possibility of releasing the enormous forces locked in the atom had burst upon the world in August 1945. Since then the USSR had exploded its own atomic weapon in August 1949 and Britain had followed in October 1952. Atomic power, however, had another side; the scientists were claiming that it held the key to a cheap and inexhaustible supply of energy once the technical problems of taming it had been resolved.

There was little general appreciation of what was entailed but the bare essentials were understood. The heaviest element of all, uranium, it seemed existed in nature as a mixture of two main groups of atoms (isotopes), chemically identical but of slightly differing weights. In certain circumstances, the lighter atoms could be made to split into a small number of fragments in a process termed fission. These fragments would then fly apart at great speed and as they slowed down through collisions with the matter around them, generate heat that might be put to use. Produced in the fission process, apart from two major fragments called fission products (which remained radioactive for a long time) were particles called neutrons, and these could induce more fissions in the uranium mass and establish a chain reaction. Other neutrons could be absorbed by the heavier uranium isotope, which through successive changes would turn into a substance called plutonium, chemically different from uranium, and also capable of fission (fissile). To produce an atomic bomb it was necessary to start with nearly pure fissile material and so two routes were open – separation (or enrichment) by highly expensive physical techniques of the lighter uranium isotope, or production of plutonium from the heavier isotope in a nuclear* reactor, in which a controlled chain reaction was maintained, followed by chemical separation.

*The terms 'atomic' and 'nuclear' have been and still are used indiscriminately to describe processes involving fission. Although bombs are usually 'atomic', it is more usual to use the word 'nuclear' when referring to devices designed to produce energy in a controlled manner.

The USA had followed both routes, while the UK, when the decision was taken in January 1947 to develop an independent nuclear deterrent, gave first priority to the second, and built plutonium production reactors at Windscale in the North of England. The heat generated in these reactors by the fission process was dispersed into the atmosphere. When, however, the call was made for an increase in the rate of plutonium production, those more interested in power succeeded in getting agreement in March 1953 for the construction of reactors (at Calder Hall) from which the heat could be extracted at a temperature sufficiently high for it to be used to generate electricity. Britain, thereby, (in conjunction with its military programme) became the first country to embark upon a civil power programme. This was placed under the direct control of a government agency that in August 1954 was constituted as the UK Atomic Energy Authority.

In the early fifties, all countries were preoccupied by the need to supplement existing sources of energy. Energy was regarded as the crucial factor in economic growth and the mounting gap between consumption and cheap indigenous supplies was a growing cause of alarm **(2)**.

Until 1929, Europe (as defined by the OEEC countries) was a net exporter of primary energy and until the war had been able to meet a large part of its demand from coal and hydro resources. Since the war, supply had not been able to keep pace, and while oil – practically non-existent in Europe – had begun to assume an ever increasing role, coal too was being imported. Price was an important factor, as although coal prices in the major producing countries had stabilised, they were still two to three times those current in the USA. Only from new hydro works could a significant increase in output be expected and suitable sites were becoming increasingly difficult to find. With Europe's overall energy needs increasing at the rate of about 4.8%/a, it could be expected that over 30% would be imported by 1965 as against 10% in 1953.

Within the energy framework, the role of electric power generation was seen to have special significance. Demand was on average doubling every ten years and, even more than gross energy, the per capita consumption of electricity was regarded as the yardstick by which a country's economic strength could be measured. Electrification was made synonymous with modernisation.

Against this background, a nuclear source which could provide from a few kilograms of fissile material, as much power as a million tons of coal, seemed heaven-sent. But nuclear energy was more than just an alternative way of generating heat; it was a talisman which conferred on its possessors, status in the present and prosperity in the future. It was almost impossible not to be affected by the excitement and the drama the subject generated. Those working in the field were, for the most part young, stimulated by the scientific and technical challenge, revelling in the priorities they were granted for the procurement of equipment, generously financed and encouraged by career prospects difficult to match in any other area. The opportunities seemed boundless and the internal vitality and enthusiasm spilled over into the consciousness of government and the public.

OEEC ATTACKS THE ENERGY PROBLEM

The problem of energy supplies at the European level was recognised in the OEEC in a memorandum submitted by the Secretary-General to the Organisation's Council in December 1953, which led to a study being undertaken by Louis Armand, Chairman of the Board of the Société Nationale des Chemins de Fer Français. His report was submitted to Council in May 1955 **(3)**, three months after the UK had startled the world by announcing on 15 February, a programme for the construction of nuclear power stations generating between 1000 and 2000MW (1.5 to 2 million kilowatts) of electricity **(4)**, to be available by 1965. Emboldened by the success of the construction programme at Calder Hall, Britain was prepared to gamble that, given a substantial programme of work – 12 stations were projected – industry would be able to build stations which generated electricity at a cost comparable with that associated with modern coal-fired stations. It will be noted that the size of station envisaged was around 100MW capacity for two reactors, a size quite in keeping with conventional practice at that time.

In his report, which largely equated energy supply to the availability of electric power, Armand made a direct correlation between the per capita consumption of energy and the standard of living prevailing in the various countries of the world – at least expressed in economic terms. He declared that the problem in the future will be to "extend the maximum kilowatts from a given amount of investment in power supply schemes, keeping in mind their respective running costs". Shortage of energy was not his prime concern, so much as the cost. If sufficient means were not devoted to the development of Europe's power supplies, expansion of the economy would be inhibited, which was of particular concern in view of the favourable energy position of the United States, where the per capita consumption of power was $3\frac{1}{2}$ times the European average. It was timely to pose the question whether the advent of the new primary source – atomic energy – did not mark a turning point in world supplies, similar to the introduction of oil, and whether, in consequence, steps should not now be taken internationally to secure its rational development.

It was well understood that the technical problems were immense, involving a great deal of research and development and that alone made "a very clear case for international co-operation between member countries of the OEEC".

Whilst Armand was not unsympathetic to the political implications, it was essentially from the economic standpoint that he assembled his arguments and presented them to Council. He was nevertheless cautious . . . "for many years to come atomic energy can hardly account for more than a small proportion of the total power supplies". His observation that the maximum speed of introduction of nuclear electricity generation would probably not be faster than that foreseen by the UK where all increase in demand could be met from atomic power stations by 1970 or 1975, can still be seen as perfectly sound.

Armand argued that the UK nuclear power programme showed that the price of nuclear power could already be regarded as economically competitive in regions where conventional fuel prices were not exceptionally high. Moreover, it could confidently be expected that technical development would reduce the costs of nuclear power more rapidly than the costs of power from conventional sources. Given, in addition, the current estimate that the total reserves of nuclear fuel were some twenty times those of fossil fuel it was not unreasonable to predict for nuclear energy a rosy future.

It could be hoped that Europe's most advanced nuclear nation would be prepared to share its experience, as the White Paper announcing the British programme included the phrase "Her Majesty's Government have always been in favour of the greatest possible international collaboration in the peaceful uses of nuclear energy". Only later was this statement seen to omit two important limitations: the determination to preserve a special relationship with the USA (a relationship which was controlled by international treaties that severely limited the freedom of action of the UK **(5)** without conferring full partnership status), and the desire to preserve acquired commercial advantages. At the time, however, the declaration of the UK provided a powerful incentive to pursue further the exchange of technical information and the joint construction and financing of plants as a European co-operation, considered by Armand to be "an essential prerequisite for the success of Europe in nuclear power [as] only in this way will possibilities in Europe bear comparison with those in the United States".

PARALLEL INITIATIVES IN NUCLEAR COLLABORATION

A further encouragement was the existence of two autonomous European organisations that had recently been set up to stimulate co-operation in the nuclear field and which included as full members, both the UK and the country with the second largest nuclear effort in Europe – France. These were the European Organisation for Nuclear Research (CERN), formally established in September 1954 **(6)**, and the European Atomic Energy Society

(EAES) which held its first meeting in March 1955. In practice, CERN was essentially academic in character, engaged on fundamental studies into the nature of matter. As the years passed, its activities became less and less relevant to the exploitation of nuclear energy, although at the beginning, the distinction between pure and applied research was less marked, and a small effort on nuclear fusion suggested that there might be applications emerging.

The EAES was an intimate gathering of the leaders of the national atomic energy commissions which most governments in Europe had set up to study and promote the new power source. Frequently referred to as a Club – the Cockcroft (or Dunworth) Club in England, the Goldschmidt Club in France – it provided a forum for the scientific leaders to meet together privately, at about two-monthly intervals, away from the glare of a learned society conference, and without the restrictions and controls that a formal intergovernmental meeting would have implied. It opened the way to an exchange of scientific and technical information at a high level, whilst preserving the privileged positions won by atomic energy commissions, notably the Atomic Energy Authority in the UK and the Commissariat à l'Energie Atomique in France, whereby policies were evolved and executed without parliamentary supervision and with minimum ministerial control.

The political, social and economic significance of atomic energy had not escaped the notice of the federalists. Under the drive of Monnet, who, on the collapse of the European Defence Community, had resigned from his position as first President of the European Coal and Steel Community in order to devote all his efforts to the promotion of European unity, the six members of the ECSC had been studying alternative areas, where co-operation might be accepted. As a result, in the same month that the OEEC Council resolved to take action in the light of Armand's report (June 1955) these six countries agreed at Messina in Sicily, to set up an intergovernmental committee to formulate proposals for an economic and political union. Chairman of the committee was the dedicated European, Paul-Henri Spaak of Belgium.

Nuclear energy was singled out for special attention as being a field in which this drive towards co-operation could be given concrete expression. The crucial differences between the OEEC and the Monnet initiatives hinged on the fundamental issue of federalism and related to:

(a) the motivation, which was primarily political among the Messina countries, economic within OEEC;

(b) the attitude of the UK which refused to take part in the Messina discussions.

On the broader international plane, co-operation in nuclear energy was also gathering momentum. Motivated by its concern over the proliferation of nuclear weapons, the United States had proposed in December 1953, the setting up of an International Atomic Energy Agency (IAEA) to be closely associated with the United Nations. Having endorsed this proposal in December 1954, the UN General Assembly also resolved to promote the non-military applications of nuclear energy by organising an international technical conference. Held in Geneva in August 1955, the "International Conference on the Peaceful Uses of Atomic Energy" was the occasion for a world-wide review of forecasts of power requirements and for the release of a huge amount of scientific and technical data on nuclear technology. One result was a great deal of publicity for the new source of energy and the promise it held of cheap, illimitable power, everywhere.

Nuclear power became a household phrase and nuclear energy was recognized as a vast source of heat. Reactors, in which the heat was generated in a controlled manner could replace the familiar boilers fuelled by coal, oil or gas in conventional power installations and were already replacing diesel engines for submarine propulsion in the US Navy.

BASIC ELEMENTS OF A POWER REACTOR

In essence, a reactor consists of an agglomeration of fissile material – the fuel – which is allowed to 'burn' and is kept at a steady temperature by passing through it a cooling fluid that carries the heat away. If this heat is to be used, then in most power systems it is made to raise steam which drives a turbo generator. Alternatively, the cooling fluid may be a gas which directly drives a gas turbine that is coupled to a generator.

Only one fissile fuel exists in Nature, the lighter isotope of uranium – uranium 235 – that occurs in natural uranium with a concentration of 0.7%. A chain reaction cannot be sustained in pure natural uranium, however large the mass, because the heavier isotope – uranium 238 – absorbs too large a proportion of any neutrons present, leaving too small a fraction to induce fission and so create more neutrons. The competition between fission and absorption can be made more equal, by introducing a 'moderator' which slows down the neutrons to thermal speeds before they are captured by the uranium 238. A reactor which exploits this effect is termed a thermal reactor. If the fuel can also be enriched in uranium 235 the choice of moderating and cooling materials is widened and the size of reacting mass, needed to cut neutron losses to the point that a chain reaction can be sustained, reduced. In the extreme case, the moderator can be dispensed with, and the fission produced by fast neutrons. The reactor is then termed a fast reactor.

Of the spare neutrons produced in the fission process, some are absorbed in control rods used to regulate the rate of reaction, some in construction materials and the shielding around the reactor needed for the protection of the surroundings, and some may be absorbed in fertile material, i.e. material which is converted into new fuel. As already noted, uranium 238 after capturing a neutron will eventually turn into plutonium. On the other hand, thorium, if present, may be transmuted into a fissile isotope of uranium not occurring in Nature, namely uranium 233. In both cases, the new fissile material produced can be separated from the fertile by relatively straightforward chemical processing techniques.

Once a nuclear capability has been developed, therefore, the reactor designer can select from three fuels, say four moderators (or none) and half a dozen coolants. Taking into account the different chemical forms in which the materials may be made up and the wide variety of engineering configurations, the range of systems he can imagine is clearly immense. Alvin Weinberg in his paper to the first Geneva 'Atoms for Peace' Conference **(7)**, spoke of 900 conceivable reactor types and 100 combinations not altogether unfeasible. Even though many of the combinations could be excluded by a cursory examination of either the physics or the engineering, the number that appeared to merit study in the second half of the 1950s remained impressively large.

What was so tantalising, was that so long as natural uranium only was available as fuel, the only practicable moderators were graphite and heavy water, the second of which was costly and difficult to separate from ordinary water; the only practicable coolants were certain gases and heavy water. However, once the more concentrated fuels came onto the market, the developers of civil nuclear power systems would be presented with an embarrassment of choices of which perhaps just one might prove to be superior to the others and command a global market. Only dimly was the magnitude of the effort needed to translate conception into a commercially viable product understood, and only slowly was the extent to which scientific nicety would become subordinate to industrial strength appreciated.

CHAPTER 3

COMPETITIVE FORMS OF CO-OPERATION IN NUCLEAR ENERGY

COUNCIL'S reaction to Armand's report was to agree "to explore the possibilities of economic and financial co-operation within the Organisation in the field of nuclear power. For this purpose it instructs a Working Party . . . to examine and report as soon as possible upon the possible scope, form and methods of such co-operation."

The OEEC Working Party (denoted WP10) consisted of Professor Leander Nikolaidis of Greece as Chairman, R. Ockrent of Belgium and W. Harpham of the UK. They began their work by sending a memorandum at the beginning of July 1955 to all the member states recalling that the object of their enquiry was "to secure the greatest benefits from nuclear power as a new factor of economic expansion". At this time "no country had yet brought into operation a generating station using nuclear power with a capacity of some 50MW or so, i.e. the size of the normal small coal or oil-fired station". Calder Hall had not yet become operational and the first land-based power demonstration reactor in the USA, Shippingport, was still in the constructional phase.

WP10's mandate was not confined to nuclear power in the narrow sense of nuclear power stations, and early in its studies, the Working Party identified 15 areas for consideration which can be summarized under the following headings:

1. Training of technicians
2. Exchange of information and patents
3. Research facilities
4. Nuclear fuel procurement, treatment and reprocessing
5. Joint construction of prototype and power reactors
6. Legislation covering trade exchanges, safety and waste
7. Finance

The first of these headings indicates how thinly was knowledge of the new technology spread in the majority of the OEEC countries. Only in Britain and in France was there an extensive body of experience which spanned all the facets of the subject, and included not just the academic physics and chemistry but also the engineering in depth. It was not unnatural, therefore, that the Working Party should begin its enquiries in the UK where it was able to lean particularly on the insight of Michael Michaels in the UK Atomic Energy Office (an extra-departmental office reporting to the Lord President of the Council). Standing between the politicans and the scientists, Michaels was one of the few people to have an appreciation of both the essentials of the technology and its political and economic implications. His influence was to be felt not only at this time, in the studies of the Working Party, but also subsequently in the general evolution of Europe's nuclear institutions, and more especially in the work of OEEC.

Altogether, WP10 visited 12 OEEC member countries, learning of national plans and gathering reactions to the idea of collaboration within the different areas they had identified. However, before their visits had been completed, the report of the Nuclear Energy Commission set up by the six ECSC countries at Messina had been published (5 November, 1955). Its principal recommendation was the creation of a series of nuclear activities under a strong international authority.

As the Commission was chaired by Armand and included Michaels from the UK, it is not surprising that on the technical plane, the main areas of interest identified were largely identical to those finally adopted by WP10. On the organisational plane, it was otherwise and the divergencies in motivation were evident. Fundamental differences between the Commission's recommendations and the ideas emerging in OEEC, concerned for example: the supply of fissile materials, with the Commission envisaging a monopoly, OEEC free trade; the degree of centralisation of research facilities, with the Commission favouring an international research centre and OEEC the co-ordination of work in national centres.

ECSC COUNTRIES

Attitudes in France towards the Commission's proposals and the competitive moves of OEEC were far from homogeneous, not least because of the country's unique strength in comparison with the other ECSC members **(8)**. French research had made important contributions to the basic science of nuclear energy, and the wartime Anglo-Canadian project for the construction of a reactor moderated by heavy water, at Chalk River, 200 km west of Ottawa, included a number of French scientists, and was initially directed by Hans Halban, one of the pioneers of nuclear energy in France. When the tripartite agreement between the USA, Britain and Canada of November 1945 then excluded France from further access to work in the three countries, France set out to create its own nuclear competence from its own resources, with effectively no outside help. The body entrusted with this task was the Commissariat à l'Energie Atomique (CEA), which was set up in January 1946, initially with the objective of studying and developing nuclear science and engineering for peaceful applications. By 1955 impressive progress had been made. Its installations included a research establishment at Saclay near Paris, and a reactor site at Marcoule in the south east. The foundations for the civil nuclear power programme were already laid, based on natural uranium fuel and graphite moderation with heavy water as a second possibility.

Government policy in regard to an independent nuclear deterrent vacillated for almost a decade, and it was not until April 1958 that the decision to develop a bomb was formally confirmed, although parliamentary approval for the creation of a military division of the CEA had been given two years earlier and within the CEA itself, the first explicit steps towards a bomb had been taken at the end of 1954.

At the end of 1955, the government in power was firmly committed to European integration, but its position was weak and events in Algeria were to result in a progressive disintegration of its authority until Général de Gaulle came to power in May 1958. Well before this, his views on the need to re-establish a sense of national purpose in France and his hostility to any actions which might materially restrict the country's powers of self-determination were exerting an influence on negotiations, particularly within the Six. OEEC actions were less objectionable, being largely devoid of political overtones.

For Germany, the issues were quite different. Only a few months previously had the country been allowed to begin work in the nuclear energy field and to make up for lost time, co-operation at the international level was seen as essential. Nevertheless, being traditionally bound to the principle of private enterprise, Germany was inimical to the idea of a supranational technical body. Industry expected to take upon itself the responsibility for nuclear power development, providing the finance and taking the technical decisions. Technical information was an industrial property that could be marketed or shared like any other commodity provided there was no state monopoly, either national or international, interfering with the natural flow of trade. Industry felt much more comfortable with an OEEC type collaboration where there need be no total package that participating members would have to accept, but a series of joint actions involving only those who were interested.

Politically, however, the German government was strongly in favour of the protective alliance that could come from the agreement at Messina and was nervous of the apparent competition OEEC collaboration might make with the smooth evolution of negotiations. Also there was a feeling that a close knit group could be much more dynamic than OEEC and develop atomic energy at a much faster pace, offering thereby a more rapid means of making up for the lost years of enforced inactivity.

Speed was everywhere something of an obsession at this time. Armand in his report to the OEEC was concerned with "policy with regard to atomic energy which will rapidly lead to satisfactory production costs". The newspapers were fond of 'races', the technical people were keen to light the first lamp from nuclear power. Under the influence of the talisman, there was everywhere a feeling of urgency that went beyond the simple need to plan for future energy supplies.

Feeling in Germany then was divided. Both industry and government were keen to participate in international programmes, but the very factors in the Commission's ideas which attracted the political arm, calling as it did for strongly centralised direction, alarmed local government in the Länder and industry. On balance, Germany wished to remain a faithful member of the OEEC but it would not jeopardise the collaboration efforts of the Six.

Belgium expressed itself in favour of OEEC collaboration while supporting vigorously the proposals of the Commission to which it was host. No conflict was seen between the two groups, and the country as a whole was against the evolution of rigid blocs which would favour protectionism and limit the flow of international trade. As a major producer of uranium (in the Congo) it was in the privileged position of raw material supplier and had long-term agreements with both the USA and the UK. It had a relatively ambitious nuclear research programme, pushed forward by a government institution, the Centre d'Etudes et d'Applications pour l'Energie Nucléaire (CEAEN) that had been set up in the Autumn of 1952 out of an initial national nuclear commission. A national research centre was being built at Mol, which would house a small natural uranium graphite reactor to begin with and other testing and prototype reactors in the future. Its research could, nevertheless, embrace only a fraction of the total field and cooperation with bigger countries through both multilateral and bilateral agreements was seen to be indispensable.

The Netherlands was more wary of the possible disruptions that the OEEC initiatives might provoke in the negotiations on the formation of the new communities of the Six and was worried by the multiplication of bodies dealing with nuclear energy. In addition to these already mentioned The Netherlands was also in partnership with Norway through the joint organisation JENER that had formally been set up in 1951. At home, research in nuclear energy begun within the Foundation for Fundamental Research on Matter (FOM), had just been given a new impetus by the creation of the Reactor Centrum Nederland (RCN) which brought together with FOM, the electricity producers and industry. Furthermore, the electricity companies' independent research organisation KEMA was intent on following its own line of research leading to the eventual development of a power reactor. It was a question of priorities. Apart from JENER, the Government saw its international commitments as first, the United Nations International Atomic Energy Agency, and second, a federal group within Europe. Nothing should be allowed to interfere with these. Indeed the Netherlands considered the formation of the nuclear community as a *fait accompli* and intimated that any discussions on a wider collaboration with OEEC should be conducted by the ECSC countries as a unit.

In Rome also, the Government was worried by the political conflict that resulted from the dual approach to European co-operation in nuclear energy, but was anxious not to adopt a stance that might result in the country's exclusion from an important development. Italy was heavily dependent upon external fuel supplies and with its expanding demand for electric power (averaging 9% in the years 1950-1955), was particularly interested in the exploitation of nuclear energy for power generation. Italy, however, had its own internal conflicts in the power field, as the private utilities struggled against the threat of nationalisation and the efforts being made to create a fuel and power monopoly. Successive governments were unable to resolve the problem and the national research organisation, CNRN, the Comitato Nazionale per le Ricerche Nucleari, that had been set up in 1952, was operating with no clear terms of reference and at times without even a budget.

Dominant industrial voice at this time was that of Edisonvolta. The company had just been the instigator and main contributor to the formation of a special nuclear company (SELNI) which grouped the electrical firms in the north and centre of Italy, and it had every intention of going ahead with a nuclear power programme based upon imported technology, in which the Italian state organisations had no part. The manufacturing industry was also preparing itself for whatever markets emerged. The industry-driven sectors were much more sympathetic to an OEEC style of collaboration that could allow direct participation, than the one proposed by the Commission which would be implemented through the State. There was, therefore, a tendency for the country to divide, with the State backing the ideas of the ECSC countries and industry seeking to impose its views through its presence in OEEC. The situation was, however, more blurred than such an analysis might imply as attitudes were strongly conditioned by the sense of dependence on outside sources of both fuel and technology, and the need, therefore, to keep as many options open as possible.

NON-ECSC COUNTRIES

For the countries not involved in the Commission's negotiations there was little ambiguity. Even those with active nuclear research and development programmes recognised that the technology was too broad to be covered individually and there were many aspects that could only be dealt with in an international context, not least of which was the procurement of special materials. None had military ambitions and their chief concern was to see that they could benefit in due time from the advent of nuclear power and that their industries could profit from the markets that would open up.

Prominent amongst the smaller countries for its early activity in nuclear research was Norway which, as Europe's principal producer of heavy water through Norsk Hydro, was well placed to initiate work on reactor systems that employed heavy water as moderator. In 1948, a joint government/ Norsk Hydro foundation for nuclear research – the Institutt for Atomenergi (IFA) – was established and construction started at Kjeller of a research reactor. Norway, however, had no uranium, whilst the Netherlands had a stock of ore bought in before the war. This led to the JENER collaboration and following the start up of the reactor in July 1951, work was begun on the building of an experimental power reactor for producing electricity and process steam at Halden. Norway had already made its attitude to collaboration clear by adopting an open policy towards other countries, apart from the Netherlands, and it was Gunnar Randers, Managing Director of the Kjeller centre, who had been the driving force in the formation of the European Atomic Energy Society. Norway had no immediate need for nuclear power but the scientific community was convinced of its importance, and keen to be among the first to apply it to merchant ship propulsion. OEEC initiatives were fully in line with political and technical thinking.

Sweden and Switzerland were no less positive. Convinced of the necessity for international collaboration, both were anxious to avoid associations that looked like political alliances or organisations that possessed any supranational powers. In Sweden, a 4/7 government 3/7 industry-owned national research organisation, AB Atomenergi, was set up as early as 1947 and had been pushing ahead with uranium prospection and research on a heavy water moderated reactor system, that could be used for power production when hydro resources had reached their limit. Its first research reactor had begun operation at a centre on the outskirts of the city of Stockholm in the Summer of 1954. Co-operation was seen as the means by which other systems could be covered, organized between national centres. Each country should be left free to arrange its own commercial links when required, as in other fields of engineering.

Switzerland was even more concerned about the need to avoid protectionist practices that would limit industry's freedom of action. National policy was based on using federal aid to encourage industrial development, but the country's efforts had been seriously hampered by its inability to procure even a modest amount of uranium. Not until the Spring of 1955 could this be assured and a consortium of industrial companies had then been formed to build a research reactor. More recently, the country had been able to acquire the reactor built in the grounds of the United Nations in Geneva for the Atoms for Peace Conference. In the light of their experience, both government and industry were anxious to gain access to the knowledge and facilities of an international community, but resolutely against any form of dirigisme, in any case unacceptable under the Helvetic constitution. They were only prepared to consider a project by project type of collaboration that allowed countries to choose to participate or not according to their interests and the finance available. Less sanguine than Sweden about the difficulties that could arise when commercially valuable information was at stake, the general feeling was that the most suitable areas for collaboration were the scientific. Basic research could be conducted in common, but industry on its own must take the responsibility for exploitation.

Other countries such as Austria and Denmark were less advanced in their national planning and therefore keen to participate in an international activity which would allow their technologists to bring themselves up-to-date and their industries to remain in touch with markets. But for all countries on the continent of Europe, the value of an international collaboration would depend upon the future policy of the strong nuclear countries – the UK and the USA.

UNITED KINGDOM

In atomic energy at this time, Britain's pre-eminence in Europe was acknowledged. British contributions to the Geneva Conference bore witness to this and the senior scientists, even from France, who had been shown the installations at the United Kingdom Atomic Energy Authority's research centre at Harwell and the production and power centres in the north of England, returned impressed by the scope, the scale and the quality alike.

The Authority (and its antecedents) had been set up, first and foremost, to build an atomic bomb and to establish the country as a nuclear power in the military sense. No secret had been made of this objective, nor that civil applications would have second priority until the primary objective had been achieved. The weapons programme led to the construction of a series of gas-cooled graphite-moderated reactors, an enrichment plant for uranium, fabrication plants for the uranium with which the reactors were fuelled and a processing plant for the extraction of plutonium. In parallel, an extensive programme of research on reactor technology was instituted intended to cover a wide variety of combinations of fuel, moderator and coolant, with a view to civil application.

Military and civil interests, as already noted, came together with the construction of the dual purpose reactors at Calder Hall and work on these progressed so well that the decision was taken to launch a truly civil power programme based on the same type of reactor. This was characterised by having natural uranium as fuel, graphite as moderator and carbon dioxide as the coolant gas. In due course the civil version became known as the Magnox reactor, on account of the magnesium alloy used to sheathe the fuel and isolate it from the coolant. Industrial interest was awakened and consortia of companies were formed to compete with each other for the contracts for complete power stations placed by the nationalized electricity undertakings. The Authority provided the basic technology and acted as consultant to both supplier and customer while retaining direct control over fuel supply, reprocessing and waste disposal.

A part of Britain's technology had been learned from America under bilateral agreements which excluded the UK from passing on either know-how or concentrated fissile materials to a third party. Moreover a key-stone of government policy had been to re-establish a full military partnership with the United States that had been cut off by the McMahon Act of August 1946. Amendments in October 1950 had gone some way to improving the situation but it was only in June 1955 that satisfactory co-operation in the military field had been forthcoming. Nothing in the civil field was to be allowed to prejudice the new relationship. Various security lapses on the UK side had severely strained relationships and the UK was anxious not to appear as a source of proliferation of US information into Europe. On the other hand, the moves towards European integration were changes that could not be ignored. The problem for the UK was the conflict of being tied to the United States, yet regarded as the leader of Europe.

Although it had firmly declined to sign the Messina agreement, the UK participated in the Brussels talks of the Commission and its full membership of an atomic energy community was theoretically not excluded. Such an eventuality was, however, improbable. Although the Chairman of the Authority, Sir Edwin (now Lord) Plowden was personally in favour of European union, his main task was the development of weapons systems. If the UK became member of a European nuclear community it would be impossible to disentangle the information rights whereas collaboration through OEEC was something that could be done without impairing relations with America.

Broad as the UK reactor research programme had become, it could not cover all eventualities, so there was an incentive for some co-operation in many of the areas considered by WP10, the principal exceptions being uranium enrichment and reprocessing, which touched too closely the military side. It was also understood that in those areas where the UK had already established a commercial 'lead', then business considerations would control its actions.

NORTH AMERICA

Although not members of the OEEC, the United States and Canada participated in the work of the Organisation and there was no intrinsic reason why these two countries also should not participate in some joint activities that were set up under the OEEC umbrella. There was no time, however, for WP10 to make a visit to the North American continent to explore the possibilities of North American involvement; instead discussions were held with representatives in Paris and Geneva where the Atoms for Peace Conference provided an opportunity for both formal and informal exchanges.

The United States had, by far, the biggest nuclear activity in the western world with its military installations, its submarine programme and an extensive civil programme that was being pursued by industry with massive government aid channelled through the US Atomic Energy Commission (AEC). Moreover just prior to the Conference, in June, President Eisenhower had proposed to grant financial and technical aid to any non-communist country developing nuclear energy. The full implications of this were still unknown. Legislation in the USA was both restrictive and complicated (at that time small quantities only of enriched material could be exported for explicit research projects) and it remained to be seen whether it was legally possible for the Government to deal with a group of European countries. Also, the influence of the creation of the International Atomic Energy Agency on events had still to be evaluated.

Canada could be more positive, expressing a decided preference for an OEEC style collaboration than for any supranational organisation.

CHAPTER 4

EUROPEAN NUCLEAR ENERGY AGENCY

ALTOGETHER, Nicolaidis and his colleagues were able to take encouragement from the reactions they had encountered to their 15 points, and the timeliness of their study was commended by Francis Perrin, High Commissioner of the French CEA. He pointed out that collaboration was currently not only necessary but easy, whereas in three years' time, it would be very difficult. Certain broad principles had become clear, the most important of which was the dislike of supranationalism and the attractions of the opportunities afforded under Article XIV of the OEEC Constitution for collaborating project by project without any obligation at the same time to subscribe to an overall programme. No country, moreover, would have powers of veto if a project had its supporters.

In the report of WP10, presented to the 307th session of the OEEC Council in January 1956 **(9)**, three types of institution were envisaged:

1. A Steering Committee for Nuclear Energy set up by the OEEC Council to: organize a confrontation of national programmes with the object of preventing duplication and identifying gaps in overall development; promote joint undertakings; harmonise national legislation; promote the training of scientists and engineers and standardisation within the field; study proposals yet to be made on international trade.
2. A Control Bureau set up by Council in association with the Steering Committee to exercise control over fissile materials, and prevent those used in the joint undertakings or produced by them from being ultimately diverted to military purposes.
3. Companies independent of the OEEC which would carry out joint projects concerning production and applied research. These would not be subordinate in any way to the Steering Committee but would establish relations with it for the mutual exchange of information.

In the list of joint undertakings which might immediately be considered, pride of place was given to fuel, the production of heavy water and the construction of power stations. The "keenest interest expressed in a great number of member countries" was in an isotope separation plant for the enrichment of uranium. A chemical separation plant for the treatment of irradiated fuels also seemed likely to attract a good number of members.

Interest in the joint construction of nuclear power stations arose from the chance it would give interconnected countries of gaining, with the minimum of investment, experience in operating such stations. Prototype reactors were grouped with reactors for the study of materials while centres for testing and basic metallurgical research, figured in a second group of joint undertakings which the working party recommended should be studied only when existing installations were being used to full capacity.

Recognition was given to the problems that would arise in establishing the status of any joint undertaking, but there were precedents in the international fields, and it would be necessary to choose the best possible arrangement that fitted each project and type of participant, bearing in mind that these would not necessarily be governments alone or even at all.

Preliminary reactions of Council and the permanent delegates to the OEEC were favourable and, together with the appropriate resolutions, the report was passed on to the inter-ministerial Council meeting held on 28 - 29 February, 1956, presided over by the Chancellor of the British Exchequer, Harold Macmillan. The meeting took place two weeks after the six ECSC countries had agreed the basis of a Treaty setting up a Common Market and the establishment of a European Community of Atomic Energy that was to be known as Euratom.

EURATOM'S EXPECTATIONS

Fresh from this diplomatic success, Paul Henri Spaak of Belgium spoke for the Six. Without wishing to denigrate the quality of the Nicolaidis report which was unanimously regarded as an excellent study, Spaak was obliged to declare that Euratom was a more positive approach. He regretted that European solidarity had not been emphasised and pointed out that Euratom was not a closed organisation but one in which countries outside the Six could participate. Co-operation in the OEEC manner was but "a bonus to national egotism". The full report of the Euratom study would be published the following month and "all OEEC countries would be invited to study the more ambitious proposals of the Six countries in the hope that they would see their way to joining them".

Despite the fact that the principle of jointly producing nuclear power under the auspices of a supranational body had already been abandoned, as well as any restrictions on national military activities, the delegates from the Euratom countries were full of confidence that Euratom would be the main international instrument for nuclear power development in Europe. Spaak, for example, compared the timidity of the Nicolaidis suggestions regarding the control of nuclear materials with the more sweeping plans of the Euratom countries, and other delegates underlined the greater scope and scale of the intended Euratom collaboration. Christian Pineau of France spoke of Euratom's aim as the "highest common multiple" contrasting it with that of OEEC where one had to look for the "lowest common denominator". However, the delegates from France, Germany and Italy all agreed that Euratom and OEEC were complementary and were prepared to acknowledge that there was no essential conflict between the two.

SPECIAL COMMITTEE FORMED

The UK from the Chair implied that it was very willing to co-operate and the USA seemed to be adopting a positive stand towards European co-operation, President Eisenhower having made known the previous week that 2500kg of uranium 235 would be made available to Europe for civil purposes. (This was, in due course, to have a strongly negative effect on the principal joint undertaking that was proposed but it can be argued that Europe thereby avoided a costly investment out of step with market requirements.) The remaining countries also being favourable, the Council resolved to set up a Special Committee* of the Council for Nuclear Energy including representatives of all member countries, together with the United States and Canada in their capacity as Associate Members of the OEEC. Instructions given to the Committee were to submit concrete proposals to the Council within three months, with a view to implementing the suggestions of the Working Party.

The Special Committee for Nuclear Energy was duly set up with Nicolaidis as Chairman and Pierre Guillaumat of France and Friston How of the UK as Vice-Chairmen. The Secretariat was composed of Guido Colonna, Deputy Secretary-General of the OEEC and Pierre Huet, a Director of OEEC and legal Counsellor to WP10. Four Working Parties were quickly formed to study Joint Undertakings, Security Control, Harmonisation of Legislation, and Co-operation in the Field of Training.

Each of these was served by a secretariat, Huet serving on the first and second together with Einar Saeland of Norway. Other relevant committees of the OEEC were called upon to give assistance, in particular, the Electricity Committee.

---

*Special Committees within the OEEC (and later the Organisation for Economic Co-operation and Development – OECD) were ad hoc committees set up to deal with specific programmes of work. Steering Committees are permanent committees covering evolving fields of activity.

In the terms of reference given to the first Working Party WP1, the emphasis was strongly on fuel enrichment and reprocessing and the production of heavy water. Prototype and experimental reactors were lumped together in a few lines with the explicit requirement that "the Working Party should examine the question as to the types of reactor member countries do not propose to construct individually and which might usefully form the subject of studies to be initiated or pursued jointly". At the first meeting of WP1 in April 1956, rapporteurs for the various sectors were appointed including for the reactor sector A. H. W. Aten, Jnr., of the Dutch Institute for Physics Research (IKO).

OPPORTUNITIES FOR COLLABORATION ON REACTOR DEVELOPMENT

When Aten made his first report to the Working Party he was forced to acknowledge that suggestions for reactor development by an international team had not proved very satisfactory. "It is evident", he continued, "that every country wants to develop its most attractive ideas by itself." It might be possible to obtain support for a high temperature reactor, presumably fuelled by a liquid metal, or propulsion reactors, or a very high flux materials testing reactor. Fast reactors and boiling water reactors were interesting but the UK was heavily engaged on the first and Norway on the second, and until these two countries had made their positions known, it was difficult to make a judgement. He concluded that "a field in which technical developments are unlikely to take place in the near future but which may play an important role at a remoter stage is likely to be most acceptable".

An attempt, nevertheless, was made to compare the costs of developing various reactor types and, for the first time, brief mention was made of high temperature reactor systems based on gas cooling and fuelled by uranium oxide or carbide. The American journal *Nucleonics* had carried two articles analysing the system and the notional costs of electricity from a station based on it **(10)**. Largely on this evidence, the system figured in the note attached to the Aten summary which indicated that the cost of setting up a research institute for the development of a boiling water, propulsion or high temperature reactor was about the same, whilst a fast reactor institute would come out at about double.

Meanwhile a Working Party of the Electricity Committee under the chairmanship of Franco Castelli of Edisonvolta had been considering the question of joint undertakings from the viewpoint of the producers and suppliers of electricity. Starting from the advanced position of a permanent committee with established communication lines, it was able to produce a thoughtful and pertinent report in the short time available. This deserved more serious consideration than it appears to have been given in the compilation of the main report of the Special Committee to Council, even allowing for the fact that the Electricity Committee's comments were attached as an annexe and its recommendation that a Study Group be formed was endorsed.

The Electricity Committee foresaw three possibilities for common action:

1. the joint construction of an already tested type to share experience;
2. the joint development of small novel types of reactor;
3. a series of autonomous developments in a co-ordinated programme of work in which experience and perhaps even the commercial risks would be shared.

These were not to be considered as mutually exclusive. For the type of collaboration envisaged in 1 and 2 a mixed company with private and government participation might be the best structure but, as a first stage, it was suggested that a pilot company be set up to draw up plans for one or more stations without prejudice to any solution ultimately adopted. In its summary, the Committee stated that "the experts of the member countries as a body attach great interest to the making of co-operative arrangements and the drawing up of a joint programme between the interested countries for the development of electricity production based on nuclear energy".

In the Special Committee's final draft **(11)**, bringing together the conclusions of the Working Parties, the section dealing with reactors was rather hesitant, and no attempt was made to define a policy. The assumption was widespread amongst the reactor experts that joint reactor development necessitated the setting up of a new joint institute and only in the annex on Administrative and Legal Problems treating joint undertakings generally, were alternative procedures considered, including the notion of confiding a project to the care of a national organisation. Here, the model was put forward with some firmness. Most probably the joint undertaking that was chiefly in mind was either the fuel enrichment or fuel reprocessing plant and for the time being, the idea did not spill over into the reactor area.

Three months was a very short time for the Special Committee to produce a report covering all the sectors specified, of which joint undertakings was but one. Yet this was done and presented to the 3rd meeting of the Special Committee at the end of June. Inevitably, ideas were blurred and when the UK delegate made the distinction between joint projects for generating power as such and joint projects for developing reactor systems, even he declared that if it was the first under consideration, then the only choice was a gas-cooled graphite-moderated reactor fuelled by slightly enriched uranium – ostensibly the UK system but really a novel type that no one had yet tried.

CREATING THE AGENCY

The essential task of the Special Committee was to make recommendations concerning the establishment of a Steering Committee and a control system. Huet, the driving force in the Special Committee, had become convinced that not only was a Steering Committee essential but all its different activities should be grouped under an Agency. This would give greater cohesion and would introduce a degree of insulation from the OEEC administration. He saw the atomic world as a closed world; senior scientific staff were not comfortable in the political atmosphere of the OEEC. At an Agency they could be made to feel that they were with their own kind. Moreover, the OEEC was becoming structurally a little heavy and an Agency with its own individuality would allow a new liberty of action with regard to the central services such as personnel administration and even budgeting. Also in its relations with the outside world it would have a certain autonomy which could make for simplification when it came to third parties.

Huet's prime objective became the creation of an Agency with real projects, not just administrative functions "Il me faut du béton" he recalls saying at a meeting in Strasbourg and it became a guiding principle to which he devoted his considerable skills and energies.

When drafting the chapter on General Organisation, in the section covering the Steering Committee, he added to the paragraph relating to its subsidiary bodies the word "the Steering Committee with its subsidiary bodies together with the control body shall be designated under the name of the European Nuclear Energy Agency". Curiously this passed without comment at the Committee's final meeting, as Euratom's defenders were more concerned with trying to stop, or at least delay, the creation of the Steering Committee. (Within the Commission, on the other hand, there was vigorous reaction to the proposal.)

The Netherlands' delegate representing the Six had tried to persuade the rest that there was no reason to form a Steering Committee yet; studies could continue perfectly well under the Special Committee and it would give an opportunity for the others to appreciate the full scope of the Euratom agreement when this was finally approved. Support from other members of the Six for this view-point was less than whole-hearted. France had indicated that she would not go against Euratom interests but was ready to approve the report. Germany also was ready to approve the report and participate. With this backing added to that of all the other members of OEEC, the Euratom case for delay was lost and, with a few modifications to the drafting, the report could be presented to Council with just the reservation that the agreement of the Six was conditional upon the new organisation not impeding in any way their closer collaboration.

At the Ministerial council meeting in the middle of July 1956, the report of the Special Committee was approved and decisions taken to implement its recommendations. These included the setting up of the Steering Committee for Nuclear Energy which in turn, as a result of Huet's persistence, was instructed to draw up within six months its proposals for the statutes of the European Nuclear Energy Agency.

On subsequent occasions, attempts were made by Euratom's promoters to delay acceptance of the Agency's statutes, but the speaker delegated to represent the views of the Six in OEEC Meetings received, as a rule, only meagre support from his colleagues and the opposition was largely ineffective.

Euratom was not alone in being unhappy with the idea of an Agency. Washington saw in it a competition with the International Atomic Energy Agency and suspected that OEEC was acting politically. Another name would have been preferred. This was not, however, the reason for the cool reception in political terms that the USA reserved for the Agency when it came into being, even though technical relationships were cordial. The USA adopted the policy that nothing was to prejudice its political relationship with Euratom and Washington could not be persuaded that the success or otherwise of Euratom was entirely its own affair and quite independent of what went on within OEEC. Even so, Washington would not go so far as to oppose the OEEC moves, as was learned when Colonna, accompanied by Huet and Saeland, made a quick visit to the USA in July to sound out the country's intentions.

A lasting impression gained by the group was the realism to be found in American industry which saw clearly that nuclear energy was not ready for immediate exploitation. This contrasted with the more enthusiastic influences in Paris which included that of Walker Cisler, chairman of the Detroit Edison Company and member of the US delegation on the Steering Committee. Cisler was involved in the decision to build many years ahead of its time the Enrico Fermi fast reactor at Lagoona Beach on the shores of Lake Erie.

THE AGENCY PEOPLE

Huet was unquestionably the man to head the new Agency. Armed with a legal background and a natural faculty for putting order into a complicated array of information, he was both ambitious and hard working. His style was not the drama of the impassioned barrister but the dry analytical approach of the judge. On the few occasions when he intervened at meetings, it was to smooth away the rising resentment of some delegate or succinctly to rearrange the essential data on a topic that was going astray.

Of aristocratic bearing, immensely cultivated, with all the polish of the French Hautes Ecoles, equally at home in French and English, he made a most effective diplomat, gently but firmly moulding the base on which the Agency could be founded.

A primary requirement was staff to concentrate on the new project, and one of his first acquisitions was the consultant who had been a member of the Secretariat of the Working Party on the Harmonisation of Legislation, Jerry Weinstein. A barrister from the UK, Weinstein was both brilliant and industrious, although so down to earth, one surprised visitor was incautious enough to ask in supercilious tones which University he came from. He received the crushing (and truthful) answer, "Oxford, Cambridge and Yale". Weinstein was to become one of the world's leading experts on nuclear law. His interests were wide and though burdened with a crippled leg resulting from infantile paralysis contracted in Singapore in 1946, he was a passionate follower of Association Football and the co-author of a best seller on the subject as well as a contributor to a number of magazines. In addition he was a connoisseur of the arts, again contributing to the literature. His languages (including Japanese) were excellent and he had a real feel for the meaning of words. His contributions to the Agency's development were immense and he was renowned for his total devotion to his work and the unswerving loyalty he gave his superiors.

During the meeting of the Working Groups of the Special Committee and on his visit to the USA, Huet had had the opportunity of working with Saeland, a quiet physical chemist from Norway, who *inter alia* was responsible for the heavy water programme at Norsk Hydro. Little interested by administrative problems, he preferred to slip away to read the scientific literature, ruminate on a technical problem and make his own calculations. Endowed with a clear scientific mind, his judgement of matters of scientific policy was respected. Never pushing himself forward, he would offer comment if it was asked for; if not he was able to keep his own counsel. Huet appreciated these qualities, so complementary to his own drive and non-scientific background. On his return from the USA, Saeland became permanently established in Paris, where his specialist knowledge was particularly relevant to the enrichment, reprocessing and heavy water projects. He was not, however, an expert on reactors, and Huet turned for suggestions to Roland Perret, a young ebullient Swiss whom he had taken on as technical consultant to serve the Special Committee.

Perret remembered having read the lengthy accounts of a lecture given at the Sorbonne on co-operation at CERN, the lessons this held for the reactor field and the areas of research that might be pursued as international ventures. The lecturer was Lew Kowarski, one of the original research workers on atomic energy problems in France, famous for his flight to England with the French stock of heavy water when France was over-run, member of the select team at the Cavendish in Cambridge, then at Chalk River. He returned from Chalk River to France to build the CEA's first research reactor that was fuelled with natural uranium and moderated with heavy water. Later, Kowarski found himself at odds with the directorate of the CEA which, because of his habit of abandoning to others the hard middle ground of routine work in order to concentrate at one extreme on tedious questions of status, and at the other, on penetrating analyses of policy to the discomfiture of those in authority, had been relieved to see him turn his attention to the establishment of CERN. The attendance at the Sorbonne lecture had been abysmal, but Kowarski had met the reporters in advance and consequently their accounts reflected only the contents of his lecture and not the lack of popular support.

Perret put his name before Huet and soon after, Kowarski was invited to become a part-time consultant to the new organisation. He had unique qualifications. He was a reactor man, he knew the American, British and French people and programmes well, yet was not tied to the career structure of a national commission. He was also available, finding himself at rather a loose end at CERN and seeing reactions develop there surprisingly similar to those he had known at the CEA. Huet's invitation was accepted and Kowarski, rejuvenated by the prospect of helping to found another European collaboration, joined the team.

CHAPTER 5

REACTOR STUDIES

UNDER the general heading Joint Undertakings, the OEEC Council had determined to set up two Working Groups: one to consider the technical, legal and financial problems raised by the construction and operation of reactors, and to make suggestions as to types which might be adopted for joint ventures; the other to make a similar study on nuclear power stations.

As Aten in his report to WP1 had already concluded, co-operation would not be easy to organise, a view that was reinforced by the statements made to Council by the UK delegate. Whilst maintaining that "the UK intends fully to collaborate in the continuing work of the Organisation" and to assist in the separation plant and other activities "on suitable terms", he finished with the words "we cannot ourselves join in these joint undertakings". This attitude was not unexpected in the Secretariat. Statements made at the last meeting of the Special Committee had evoked the somewhat acid comment from the French delegation that aiding was not participating and those who paid must expect to get the results. Still Huet did not abandon hope of some useful co-operation with the UK.

During the Autumn, the names of the representatives to sit on the two Working Groups were assembled under the abbreviated titles of REP and REX (Réacteurs de Puissance; Réacteurs Expérimentaux). The press release issued on the occasion of their first meeting in October 1956 spoke of their respective functions as: "to study the means for speeding up the construction of nuclear power stations; and to study the establishment of a joint research centre". This was a very free interpretation of the Council's decision and the report of the Special Committee, but it was justified by the terms of reference that had been prepared for the group in their initial working documents.

EXIT THE ELECTRICITY UNDERTAKINGS

Castelli was made chairman of REP and with him as members, both of the main party and its committee of experts, were several others who had served on the Working Party set up by the Electricity Committee. There was no inherent reason then for the drift away from the ideas that had been embodied in that Working Party's report, towards the much more narrow objective of considering only large 'developed' reactors.

One influence was the Suez crisis of 1956 which had the effect of lending a new urgency to the construction of nuclear power stations in Europe in order to provide as quickly as possible an alternative source of energy to oil. Another was the inauguration in October by H.M. the Queen of the combined plutonium production and power plant of Calder Hall in the UK which, by its acknowledged success, gave the impression that nuclar power had indeed arrived. Moreover, it had become known that the first industrial stations in the UK would have an output per reactor of about 150MW(e)* which would almost certainly lead to a substantial increase in the total capacity installed under the first British programme. Such an increase was confirmed on 5 March, 1957 when the target for nuclear generating capacity to be installed by the end of 1965 was raised to 5000–6000MW.

REP began its work by instituting a study of the power programmes of the member countries. This was completed in February 1957. In March, the experts were talking of the need for speed and large sizes, picking out as possible systems: gas-cooled graphite-moderated natural uranium-fuelled reactors (GCGR) as in the British and French national programmes; two types of light (as distinct from heavy) water-moderated and cooled reactors viz. the pressurised water reactor (PWR) as developed by Westinghouse for the first US submarine *Nautilus* and the first US civil demonstration power plant at Shippingport, and the boiling water reactor (BWR) under development by General Electric of America; the sodium-cooled graphite-moderated reactor (SGR) as developed by Atomics International in the USA and used to power the submarine *Sea Wolf.*

From amongst these, REP recommended either the GCGR reactor or a PWR for large capacity plant with a preference for the PWR, adding as variant, a pressurised heavy water reactor on the lines of the Swedish design for a power and district heating plant to be built at Ågesta.

For low capacity pilot units it was thought that the BWR and possibly the SGR would be suitable. This was later modified after discussions with the USAEC and the PWR and BWR were from thereon considered to be equivalent.

In drafting its report, REP still envisaged a harmonisation of construction programmes and a wide exchange of information to help electricity undertakings choose the best system. It also proposed the setting up of a Study Group to select the type of reactor for a joint undertaking, work out

*Electrical output is denoted by (e) and thermal output by (th). The ratio is the thermal efficiency which is strongly dependent upon operating temperature.

the feasibility design, identify the site and prepare the administrative structure.

Switzerland was particularly active in promoting the collaboration and Huet and Kowarski went to Zurich to narrow the field of choice. This led to the terms of reference for the Study Group on Nuclear Power Stations being drafted in the definite form that its task was to finalise one or two projects of about 150MW electrical output for which it "shall consider a Calder type and a water-cooled reactor (PWR)"**(12)**.

Just prior to the meeting of the Steering Committee on 9 May, the Secretary of REP wrote to Huet that Germany, Italy and Belgium were all in favour of the proposal but although the Steering Committee accepted the report it soon became clear that there was little hope of the Euratom countries taking part. In June, the UK made the comment that as all the information on gas-cooled and pressurised water reactors was published there was little point in forming a Study Group and at the fifth session of the Steering Committee in July, it was made known that only Switzerland, Austria and Greece had declared their willingness to take part. The activities of REP were virtually at an end.

One consequence of REP's disappearance was that the representatives of the electricity undertakings, by far the most important customers for the products of nuclear energy developments, slipped from the scene and exerted no further influence on events. Already, by setting up the Steering Committee for Nuclear Energy and the embryonic European Nuclear Energy Agency, the first wedge had been driven between the prime users and those concerned with planning the means by which their requirements would be met. By concentrating attention only on what might be built immediately, before they had any competence to act, the electricity producers forfeited the chance of acting in concert and steering developments in a direction that, in the long-term, could have proved most beneficial. They left the field to the nuclear specialists whose interests were concentrated on the technology rather than on the reliable supply of cheap electricity. Subsequently, they were to find themselves trapped in the market that had been fashioned by the few industrial corporations which had sufficient resources to weather the technical, financial and social storms that broke.

THE RESEARCH APPROACH

It would have been astonishing if the situation had developed otherwise. Nuclear energy in the leading European countries was, if not military, a scientific hegemony. The Authority in the UK and the CEA in France for example, determined needs and means alike, the role of the UK Central Electricity Authority (later the Central Electricity Generating Board) and

Electricité de France being to take on trust what the national research organisation considered to be most suitable. Research and development were the key activities in civil power and although it must be regretted that OEEC did not press for the continuing presence of the market representatives, it was only reflecting the syndrome current amongst its members.

At the centre of the budding Agency were lawyers and research-minded scientists with little industrial awareness. It was natural that they should put the accent in the reactor field on research into scientifically interesting systems. Moreover, Aten's study had indicated that only speculative projects, far from industrialization would receive support and it was evident that collaborations between electricity undertakings would be very much more difficult to mount through OEEC than research collaborations between government-controlled commissions. Speed too was important. The prime objective was to make the European Nuclear Energy Agency a success and for this it had to have projects, and quickly. It was up to REX to find some reactor, or reactors, which would be built as an Agency project, preferably with the active participation of the UK. The role of any reactor in the scheme of long-term exploitation would emerge automatically if joint agreement could be reached to build it.

When REX began its work in the Autumn of 1956 it was fortunate in finding a chairman who was both well versed in reactor technology and experienced in the international field. At the very moment when Huet and Kowarski were discussing how they could persuade Sigvaard Eklund, Head of the Physics Department in the Swedish company AB Atomenergi (subsequently Director-General of the IAEA) to take on the job, a call came from the Swedish permanent delegation intimating that they would be well pleased if someone from their country were invited. Eklund was suggested. His authority was in itself a major contribution and his nationality gave balance to the committee's structure. From amongst the national delegates a committee of experts was chosen comprising, under him, Aten, H. de Laboulaye of the French CEA, S. Gallone of the Italian CNRN and Compton Rennie from Harwell. This committee was completed by Kowarski with Perret acting as secretary.

At its first exploratory meeting in October, Kowarski outlined the programme which he had presented at the Sorbonne in May, in which he put forward four reactor types as deserving attention – a fast breeder reactor (FBR), a BWR, a liquid metal fuelled reactor (LMFR) and, in a different category, a big materials testing reactor (MTR). Agreement was quickly reached on the principle that only non-proven systems should be considered, i.e. those which would be suitable for development over a five year period, and would then serve as models for the construction of prototypes or power stations. Gallone, whilst anxious that the systems studied should not be so advanced as to be beyond the scope of the less nuclear developed nations, was ready with the others to accept as a discussion formula Kowarski's idea for an international research centre. The centre would house laboratories and, say, four reactors – three experimental and one MTR – and would work in close collaboration with national centres which might undertake some research for the international laboratory. In return, the central laboratory might make available to the national laboratories a part of its facilities. This is roughly the pattern upon which CERN has so successfully operated over more than 20 years.

In the course of their discussions, the group of experts identified in addition to Kowarski's original list, the homogeneous aqueous reactor (HAR), which was being developed in the UK and in the Netherlands. This was a reactor in which the fuel and water moderator were intimately mixed, either as a solution or slurry, and on paper at least, it seemed a simple and attractive system. Rennie agreed to assemble data on its outstanding problems as well as those of the fast reactor, while Gallone did the same for the BWR.

Consideration was also given to the various styles of organisation that could be adopted for the joint centre, estimated to need an annual budget of $30M. In the Experts' report made to the full working group on experimental reactors on 18 January, 1957, two possibilities were indicated – either a self-contained autonomous organisation like CERN, or a collaboration that formed part of a wider variety of activities within OEEC. It was also recognised that as an alternative to a jointly owned centre where all the work was concentrated in one place, it was feasible to locate different projects on separate sites adjacent to existing national laboratories.

BRITAIN OFFERS A SITE

Before REX had had time to digest its Experts' proposals, the Steering Committee was given a preview of the main points and quickly showed a marked preference for the idea of decentralisation and the use of existing national facilities. Moreover, three countries indicated that they would be ready to offer a site for a joint project – France, Germany and the UK.

The private message that Bertrand Goldschmidt* had transmitted to Huet after a talk with Sir John Cockcroft, to the effect that Britain was ready to offer a reactor for joint development ("Votre affaire est gagnée, les anglais proposent un réacteur") had been well founded after all. The news, so crucial to the future of the Agency, had until then been greeted with scepticism, as active participation by the UK had not really been expected.

The next move was to ask Rennie to prepare for the REX Committee of Experts, a note on the conditions that would have to be met for an inter-

*Original member of the CEA following his return from Chalk River and from 1958, Director of External Relations and Programmes.

national project to be attached to a national centre, with firmly in mind the idea that national meant British.

During the early fifties, Rennie had had the delicate job of smoothing relations between the research and industrial arms of the UK Atomic Energy Authority at a time when the mutual antagonism between Cockcroft and Sir Christopher (now Lord) Hinton, the Members responsible*, was at its height. It already said much for Rennie's grasp of scientific and engineering principles as well as his personal tact that he was able to retain the respect of both these men. When the necessity for a permanent link-man diminished, he returned to reactor work in the Reactor Division at Harwell and was soon appointed overseas liaison officer to process the various bilateral agreements which the Authority was negotiating. Requests for co-operation were handled by the Overseas Relations Committee, chaired by the Authority's Secretary, David E. H. Peirson.

In Rennie's mind was still the idea of a joint undertaking consisting of a main centre with a staff of several hundred people grouped round one or more reactors, with out-stations located at other national centres. He sketched out a structure which comprised, at the top, a board of management (responsible also for the appointment of the director and senior staff) aided by subsidiary committees including a Scientific and Technical Committee and a Budget Committee which were substantially common to all the projects in so far as the distribution of the participants permitted. He proposed that the management of the international project should be distinct from the management of the national site to which it was attached, and that services provided by the national to the international centre, should be paid for. He recognised the need for harmonising the salary structure of the staff within the various projects and with those in the associated national centres. The question of safety regulations was also raised. These should be at least as stringent as those in force on the national site and it was implied that supervision by the host authority was the only practical solution.

For a project located in Britain, the question of diplomatic immunities could create difficulties, as Whitehall was strongly opposed to the invasion of foreign scientists armed with special privileges. At the Steering Committee, the British delegate had stressed that only those privileges which were essential to the well-being of the project should be granted, "not those it would be nice to have".

Rennie's paper was incorporated into the Expert's report which the REX study group was able to accept with only minor amendments, and the full

*Cockcroft was the Authority's Member for Research and, until 1958 Director of Harwell. Hinton was Member for Engineering and Production, and Managing Director of the Industrial Group which had factories dispersed in the N.W. of England and Scotland, and a headquarters at Risley near Warrington in Lancashire.

report of the Working Group, compiled very largely by Kowarski and Rennie was presented to the Steering Committee at its 3rd session on 21 March, 1957, where it was well received. For immediate construction, an MTR and BWR were put forward, while technological studies were proposed on an HAR, LMFR and FBR. The UK stated explicitly that it would be prepared to participate in the development and construction of one of the reactors and was able to announce that it was prepared to receive the project on a new research site that was to be opened up on Winfrith Heath on the south coast of England between Bournemouth and Weymouth. Other countries too reiterated their willingness to be hosts to a joint undertaking.

Finally the proposal of Sweden that a Study Group be set up to prepare proposals for the construction of an HAR, an MTR and an FBR was adopted in spite of an intervention from the Belgian delegate who pointed out that an MTR was superfluous in view of their reactor that was being built at Mol and which would be ready in two years' time. It was well understood that the HAR was destined for Winfrith and Huet and Kowarski lost no time in going to Harwell (the parent centre) to work out with the UK authorities the framework for the enterprise; this they hoped to complete by June. Meanwhile in Paris, Weinstein began the drafting of the convention, the first in a long series which attempted to reconcile the different demands of all the potential participants. The speed with which new documents were produced, modified, translated and re-issued, is a tribute to Huet's drive and to the energy and enthusiasm of his staff. Apart from reactors, plans were being pushed forward for the formation of the Agency, a joint reprocessing plant, a heavy water plant, co-ordination of national legislation on trade, security and insurance, and training of nuclear engineers. Huet and his team quickly established a reputation for efficiency unhampered by bureaucratic formalism. Their terms of reference were only broadly defined, they had no limiting treaty to analyse clause by clause and although Huet's own determination to establish and head the Agency was evident, he was a good listener. Ready to travel anywhere to discuss his project, he was tireless in his search for formulae that would take into account the most important demands of the potential member countries.

The statutes of the new Study Group were approved by the OEEC Council in May and following the proposal by Norway, its boiling heavy water reactor at Halden was added to the list of possible joint undertakings. Thirteen countries indicated their interest in at least one of the reactors cited: Austria, Belgium, Denmark, France, Germany, Italy, the Netherlands, Norway, Portugal, Sweden, Switzerland, Turkey and the UK. The USA indicated that it was favourable to the HAR and promised to make available all relevant information on similar developments funded by the AEC.

BACKLASH FROM THE EURATOM TREATY

Despite the strains apparent at the previous Steering Committee meeting, it seemed that the Agency and Euratom would be able to exist side by side without conflict. Returning to a meeting of REX from the Intergovernmental Conference for the Common Market and Euratom, for example, the French delegate on behalf of the Six had this to say: "Conscious of the absolute necessity for the European countries to multiply the methods of approach for achieving the numerous possible types of power reactor, the six countries of Euratom are prepared to make a positive contribution to the proposal of the OEEC for the joint building of a number of experimental reactors".

Of the five reactors under discussion, the Euratom countries had plans for an MTR in which other OEEC countries could participate; no definite ideas had so far been put forward for a BWR so there was an interest in co-operating on this system and, less urgently, also on an LMFR. The UK was considered to be the best equipped to take the lead in building an HAR. Moreover, it was understood that "for any OEEC project regarding experimental reactors, each of the six countries would be absolutely free to participate in its own name and with its own funds, whatever the participation in this project of Euratom as such".

If the way was clear for extensive collaboration in reactor development, there was no rush of countries anxious to take the lead. The Treaty of Rome, setting up the European Economic Community and the European Community of Atomic Energy, had been signed on 25 March, 1957, and in May the report of the 'Three Wise Men' was published **(13)**. Prepared by Armand with Franz Etzel of Germany and Francesco Giordani of Italy, in the shadow of the closure of the Suez Canal and the Middle East war, it set a target for the Euratom countries of 15 000MW installed nuclear power by 1967 – a highly ambitious programme which nevertheless at the time did not seem unattainable. It fell into a vacuum. The Euratom Treaty, following ratification by the six members was to come into force in January 1958. In the interval, the Brussels steering committee met only twice, concluding each time that nothing much could be done until the Commission had been formed. According to Jules Guéron (the architect of 'Annex 5' to the Euratom Treaty which set down an initial programme of research and training similar to that appearing in the REX report, and subsequently Euratom's Director General of Research and Development), 1957 was a year characterised by the governments of the Six, "planting flags", so leaving Euratom very little room for manoeuvre when finally it came into being. It is not the intention here to embark on an analysis of Euratom's evolution, but it must be noted that in 1957, the introspection evident during the wait for ratification of the Treaty of Rome, carried over into the negotiations on the joint enterprises of OEEC; progress was slow in most areas.

FIRST REACTOR PROJECT FOUNDERS

Mid-June 1957 was the date set down by the Steering Committee for completion of the constitution of the joint establishment where the HAR would be developed, and for a short time this remarkably early deadline did not seem impossible. While Rennie was preparing a detailed technical report on the HAR, Huet and Peirson were coming to an understanding on the essential features of a project located in Britain.

In the Secretary of the Authority, Huet found a man of his own stature, of outstanding intelligence and wide experience. During the war, he had been Lord Beaverbrook's private secretary, first in the Ministry of Aircraft Production and later in the Ministry of Supply, and when a Division of Atomic Energy was formed, he was one of the first to be appointed. Peirson was the ideal civil servant, never putting his own interests before those of his country, tireless in the search for optimum solutions, ready to invent novel procedures for novel situations and willing to take maximum responsibility on his own shoulders. Rennie had enormous respect for him – "he always applied his mind to the problem rather than going back to pre-established positions". His integrity was absolute and he was also aware that organisations were composed of people. In his private as well as his public life, he was motivated by a sense of duty towards his fellow men. Peirson was a convinced European and did all that he could while he remained with the Authority, to promote the Winfrith collaboration and make it a success.

For the joint reactor projects, it was envisaged that supreme direction would be in the hands of a Board of Management made up of one delegate per participating country. The Board would appoint an executive committee from amongst its members, the project director and administrative secretary. Development tasks would be carried out by, and in, the various national facilities and the cost recovered from the Project's budget. The design of any reactor would be done at the centre by a team consisting of the project director, seconded staff and staff recruited by the host country on behalf of OEEC if needed. Its construction would be undertaken by the host country for a determined fee, which not only assured competence but also simplified the problems of reactor safety. A surprising feature of the proposals was that the question of information handling was left open. It was intimated that a policy of open publication could be adopted rather than restriction to the contributing countries – a liberal approach that was to find little favour amongst the other members of the Study Group.

Whereas the Steering Committee in June was able to approve the technical document setting out a programme (dubbed Eugene) costing £10M, the Study Group had to report that it was not yet in a position to recommend a constitution. Again at the meeting at the end of November it had to be explained that there were divergencies of view on even the basic form of the establishment, as some countries were now in favour of an autonomous research centre rather than one that came within the framework of the OEEC Council. There was also disagreement over the powers to be attributed to the various management committees and groups, some countries favouring direct international management of the project through a committee whilst the UK, in particular, insisted that a director should be put in charge of the project and be solely answerable to the Board. There was a question too of ownership of installations and information.

It was evident that these problems would take a lot of time to resolve and Kowarski urged that whilst they were being sorted out, the technical programme be started with a limit on the budget commitment of 1% of the programme estimate. However, Cockcroft had sent a personal message stating that it would be premature to engage staff at this stage though he promised that Harwell experts would be made available for continued studies. With some reluctance the Steering Committee authorised that studies should go on, making the proviso that this was done on the basis of voluntary effort only.

Little progress was made in the ensuing weeks and Huet became convinced that the Study Group was not competent to discuss the statutes while the jurists were only making the problems more difficult than they really were. He proposed an unofficial meeting at a high level to settle the basic elements. The date slipped from November to December and into the New Year. At the December session of the Steering Committee, Halden took first place on the Agenda in the reactor discussion and Eugene was not even mentioned.

Meanwhile, although the Study Group was pressing ahead with the formation of a new group of experts there came the disturbing news that a joint study made in the USA by the Babcock & Wilcox Company and the Nuclear Power Group (a grouping of electricity companies) forecast power costs from a HAR some three times those to be expected from conventional plant **(14)**. Even if the essential technical problems were solved, it was believed that there would inevitably be heavy maintenance costs for the regular replacement of highly radioactive circuit components. This was no longer a side issue. The insouciance evident at the time of the first Geneva Conference was disappearing and nuclear power was being regarded in more analytical terms. Factors of that size could not be ignored and Saeland warned Huet that this report could have serious repercussions on Eugene, and would provide heavy ammunition for anyone trying to make difficulties.

Moreover, little encouragement could be gained from the information emerging from the main HAR experiment in the USA. When commissioning tests had been started in May 1956 on HRE-2 at Oak Ridge, contamination of the piping had been detected **(15)**. After extensive cleaning, the circuit was run with external heaters, at which point leaks were discovered that would have been disastrous had the circuit been radioactive. As a result, much of the piping had to be replaced and it was only in December 1957 that the reactor could finally run under its own power. Corrosion rates much in excess of those experienced with an inactive circuit were observed and the uranium would not stay in solution.

By this time, it was being strongly rumoured that the UK had become disenchanted with the system as a result of its own experiments and the news from America, and when at the Steering Committee meeting in February 1958, the British delegate proposed that there should be a high level technical meeting of those with knowledge and authority, it was believed that this would mark the end of the Eugene collaboration.

There were compensations. The Agency had come formally into existence on 1 February (for diplomatic reasons one month after Euratom) with Huet as Director and Saeland as Deputy Director; the Eurochemic Convention for a fuel reprocessing plant to be built at Mol was signed, progress was being made on the agreement for a joint undertaking at Halden, and in other areas of co-operation; the Agency could feel pleased with its work **(16)**. Nevertheless, it was generally assumed that the Winfrith project was lost, and that this would be officially announced at the meeting of top level experts called for 24, March, 1958.

CHAPTER 6

ORIGINS OF THE HIGH TEMPERATURE REACTOR

IN THE complex world of nuclear energy, many people must contribute if a given basic idea is to be brought to practical realisation. Reference has already been made to the large number of different reactor systems that can be conceived and it is almost certain that every one, even remotely feasible, was conjured up at some time in the middle forties over the coffee cups in the various nuclear research centres. There was no secret about the attributes that should be looked for. In addition to being self-sustaining as far as the basic chain reaction is concerned, a good power reactor should have the following characteristics: a high temperature outlet fluid, good conversion by the spare neutrons of fertile material into new fuel; long residence times of the fuel in the reactor to minimise handling costs (in the jargon high burn-up), compact and simple engineering; stable materials; no supply problems. As for the safety criteria, these can be summarised as demanding docility in normal operation and a slow and non-dangerous progression under fault conditions.

Once the initial period in the establishment of a nuclear power programme (when only uranium 235 is available as fuel) has been completed, basically two fuel cycles offer the prospect of high utilisation of raw materials. These are the uranium 238/plutonium cycle and the thorium/uranium 233 cycle. Elementary physics considerations suggest that if maximum conversion is the primary criterion, the most effective fuel is plutonium, burning and breeding in a fast reactor, whereas if a thermal reactor is preferred, the best fuel is uranium 233 mixed with thorium. For maximum efficiency of conversion of the thorium into uranium 233, two conditions should be fulfilled. The fuel and moderator in the reactor should be homogeneously mixed and the fission products produced in the fission process removed from the reactor as soon as possible after they are formed, as they act as poisons, wastefully absorbing neutrons. (Some of them are, in fact, also 'delayed' neutron producers and their presence gives the reactor a response time measured in seconds rather than in fractions of a microsecond which is very important for control).

No system has yet been devised which earns maximum marks for all the criteria that can be put forward; each system had its own strengths and weaknesses and the comparison between different systems is more subjective than scientific until a great deal of research and development has been done. Even then a large subjective element remains, opinions being strongly conditioned by background and experience and, it must be recognised, by career prospects. So, in the early days of nuclear energy, it was usual to find groups of scientists promoting a given system against the opposition of other groups pushing other systems with equal vigour and equally cogent arguments.

The first serious thinking about achieving high temperatures in a gas-cooled reactor can be attributed at Harwell to Stefan Bauer, who, in the years 1948/49, was considering as fuel, a bed of beryllium oxide balls coated with uranium and recoated with beryllium oxide, whereas on more conventional lines, Jack Diamond was evaluating for the Navy, a helium-cooled graphite-moderated reactor as a possible submarine propulsion unit. When Diamond left Harwell to become professor of mechanical engineering at Manchester University, the initiative passed to R. V. (Dick) Moore and B. L. Goodlet, who concentrated on natural uranium fuelling, carbon dioxide cooling and more modest outlet temperatures. Their cost analysis led to the decision to build the dual purpose, carbon dioxide-cooled, graphite-moderated, natural uranium-fuelled reactors at Calder Hall which led directly to the launching of the British nuclear power programme.

Interest in higher temperatures revived in 1955 as a result of the publication of a paper by B. Terry Price and John Kay on self-sustaining fuels, and in particular the merits of a homogeneous uranium/thorium system. The idea of continuous de-poisoning by removing the fission products from the reactor core was also pressed. This prepared the ground for the more specific proposals of an engineer of great virtuosity – Peter Fortescue. He was well-known in the Authority for the rate at which he generated new ideas and became totally committed to them (six months before, it had been an inverted pebble bed fast reactor) but he was nonetheless highly respected both at Harwell and Risley. At the end of 1955, he extolled the virtues of gas cooling and urged the Reactor Division at Harwell, under John V. Dunworth, to allocate effort to the design of a homogeneous reactor with ceramic fuel and moderator which would withstand temperatures approaching 1000°C. The cooling gas could then be made to drive a gas turbine and thermal efficiencies could be obtained better than with any system involving steam. Fortescue had already worked on a gas reactor concept, dubbed Felix, and made experiments on circulating uranium fluoride through a gas turbine.

The new proposal was attractive for a number of reasons. Following the decision of February of that year to launch the civil programme of Magnox reactors, it was necessary to consider what to do next. Even though Hinton was urging the engineering view-point of limiting the number of novel approaches and going forward in modest steps, at Harwell the spirit was one of adventure where systems were promoted as much for their scientific interest as for any commercial reason; for example, the Metallurgical Division under (Sir) Monty Finniston was promoting the liquid metal-fuelled reactor – the metallurgists' reactor, while the Chemistry Division under Robert Spence was promoting the homogeneous aqueous reactor – the chemists' system. The Reactor Division was needing a system to take the place of the fast reactor that was being taken over by Risley and Fortescue's reactor was a physicists' and engineers' reactor, not lacking interest for the other Divisions and sufficiently futuristic to stay at Harwell.

It had the merit also of being a natural follow-on to the Magnox reactors (though the relationship at that time was far less obvious); it held the promise of high temperatures and thus good thermal efficiencies; it would exploit the thorium cycle in contrast to the uranium/plutonium cycle of Magnox; by removing fission products soon after they were produced it would have exceptionally good fuel utilisation; and it could be built to a very compact design and so would be attractive for special applications such as ship propulsion or even spacecraft drive.

It was agreed then to follow tradition and create a special working party. A physicist, Leslie Shepherd, was made deputy to Fortescue, responsible for research and development and, incidentally, for keeping Fortescue's feet on the ground. Not that Shepherd was altogether conventional, as outside his nuclear interests, he had a passion for all matters connected with interplanetary travel. It was this last characteristic – not considered to be a mark of seriousness by some – which had led to his nomination as Harwell's representative on the Ministry of Supply's Working Party on Nuclear Propulsion of Aircraft. He had written a number of papers on hydrogen cooled high temperature reactors for rocket propulsion and consequently became identified with high temperatures although his main line was fast reactors. Also in this study group at the beginning were Price, J. Malcolm Hutcheon, J. Archibald Robertson concerned with fuel irradiation, and Jack Williams for fuel fabrication. They met for the first time as a group on 30 December, 1955. Soon the group was joined by George Lockett "the best intuitive designer at Harwell" (Shepherd), "in the World" (Fortescue). An engineer of great experience and thoroughness, easy to work with, unworried by considerations of status position, he became the engineering sheet anchor of the project. Other names to be found in those days were L. A. (Mike) Husain, Frank Sterry and Derek Wordsworth.

Work centred on the design of a so-called fuel testing reactor of 30 MW(th) which had a core (i.e. the central part of the reactor containing the fuel) about 4 m in diameter and 2½ m high enclosed in a steel pressure vessel. Coolant inlet temperature was 350°C and outlet temperature 850°C. Ideas changed regularly and many of the fuel designs now seem bizarre in view of the enormous present-day concern over integrity-plates of graphite soaked in a uranium salt solution, dried off and then canned in beryllium or zirconium, uranium drawn into wires and trapped between a centre rod and an impermeable sleeve. . . . Progress was, however, evident and additional motivation was provided by a report of David Griffiths, a visiting Australian scientist, on the economics of the later designs of reactors that might develop from Magnox. The progression was seen as passing through a series of improvements to the point where there were no longer any metallic components in the core of the reactor, the fertile material was thorium and power was probably generated by closed cycle gas turbines. The prospect was sufficiently promising for a research committee to be formed under Dunworth, comprising members of interested groups at Harwell and representatives from the Authority's Industrial and Weapons Groups.

One problem which was to embarrass the project throughout its life was the number of variants that could be devised. A ceramic core and gas cooling left the actual choice of fuel, moderator, fertile material and coolant still open, and each modification meant really a different reactor system. For the present, it was decided to concentrate on graphite as moderator and fuel matrix and to consider the system primarily as leading to the establishment of a uranium 233/thorium cycle although initially the fuel would have to be uranium highly enriched in uranium 235, there being no significant quantities of uranium 233 available. Fortescue led the group, Shepherd the scientific side, Lockett the engineering design and a tempestuous metallurgist called Roy Huddle, the fuel working group. Huddle, the inventor of Magnox, had become aware of Fortescue's activities through lending him some money for lunch in the canteen and then going back to his laboratory to collect it. Intrigued by the problems displayed before him and fresh from a blazing row with Finniston (his Division Head) he had himself attached to the project. Via Shepherd he also made contact with the Royal Aircraft Establishment at Farnborough, beginning thereby a long association on the development of special graphite processing techniques for use in moderators and fuels.

During 1956, Lockett completed his preliminary design proposals for a modest 10MW(th) helium-cooled reactor experiment, in which the core comprised an array of fuel elements, each consisting of seven rods sitting on hollow spikes. Each rod was hexagonal in outer section and pierced by a circular hole into which fuel capsules were inserted. Through the small annular space that was left, a stream of gas could be led carrying the fission products away through the hollow support. The cooling gas flowed round the outside of the rods and out to heat exchangers. Encouraging progress had been made in the fabrication of fuels and a record breaking burn-up had been achieved with cylindrical samples irradiated in one of the Harwell research reactors. Meanwhile, Shepherd had become convinced that a preliminary high temperature experiment would be needed with a mini reactor generating almost no heat itself but raised to the operating temperature of the main experimental reactor, in order to evaluate the physics data. He had therefore gone ahead with plans for such a reactor (named Zenith) and these were in an advanced state.

Late in November at a symposium arranged to inform the industry of recent developments, Fortescue's ideas were introduced. They evoked considerable interest because of the possibility they offered of attaining high thermal efficiencies and a low unit capital cost. In effect, the system received recognition and the construction of Zenith was authorized in the New Year.

So similar was the original reactor design to what was eventually built and operated so successfully, it is perhaps necessary to point out that the range of unknowns was still very large, and in terms of a power reactor the system as it stood was not at all practicable. It was perfectly reasonable to design a small experimental reactor in which some of the fission products could be expected to escape into the main coolant stream but, as Dunworth regularly emphasised, this could not be permitted in a large power reactor. Quite apart from the seal problem at the bottom of the spikes, the fuel had to be encased in an impermeable graphite sleeve to isolate the purge from the main coolant stream. No one yet knew how impermeable graphite could be made nor whether a material could be devised which would stand up to irradiation at high temperature. The more one learned about graphite the more unpredictable a material it seemed to be: sometimes shrinking, sometimes growing, hardly a structural material. Similarly, it was not too difficult to design a steel pressure vessel for a small reactor experiment working at high temperature; scaled up to a commercial size the problem was altogether different. Again helium was in short supply so could hardly be considered for a power programme. The system of the Harwell study group was still very much at the conceptual stage.

HTR BECOMES AN AUTHORITY PROJECT

To take the project (very briefly called HUGO) to the next stage, the Authority again followed its usual practice and formed a special Research Committee to study what became known as the High Temperature Gas-Cooled Reactor – the HTGCR, or more shortly the HTR. With Cockcroft in the chair, the Committee brought together the top people from Harwell and Risley, Sir Leonard Owen, Director of Engineering and deputy to

Hinton leading the northern team, which examined the proposals from the two standpoints of immediate constructional difficulties and ultimate application.

At once the use of helium was questioned, as the gas at that time was only available from the USA and was being classified as a strategic material. It was argued, however, that for the reactor experiment, which was essentially designed to develop the fuel and study fission product migration, it was necessary to choose an inert gas. For a power reactor series, another coolant would have to be considered and for a time provision was made in the experiment for the substitution of nitrogen. Owen was not too happy either, about the use of coolant circulators that ran on gas bearings as against the more usual oil lubricated roller or journal bearings, but was persuaded that development on these was sufficiently advanced to justify going on. The main worry, however, was the control of the escaping fission products even in the experiment, and reliance on graphite as a barrier and a strucural material. Even so, there was general agreement that the system presented a fascinating challenge.

Cockcroft quietly kept the Committee moving. He had a strong interest in the project, paying it far more attention than was usual in one so distant from his staff and he was determined to see it survive the cutbacks that were threatened in the number of reactors studied as the British Government began to count the cost of the broad programme of research that had been mounted. He had been successful in putting through the new site at Winfrith as an out-station of Harwell and it had been agreed that if the fast reactor went north, the HTR would stay south. Zenith and the HTR experiment were to be built at Winfrith as one group in a set of new reactors that Harwell would control. Money was tight and the HAR, in which Cockcroft was beginning to have less faith (combining as Dunworth expressed it, all the problems of a nuclear reactor with those of a highly corrosive chemical plant) was temporarily saved by offering it to Europe.

Actual construction and so the detailed design of the HTR experiment would normally be in the hands of Risley and at the beginning of May 1957, Cockcroft stated that as soon as financial approval had been obtained in the Summer, design, construction and civil engineering would become the responsibility of the Industrial Group.

Some of the team at Harwell had hoped that this eventuality could be avoided and, when Fortescue left soon after to go to the USA it was popularly believed that resentment was the cause. This was not so. In May 1956, Cockcroft had introduced him to Freddy de Hoffman at the meeting of the European Atomic Energy Society at Monte Faito near Naples, when Cockcroft announced the operation of Calder Hall and some discussion took place on advanced gas graphite concepts. De Hoffman had been appointed by John J. Hopkins of General Dynamics as a one-man think-tank to determine what system the company's nuclear division, General Atomic (GA), should specialise in for power production. He had toured Europe making contact with the various atomic energy commissions and on his return he organised in San Diego in the Summer of 1956, a three-week symposium to which he invited a wide gathering of European and American scientists. Fortescue was detailed from Harwell and was one of the few who followed the symposium from beginning to end. De Hoffman saw in the HTR a reactor that fulfilled his essential requirements. It was ripe for development, it would exploit the thorium fuel cycle, largely ignored to date, and it had no industrial protagonists as yet in the USA. There was the background experience of gas cooling and graphite moderation from the US production reactors; the AEC's Oak Ridge Laboratory was continuing with graphite research; the long-term objective of generating power through closed cycle gas turbines was directly in line with the main company's business. Having taken his decision, de Hoffman began to recruit staff and he invited Fortescue to spend a sabbatical year in San Diego while the strength built up. Fortescue was delighted with the opportunity and with Cockcroft's blessing went off to the USA on 5 May, 1957 for one year – a year that turned into two and then became indefinite. Twenty years later, he was still entirely English in manner and speech, still teeming with ideas, still enthusiastic, and still with GA.

Transformed into a Design Committee under the Chairmanship of Owen, recently appointed Managing Director of the Industrial Group following Hinton's resignation on 1 August, 1957 to become Chairman of the newly constituted Central Electricity Generating Board, Risley and Harwell sat down together to define the main features of the HTR experiment. Risley was to assemble a design team under one of its senior engineers, Harry Cartwright, and he would have responsibility for the main engineering while Harwell continued with research and some detailed design work.

A 37-month programme was foreseen starting from August 1957 with construction beginning in January 1958, completion in September 1960 and first criticality* in March 1961. No time could be wasted and it was planned to seek Treasury approval in mid-November. Conscious also by then of the problems of introducing industry to a reactor system, two companies which had participated in development work with the Harwell team and were active in the gas turbine field – Ruston and Hornsby, and Rolls-Royce – were invited to detach staff to the project. Early though this was in the evolution of the HTR system, the Authority was expecting an eventual application. Other countries might be developing systems for training purposes or because of their inherent scientific interest but at Risley at least, the justification was exploitation in the near or longer term.

---

*Criticality is when a self-sustaining chain reaction is produced for the first time.

No conflict was seen with the parallel work that was going on to decrease the capital cost of the next round of power reactors by developing an Advanced Gas-cooled Reactor (AGR), fuelled by enriched uranium and more compact than the Magnox reactors, as the time-scales were so different. The AGR was seen as a single step beyond Magnox and the Authority was ready in September 1957 to begin the planning of the demonstration plant to be built at Windscale. The HTR was regarded as the step beyond.

DISRUPTION

However, on 10 October, 1957, before Risley had had time to set up its design team, one of the production reactors at Windscale caught fire and the atomic energy world was facing its first major civilian incident **(17)**. In the event, the damage to people was undetectable and the impact on property outside the reactor area of short duration only. Within the Authority on the other hand, the repercussions were far-reaching. It was recognised that a complete re-organisation of safety procedures would be necessary and much effort would have to be devoted, first to the enquiry and then to clearing up the mess. One of the first victims was Harwell's HTR and at only the 3rd meeting of the Design Committee, held on 7 November (the only Authority meeting on the HTR which Cockcroft did not personally attend), Risley announced that no design effort was available and the project would have to be put off for several months. The working parties studying fuel elements, coolant processing, physics and a new one on safety, were continued and members were told that the Design Committee would be reconstituted as a Research Committee.

Cockcroft was not prepared to let the programme drift and Shepherd was instructed to prepare a paper on the HTR for the next meeting of the European Atomic Energy Society planned for Rome on 27-29 November, when the subject would be gas-cooled reactors, including a session at the end on "Other Types of Inactive Gas Reactors'. (At the same meeting, Huddle was reading a paper on materials problems in advanced gas-cooled reactor designs.) None of the information assembled on the HTR had been declassified, but Shepherd was instructed to go ahead and assume that it would be released in time. Already in Cockcroft's mind it seems was the thought that the only way to get the experiment built quickly was to bring in external funds and spread the research load. Cockcroft had also a strongly positive attitude towards collaborating with the Continent in a research programme that had genuine merit, and by this time he no longer had any faith left in the HAR.

Rennie was put in the picture. He was still working hard on trying to get something started at the international level and was turning his attention more and more to Halden which he was hoping to persuade the Authority to join, notwithstanding the bilateral agreement between the UK and Norway which suggested that any further commitment would only cost money and would bring nothing extra in return.

At Rome, then, the HTR was unobtrusively presented to the senior nuclear scientists of Europe and the point registered that the UK had already put in a great deal of effort on this system and foresaw for it a promising future.

CHAPTER 7

BIRTH OF DRAGON

FEW OF the top level experts from Europe and North America who assembled in Paris on 24 March, 1958 under the presidency of Francis Perrin were prepared for the proposition that was to be made by the British. Cockcroft requested a change in the agenda in order to discuss first the technical aspects of reactors which could be the object of an international co-operation within OEEC. Then with characteristic flair for the dramatic, he went on to announce that the UK was no longer prepared to sponsor the construction of a homogeneous aqueous reactor in Britain; if this project was to go ahead another host country would have to be found. In its stead, the Authority would like to propose that a high temperature, gas-cooled reactor project be established at Winfrith.

Certain administrative conditions were laid down. The project would be under the control of a chief executive who, though appointed by an international board of management, would be a national of the host country. This was considered necessary in view of the legal and safety responsibilities that the host country would have to assume. It was also indicated that the construction of the reactor would be undertaken by a contractor of the host country and it was assumed by the UK delegates that Risley would be the chosen organisation. International supervision would be ensured by the type of management structure instituted at CERN, where supreme control of policy and budgets was in the hands of an international Council, assisted by a Scientific Policy Committee to advise on technical matters and a Finance Committee which would keep watch on the financial and contractual aspects. It was expected that research contracts relevant to fuel development, for example, would be placed in laboratories of the participating countries.

Not all the countries represented were happy with this sudden change in policy on the part of the British, as for many months their national programmes had been developing on the assumption that there would be an international HAR. The Netherlands had a particular interest in the system and a small experiment was being built at Arnhem. Sweden, also, was anxious not to abandon it and urged that the two projects be studied in parallel. So great, however, was the reputation of the British delegation – Sir John

Cockcroft, Sir William Cook* and Sir Leonard Owen – and so respected the British effort for its thoroughness and its down-to-earth policies, the majority of countries were prepared to go along with the new project on these grounds alone.

In his presentation, Cockcroft made the point that once the experiment had been constructed and operated, each country would be free to construct a reactor of similar type on its own. There was no suggestion that there might be an international follow up. Within the Authority it had already been decided that the project should not be open-ended but should be of limited duration and confined to the experimental phase. Basic knowledge only was to be exchanged.

Before the top level experts moved on to other matters, it was agreed that the UK would circulate a description of the proposed reactor project covering both its technical and administrative aspects and the Secretariat would form a panel of up to seven experts who would examine the proposal and prepare a technical report for submission to a second top level meeting **(18)**. At the same time, the Secretariat would convene a meeting of administrative and legal experts drawn from all countries interested, plus Euratom, to consider the administrative structure.

The technical panel met on 2 May in Paris and subjected the British delegation of Rennie and Shepherd to a searching questioning on the technical aspects of the proposal, Shepherd doing most of the answering and demonstrating his wide grasp of the problems and his realistic approach to the difficulties. It was also an opportunity to make a preliminary exploration of the research facilities available in other countries, particularly for irradiating components. The panel agreed that:

1. the proposal presented by the UK appeared promising from the technical point of view and could be reasonably recommended as a subject for European co-operation;

2. other countries possessed scientific and technical resources enabling them to contribute to a joint programme in the field of high temperature reactors.

Co-operation, it was proposed, should not be limited to one reactor experiment but should include research, as outlined by the UK, leading to a wider knowledge of HTRs. Many fundamental questions relating to the fuel remained to be answered and different avenues would have to be explored.

With little modification apart from strengthening the section dealing with the technical contributions from participating countries to make it *essential* that countries with relevant resources should make every effort to co-

*Cook succeeded Hinton as Member for Engineering and Production.

operate fully in the project by making qualified people available, these recommendations were passed to the second meeting of top level experts on 19 May. They in turn passed them to the Steering Committee the following day which had no difficulty in endorsing them. It may be noted that at this time also the UK made known its willingness to participate in Halden.

Huet was hopeful that it would be possible to reach agreement on the project at the Steering Committee's meeting in June, with a positive decision in September so that work could begin in December.

ORIGINS OF "DRAGON"

By the middle of April, in internal correspondence in the Agency, the project was being referred to as Dragon, a term which owed its origins indirectly to Otto Frisch. In the early days of atomic energy at Los Alamos, he devised one particularly awesome assembly that was called the Dragon Experiment, following Richard Feynman's comment that it was like tickling the tail of a sleeping Dragon **(19)**. Remembering this, Kowarski, who had a strong belief in the need to christen projects at an early date to give them a clear identity, dubbed the proposed reactor "l'Eudragon". Pronounced the same as le Dragon, the shortened form was immediately adopted and the title seemed eminently suited to a mysterious creature breathing fire, that appeared under the banner of St. George. Over the months that followed, frequent repetition of the pun that the negotiations still "drag on", resulted in the myth that this was indeed the origin of the name.

CHAPTER 8

COMPETITION

IN SEPTEMBER 1958 at the Second International Conference on the Peaceful Uses of Atomic Energy, organised by the United Nations in Geneva, the nuclear countries were able to compare national programmes and the progress that had been made since 1955. Atomic energy was no longer hailed as the panacea for the developing countries and the headline news was the release by the Soviet Union, the USA and others, of the results of their research into thermonuclear fusion, which would take over from fission when the uranium ran out. At the same time an enormous amount of information was presented on reactor developments, the underlying message of which was that a concentrated industrial effort was needed if nuclear power was to be competitive with conventional power sources. The scale of activity in the USA was impressive, and the importance of commercial weight was evident in the figures published by General Electric for the 180MW(e) Dresden station, which indicated a capital cost of less than $170/kW installed, and a generating cost of 0.75cts/kWh – comparable with that obtained from the most modern coal-fired stations.

Three papers only were given on reactors with ceramic cores; one on "The Possibilities of Achieving High Temperatures in a Gas-Cooled Reactor", by Shepherd, Huddle, Husain, Lockett, Sterry and Wordsworth and one by R. Schulten of the Nuclear Energy Group of Brown Boveri & Cie/Fried. Krupp on a 15MW High Temperature Pebble-Bed Reactor. The third was on the High Temperature Zero Energy Reactor, Zenith. (No mention was made of any OEEC negotiations.) US papers on graphite-moderated gas-cooled reactors related solely to systems with metal-clad fuel elements, notably the EGCR of Oak Ridge although in the commercial exhibition, General Atomic displayed the design of a 22MW(e) helium-cooled reactor fuelled by an intimate mixture of uranium and thorium in graphite, supported in graphite sleeves that acted also as structural elements. Fortescue's influence was evident.

THE PEBBLE-BED REACTOR

Stadtwerke Düsseldorf, a medium-sized electricity company in N.W. Germany, worried by an expected shortage of coal and by its rising cost and surrounded by bigger companies with lignite reserves, notably the Rheinish-Westfählisches Elektricitätswerk, in 1956 had brought together a group of 10 utilities to gain experience in the new field of nuclear power generation which Calder Hall had shown to be near commercial viability. They began discussions with (German) Babcock and Wilcox, a company which had direct connections with the British Babcock and Wilcox that was in partnership with English Electric and Taylor Woodrow in one of the British consortia. The outcome was an offer for a 40MW(e) natural uranium gas-cooled graphite system.

Associated with the nuclear reactor was the turbo-generation plant which the group proposed to acquire from Brown Boveri in Mannheim. When they made contact with this company, they were introduced to a new system that was being developed in association with Fried. Krupp. Innovator of this system, that came to be known as the pebble-bed reactor, was Rudolf Schulten, a scientist who had worked under Heisenberg at the Max Planck Institute at Göttingen and had been the first member of the reactor development group of the MPI to be engaged on reactor technology. Some little time before Calder Hall became operational, he had visited Harwell and had studied there the experimental natural uranium reactor BEPO. Convinced that much higher temperatures would be needed for efficient power generation and that to achieve practical reactor dimensions, enriched uranium would be necessary as fuel, he had returned to devise a system based on ceramics. He had conceived the idea of binding the fuel and moderator together in spherical pebbles, loading them into a thick bed, which would then be cooled by an inert gas. The only noble gas that seemed to be available was neon and only later was it concluded that helium could be bought in sufficient quantitites or distilled from the air if necessary.

Brown Boveri was sufficiently attracted by his proposals to begin development of the system with Krupp and together they were able to persuade W. Cautius, the head of the electricity group, to transfer his attentions to the system. In contrast to the national undertakings in Britain and France, German companies had no inhibitions about adopting a system that required highly enriched uranium, which was available only from the USA. Following the US offer to supply fuel to installations open to inspection, there was no fear that there would be any shortage, and the pebble-bed reactor appeared to be a direct route to nuclear power with steam conditions similar to those found in modern fossil fuelled plant. Reactors such as the Magnox and the American light water reactors operated at significantly lower temperatures and their thermal efficiency was only about 30% as against 40% for the latest coal-fired stations.

In April 1957, BBC/Krupp was awarded the contract for a design study of a 15MW(e) pebble-bed plant which was completed in April of the following year. This was considered to be satisfactory and a site for construction was sought. In the meantime the Government of Nord-Rhein Westphalia, with some support from industry, had decided on the establishment of a nuclear research centre at Jülich some 80km west of Düsseldorf, and had ordered a heavy water materials testing reactor and a water-cooled research reactor from England. Responsibility for installation was confided to the Stadtwerke Düsseldorf which the company was happy to accept as it provided a training ground for the more serious project they would later be undertaking. Jülich was a natural choice then for the site for the pebble-bed and a contract for its construction alongside the research centre was finally placed in August 1959. A special joint company with the abbreviated title of AVR was constituted to manage the project and operate the reactor **(20)**.

The cost of the plant was estimated at around DM40M and the Federal Government was approached for assistance in its funding. This was probably the first time that the relevant Ministry was made aware of the nature of the project and it was still some time before it was related to any British work or Euratom's activities. Communication within Germany, because of the independent status of the Länder and the industrial organisations, was restricted. Federal research funds were concentrated in the Karlsruhe centre in the south, and delegates to the OEEC Committees came either from Bonn or from Karlsruhe. While Bonn took little interest in what was going on at Jülich, the teams in the north proceeded independently and took little notice of any negotiations going on in Brussels or Paris. Ch. Marnet, then a physicist working for the Düsseldorf company, recalls that the first time he learned there was a British high temperature gas-cooled project was when he visited Harwell in mid-1958 to discuss the MTR, and saw a drawing of the reactor hanging on a wall. They were only really conscious at Jülich of the Dragon project after the Agreement had been signed and they were asked to nominate staff to work on it. Some mention of the pebble-bed concept was made at the EAES meeting in Rome in November 1957 but Schulten met Shepherd and Lockett for the first time only at the 1958 Geneva Conference.

HTR PROJECTS IN THE USA

Several projects were known to be under study in the USA from information gleaned before and at the Geneva Conference. More was divulged at a

Conference on Gas-Cooled Reactors held at the Oak Ridge National Laboratory (ORNL) on 21 and 22 October **(21)**. There, further details were given of ORNL's own EGCR, General Atomic's high temperature maritime helium-cooled reactor with metal-clad fuel elements, and Sanderson and Porter's pebble-bed design for which the Battelle Institute was investigating different methods of manufacturing fuel balls. It was also known from Geneva that General Atomic was completing a proposal for a civil high temperature reactor and a decision was imminent on whether finally metal canning would be entirely dispensed with.

On 24 November, 1958 came the announcement from the USAEC that in response to its invitation issued in September, it had received a joint proposal from the Philadelphia Electric Company and High Temperature Reactor Development Associates for the design, construction and operation of a prototype helium-cooled, graphite-moderated nuclear power plant generating 30-40MW of electricity. Prime contractors would be the Bechtel Corporation, with General Atomic sub-contractor for the nuclear portion of the plant.

CHAPTER 9

THE DRAGON AGREEMENT

EVEN though few people in the UK on the technical side were aware of the Paris discussions in March 1958, the offer that was made was not the lone effort that has, at times, been attributed to Cockcroft. Owen was a frequent speaker at the meeting and only a few days afterwards, the Agency received a document prepared by Peirson setting out the Authority's views on the administrative structure for consideration by the administrative and legal experts. In addition to reiterating the general points already made, it recommended that the legal personality of the collaboration in the UK should be the Authority. It drew attention to the fact that forming a special company similar to Eurochemic would take a great deal of time and administrative effort and the granting of a special status and immunities in the UK might involve a delay of two years. On the other hand, if the Project were under the auspices of the Authority it would automatically enjoy the same privileges, e.g. in the matter of taxation. To accommodate the international aspect, a contract could be drawn up between the Authority and an international company registered in another country. Having spent nearly a year making little progress on the statutes of a centre to house the HAR, the Authority was anxious to avoid another long debate which could delay the Project indefinitely.

As to staff, it was proposed that all personnel should be on secondment from parent organisations and paid by these organisations, the salary appropriate to their position in the Project being recouped from the Project funds. Humanitarian reasons were behind this idea of Peirson (that had figured also in the draft of the HAR agreement) as well as considerations of speed in getting started. He reasoned that as the Project was of limited duration, technical staff should not be obliged to leave the organisations in which they had career and pension interests. Their return to their former posts would be much easier if they were simply seconded.

Although under the plan put forward, the Authority would be the legal owner of the physical assets of the Project, it was foreseen that on termination, the proceeds of the realisation of the assets would be shared with the other participants. All participants would have full access to the reactor during the life of the Project and to all information arising. Patents would be owned jointly with a right to exploit on a royalty free basis.

CONTINENTAL HESITATIONS

Reasonable though the proposals seemed to the UK, on the Continent they had the appearance of conditions which related, not to an international enterprise, but to a British operation to which other countries would be allowed to contribute. Following discussions with Perrin and Goldschmidt, Kowarski analysed the problem in the following way. The UK wished to see a full internationalisation of the Project, so Europe was informed, yet imposed three 'brutal' conditions, notably: the Director was to be British, there was to be complete British control over the realisation of the reactor and the administration was also to be British.

To be clear on the second point, Peirson had indicated that the design team would be international, but the design would have to conform to national legislation and site criteria as regards safety, and the proposal that Risley should be responsible for construction was put forward as a convenience.

Kowarski saw the problem as one of presentation rather than of substance. To begin with, whilst the Authority could suggest the name of a director he would have to be appointed by the Project's international Board of Management, and if the Board put forward a name, the Authority would have right of veto. Similarly, as the organisation assuming the liability of the reactor, the Authority would have the right of veto over design or operating aspects which it deemed inadmissible. As to administration, this should be seen as a service the Authority was prepared to give in the interests of economy.

If the British could be accused of a certain arrogance, it had to be admitted that their corporate experience in reactor technology was so much greater than that of the Continent at this time, it was natural that they should consider that the success of the Project would be best assured by having it under their control. At the same time, it was equally natural that there were some suspicions of the UK's bona fides, following the distant attitude that had been adopted towards the Communities and Eurochemic and the initial hesitations over Halden.

Within the newly constituted Commission of Euratom, sharp differences were arising on the attitude to be adopted towards OEEC. Armand had been appointed President but illness prevented him from exercising his functions and responsibility devolved largely on Enrico Medi, the Vice-President and Commissioner in charge of research. Medi was a geophysicist, described often as a mystic, with almost no experience of reactors, yet anxious to exercise a personal authority in the field. He was opposed in principle to financing joint projects outside Euratom territory. Others suggested that Euratom should only agree to participate if the UK, in its turn agreed to participate in a continental project. Guéron, on the other hand, by then Director General of Research and Development, saw in Dragon the possibility of a valid reactor project for the Community. At the same time, he was strongly opposed to any independent action being taken individually by member countries, even to the point of insisting that their experts should communicate with the Agency only through Euratom. In the Netherlands (and Sweden), there were still *arrière-pensées* over the HAR.

Huet worked hard to allay the doubts and, in collaboration with Peirson, to evolve an acceptable structure. Regular meetings were held at London airport with Huet flying from Paris, Peirson coming out from London and Rennie and Denis Willson (Technical Secretary at Harwell) driving up from Harwell. Through this arrangement, a meeting could be held in the morning and they could all be back at their desks in the afternoon.

Similar though the problems were to those of the preceding year, when the statutes for an HAR collaboration were under discussion, precedents had been established as other joint undertakings had not stood still **(22)**. The agreement on Eurochemic had been signed, that on Halden was almost ready for signature, and would cover a joint experimental programme with ownership of the reactor retained by the Norwegian Institutt for Atomenergi and its operation carried out under the responsibility of the Institutt. An international committee – the Halden Committee – would be responsible for approving the joint programme and the annual budget. Within this committee, the principle had been established that decisions would be taken on the basis of a ⅔ majority of the votes cast, the Institutt and Euratom each having three votes, the UK two, and the other members of the collaboration one each, reflecting thereby the financial participation of Norway and Euratom, 27.3% each, the UK 18%, Sweden and Switzerland 9.5% each, Austria and Denmark 4.1% each*. Assisting the Halden Committee was the Halden Technical Group composed of senior technical specialists designated by the Signatories, whose task was to formulate the joint programme and supervise its implementation, as well as approve contracts above a value of 50 000 EPU U/A and the conditions of service of those working on the joint programme.

At Halden therefore, two of the brutal points had been accepted, including the general principle of running an international project under OEEC auspices without creating a special company. The first draft agreements, prepared by Peirson and modified successively following his discussions with Huet, consequently abandoned the idea of establishing a company on the Continent and followed the lines laid down for Halden. Important distinctions were also made between the Authority as effective legal personality, taking responsibility for safety and providing services,

*The first Halden agreement covered an expenditure of 3.66 MEPU U/A over a period of three years (roughly £1M).

and the Project's Chief Executive. It was nevertheless proposed that whilst he and other senior staff would be appointed by the Project's Board of Management, he would be "acting on behalf of the Authority".

The Board would have responsibilities similar to those of the Halden Committee and it would be aided by a General Purposes Committee consisting of technical specialists with duties similar to those of the Halden Technical Group. Ownership of the reactor would rest with the Authority and operation would be carried out under its responsibility. Provision was also made that, contrary to the original proposals, the reactor would be insured, premiums being paid out of the operating budget and the Authority would then assume sole responsibility in respect of any claims arising out of such operation. No mention was made of responsibility for construction, or administration; ancillary services it was suggested 'may' be performed on behalf of the Project by the Authority and the costs recovered from Project funds.

The draft agreement that was finally submitted to the Group of Experts on 16 June went one stage further. While including the clause that "all legal acts relating to the execution of the joint programme shall be performed on behalf of the Signatories by the Authority which will be the owner of any experimental reactor built in the UK, and the operation of any such reactor shall be carried out under the responsibility and control . . . of the Authority . . .", it stated explicitly in a subsequent clause that "the Chief Executive shall be responsible to the Board of Management". In this last phrase lay the essential safeguard that the Project would be truly under international direction.

With but a few minor drafting rearrangements and modifications including the substitution of the word 'supervision' for 'control', and 'determine' for 'approve' in the clause concerning the duties of the Board of Management regarding the joint programme and budgets, the essentials of the agreement were to remain unchanged. Only the exact terms of the patents provisions had still to be settled as they had to conform also with Euratom's internal rules. Over the ensuing months, a number of versions were tried but the substance of the clause that put all the Signatories on an equal footing was not a matter of dispute.

Germany was anxious that means be found of permitting private institutions to become members of the collaboration, a provision that both the UK and Euratom regarded as quite unnecessary, on the grounds that staff could be seconded not only by the Signatories themselves but also by bodies designated by them. If the fundamental difference between being a member and seconding staff was appreciated, it was no doubt considered even undesirable that there should be industrial influences at the level of the Signatories. The Project, it should be remembered, was to be a short-term research project set up between governments or their agencies. Germany did not press the point.

FINANCIAL PROVISIONS

Still to be decided was the financial basis of the agreement that was to cover the 5-year period, hopefully beginning on 1 January, 1959. During all the discussions on the HAR, the sum of £10M had been quoted as the project cost and was again used by Cockcroft at the meetings of top level experts when introducing the HTR. Rennie and Shepherd at the technical interrogation by the restricted Group of Experts, even went so far as to say that it would be difficult to spend more than this sum within the framework of the proposed collaboration. Meanwhile Risley was preparing a detailed costing of the reactor and installations, and Shepherd was asked to examine the associated research programme. The preliminary conclusions showed that taking into account recent design modifications such as complete containment of the primary system in a pressure-tight shell, the overall cost would, in fact, lie between £12M and £14M. This figure, announced to the June meeting of the Steering Committee as an order of magnitude cost, came as an uncomfortable surprise to the delegates who requested that a more definite break-down be prepared for their next meeting scheduled for 28 July, 1958.

In a preliminary note circulated to the delegates, the UK reiterated the difficulty of preparing firm estimates even for a national project, let alone one that would subsequently be subject to the decisions of an international management board. It also raised the question of the duration of the agreement, pointing out that the reactor might not be completed until some time in 1964 and it seemed more logical to maintain the collaborative agreement, therefore, for a period of seven or eight years, or at least anticipate a renewal for two or three years beyond the 5-year limit. The question of duration raised fundamental issues and it was not surprising that the meeting planned for July had to be postponed.

It was not until 13 October that firm estimates were presented to a meeting of experts. The grand total came to £13.6M and was broken down as follows:

Construction costs including the reactor in its building, the first fuel charge and graphite, administrative and service building plus an allowance for contingencies	£5.1M
Other associated facilities including facilities for the production of fuel elements, reprocessing and post-irradiation examination	£2.0M
Research and development costs including test experiments both within and outside the UK and the operation of the zero-energy experiment, Zenith (but none of its capital cost nor the capital cost of a big loop to go into one of the UK MTRs)	£6.5M

The UK emphasised that this was a best estimation, in the full sense of the term and steadfastly refused to change the figure under pressure from the other interested countries. These consisted, it would seem by then, of Euratom acting for the Six, Austria, Denmark, Norway, Sweden and Switzerland.

Increases could be explained partially by the inclusion of the costs of fuel element production and facilities for their examination when removed from the reactor (which had been omitted from the original figure) but in effect it was only in the Summer that the serious costing had been done, after which, the full scope of what had to be included within the international project (as against being available somewhere else within the Authority) was worked out.

PROPORTIONAL CONTRIBUTIONS

In March already, Huet had foreseen the problem of apportioning financial responsibility in view of the dominant role the UK would play in the realisation of the Project and was thinking in terms of 40% UK, 45% the Euratom countries and 15% the remainder. Perrin suggested that the UK should contribute 50% but Huet felt this to be a little excessive.

For the Commission of Euratom, the problem was to decide to what extent the Project came within its existing programme. Its budget of 215M EUA (roughly $215M) for the period 1958-62, which had been approved the previous year, was designed to fund a wide range of initiatives within the Community. While the general programme had still to be translated into specific action, Dragon was, in a sense, supplementary. The Commission's first reaction therefore was to consider going back to its member countries for additional funds to cover one third of the Project's costs on the assumption that the three main groups, namely Euratom, the UK and other OEEC countries would pay the same. By the end of July, however, the Commission had concluded that it would need to allocate funds from within its existing budget and 20% would be a fair contribution. This would leave a very high percentage for the UK, as on the basis of net national revenue, the five other countries likely to be involved would pay 13.2%.

Huet reasoned that as the voting power in the Board of Management and General Purposes Committee would reflect the scale of contributions, the 86.8% remaining should be shared equally by Euratom and the UK. This would leave no single bloc with a majority, yet would give the two main contributors an effective veto. It was assumed that a ⅔ majority in voting would be required as in Halden, although the Authority was quite prepared to accept a simple majority, treating all the Signatories equally.

At this stage, it was not clear whether the Euratom countries might be contributing individually as well as collectively, since representatives of both the Commission and the individual countries appeared at OEEC meetings. Euratom's position was equivocal. It was anxious to control completely the nuclear activities of its member countries other than those considered purely internal; it was being urged by its own Scientific and Technical Committee, notably in the persons of Eduardo Amaldi and Perrin, to support the Project; it wished to avoid any research centre being set up under the auspices of the OEEC; it required complete equivalence between the participating members of any international project, yet the Commission did not feel able to match this by its contributions. At the October meeting, Belgium, France, the Netherlands and Italy all made it clear that, as they had intimated within Euratom in May, they would not contemplate any contribution outside their participation through Euratom; Germany reserved its position.

As for the five independent countries,they favoured the principle of basing contributions on national income. The UK recognising its special position as host country, was prepared to accept the Agency's plan as a basis for discussion.

Little progress was made in the ensuing weeks and the UK circulated a firm note to the governments of the Euratom countries, calling on them to reconsider their position, otherwise the UK might reluctantly feel compelled to proceed on an exclusively national basis. There is some evidence that this was seen, particularly in Holland, as political pressure to which the Dutch were rather sensitive.

Euratom, was by then finding itself in the embarrassing position of having its own programme curtailed through lack of agreement on what should be done, and its first year's budget cut from 43M EUA to 28M EUA. The Commission's reaction was to instruct its delegates to the meeting of the restricted Steering Committee of OEEC, which was finally to meet on 4 December, to demand a re-estimation by the experts of the Project's costs. Delegates were thus faced not only with the known difficulty of the proportional contributions but with a demand for a new costing. This was made through the German delegate who was also by this time conscious of the BBC/Krupp development of the pebble-bed high temperature reactor which, it was intimated, could be built more cheaply. The Commission's own representative went further. He saw in the pebble-bed and in other projects which Euratom believed were being studied in its members countries, direct competition to Dragon, and so wished to reconsider the fundamental question of whether Dragon should be supported at all. Moreover, he suggested that the UK project was, in any case, inferior to the American project to which Euratom's members would no doubt have access through the co-operation agreement with the USAEC that had been signed on 8 November (against the judgement of France).

An attempt was made to clear these last objections, by assembling at lunch-time a group of experts from those present, to review the implications of the other projects. France had already been quite categoric that its own research into a (fairly) high temperature gas-cooled reactor based on beryllium-canned fuel elements was in no sense competitive and Kowarski had been told some time previously, that it was not regarded as coming within the ambit of the Dragon collaboration. The French representative again made his point and stated the country's interest in participating in Dragon. The German delegate also declared that they saw the pebble-bed and Dragon as complementary, not competitive, but were anxious to see to what extent the proposed programme could be reduced without undue detriment to its value. No one could give any more details of the General Atomic reactor which was still a proposal only. Euratom however, insisted that whilst the French and German comments were interesting, it would still like more information on their projects, exposing rather clearly the difficulty Euratom was meeting in finding out even what was going on within its member countries.

With growing exasperation, the British delegate insisted that the estimates that had been made on the Project costs resulted from the extensive UK experience in carrying out such experiments and it would be both dishonest and futile, artificially to reduce the figure just to get agreement. If the Board of Management were able to carry out its programme for a lesser sum, no one would be more pleased than the UK as the major national contributor. In any case, the savings that might be involved spread over eleven governments represented sums that were quite insignificant. As to the proportional contributions, the Five had already stated their willingness to proceed on the basis of national income and the UK in the light of the Agency's arguments was prepared to contribute up to 43.4% (although its own inclination had been to limit its share to 35%). It was up to Euratom and its members to contribute the remainder. The only ray of light in what was otherwise a gloomy meeting was the indication from France, following an initiative it would seem from Perrin, that if the margin between what Euratom would provide and the sums necessary was not too great, France would be prepared to make a separate contribution.

It is more than possible that without this gesture and the propitiatory intervention of Huet, summarising the difficulties and persuading the UK not to insist on an immediate decision but await the end of the year, the project as a full international collaboration would have died at the meeting. The HTR would then have been developed as a UK national project with possibly the participation of some other countries, as both Italy and Switzerland had indicated their wish to be involved. This could have had the ironic consequence that a full-scale power reactor would almost certainly have followed.

INDEPENDENT ACTIVITY IN THE UK

The warning that the UK would go ahead alone if the other European countries were not prepared to make an adequate contribution was not an idle threat. Risley's inactivity following the Windscale fire lasted a few months only and in February 1958, J. D. (Pat) Thorn was appointed Deputy Chief Engineer (under Jack Tatlock) with the specific task of organising an HTR design group. Thorn had previously been Chief Engineer at Ruston and Hornsby, working *inter alia,* on the development of gas-lubricated bearings for Harwell before moving to Risley where he was attached to the gas-diffusion uranium enrichment plant at Capenhurst. Working in close collaboration with Lockett at Harwell, who went on with much of the detailed design work, Thorn proceeded on the assumption that his design office would be in action by May and Treasury approval for the construction of the reactor at Winfrith would be sought the following December. Like the majority of those engaged on the project, he was unaware of any impending changes in organisation until they were disclosed to the members of the Reserch Committee in June. At that time the Authority had some 74 professional staff working on research for the HTR. Most of these were at Harwell, six people were at the National Gas Turbine Establishment at Pyestock, and some effort was going on at RAE Farnborough.

In the Autumn, a transitional group was established at Harwell with Rennie as its Head to prepare the way for internationalisation but there was no running down of effort. Indeed at its meeting at the beginning of October, the Research Committee was informed that Cockcroft had agreed with Owen that "the HTR project should proceed as fast as possible when staff became available", and it was decided to go ahead immediately with a complete review of the whole design concept and with a new design of the reactor experiment. Further stimulus came in November from Shepherd who was then President of the British Interplanetary Society and had been in the USA to read a paper on advanced propulsion systems before a meeting sponsored by the USAF School of Aviation Medicine. He had seized the opportunity to investigate the HTR work going on there and had been both surprised and impressed by the amount of effort being expended. It was not the time to let things slide.

ALTERNATIVE FUNDING

One of the compromise decisions taken by the Steering Committee was to convoke a meeting of the REX Committee under Eklund to make recommendations on the cost of a minimal programme. Not surprisingly, it could add little to what had been said before. It did, however, elicit a further offer from the UK that the Authority would be prepared to assume the capital cost of certain installations such as the administration building, charging the Project instead a rent. This would have the effect of diminishing the overall cost by a few hundred pounds. It was pointed out that the main installations would have a significant residual value of some millions of pounds which would be credited to the Project at the end of the agreement. The Project would also benefit from constructions already completed including the zero energy experiment Zenith. Otherwise, the general conclusion of REX was that it would be unwise to recommend any reduction in the scope of the programme proposed.

Relations between the Authority and Euratom were by the end of the year at a low ebb. In parallel with the Dragon discussions, negotiations had been continuing throughout the Autumn on a joint UK-Euratom agreement. On its side, Euratom was concluding that the British had no intention of entering into any real partnership and saw the USA as a much greater source of material and information, whereas the UK had become convinced that Euratom would be prevented by its own members from ever being a real force and was content to let things drift. A real break between the UK and Euratom was, however, not desired and Dragon was considered sufficiently important for the UK representative in Brussels to send a note dated 15 January, 1959 to the Commission, stating that although the UK was ready to go ahead alone "they would on political grounds deplore its collapse as a joint venture at this particular moment. They would also consider it an unfortunate omen for the future of Euratom-UK co-operation." The Agency for some time had been urging the two parties to consider Dragon as a positive example of UK-Euratom co-operation though sensitive to the danger of the Agency itself losing some of the credit for being the sponsor of the enterprise.

Within the Agency, it was seen as imperative to find an additional source of revenue within Europe and armed with the knowledge that France was prepared to contribute separately a sum of the order of £0.7M, Nicolaidis and Huet went to Bonn on 13 January to persuade the German Government to do the same. By then convinced that Dragon could be a useful complement to the pebble-bed, Germany agreed and Huet returned to Paris to devise a scheme which would meet all the objections put forward. His proposal, which he discussed with Peirson and Guéron in Brussels the following week, was to base the agreement on a total expenditure over five years of £13.3M but not to commit a sum of more than £10M over the first three years and to decide then how the Project should be financed in the light of the value of the probable assets that would remain at the end. Of this £10M, the UK would contribute 43.4%, the Five 13.2%, Euratom 29.3% (in money only a little greater than 20% of the full £13.6M) and France and

Germany each 7%, making the total from the Euratom countries 43.4% like the British. Neither Guéron nor Peirson approved the principle of leaving undecided the final £3.6M (the proposed change in total bringing it down to £13.3M was ignored) and Peirson proposed on his own account that the Authority should advance this sum in return for the acquisition of the capital assets remaining. This was approved, and Peirson returned to London to obtain the agreement of the Authority, Huet to Paris to redraft the agreement, leaving Guéron the task of convincing his Commission.

Some measure of the difficulties he had been facing can be gauged from the fact that Medi had come up with the new idea of supporting an HTR vigorously to the tune of, say 55%, dissociating it from the UK-Euratom agreement (to which it was not of course tied) and building it in Holland. In an *aide-mémoire,* written with exaggerated politeness, Guéuron reminded the Vice-President that the UK project had been judged valuable by several groups of experts; Euratom's own Scientific and Technical Committee had urged the Commission to decide in favour of the Project; the French and German projects had not been offered as international projects "in spite of my appeals"; they were not copies of Dragon; the US project was still just a proposal. He concluded with the strong recommendation that Euratom should either take an important part in Dragon or drop the whole idea.

However, the leadership problems within the Commission were coming to a head and on 14 January, 1959, Armand tendered his resignation. His replacement as both Commissioner and President was Etienne Hirsch, a vigorous and able administrator with whom Guéron would work in close co-operation. Officially, he did not take office until 2 February but by then the Commission had agreed to participate in Dragon to the tune of 29% of £10M over five years provided that:

1. the higher payment of Euratom and the UK was recognised in the management structure;
2. France and Germany each contributed an additional 7%;
3. all funds from Euratom and its members were managed by Euratom;
4. the number of personnel engaged was proportional to the contributions;
5. a reasonable part of the research was done in member countries.

Belgium came in with the 0.4% to complete the total. In practice, these separate contributions were never made. The countries concerned proposed that they should be contributed to the Euratom research budget only when the 215M EUA had all been used up and whilst there were already signs that this was unlikely to happen, a revision of the Dragon Agreement eventually intervened and superseded the arrangement. One lasting effect was the granting to both France and Germany of the right to send to the Board of Management meetings, a representative as part of the Euratom delegation, a practice that persisted to the end. Belgium was compensated by the appointment of a Belgium national as secretary to the Board of Management. An important psychological result of the arrangement of funding was that Euratom became committed to the full contribution of 43.4%, equal to that of the Authority, a principle which the Agency believed firmly to be right.

It should not be assumed that within the Five the decision to participate in Dragon had been taken without serious reflection. Nevertheless, they were spared all the inner conflicts that went on in Euratom. There was an inherent attraction of joining a project in the UK and so be in a position to keep more closely in touch with British developments. Also at an early stage it had been tacitly agreed that their contributions should be proportional to the net national revenue and this principle was never seriously questioned. Spain, the only other country which might have participated, was not prepared to accept the same terms and dropped out of the running.

AGREEMENT

On 12 February, 1959, the Steering Committee was presented with the plan that had been formulated in Brussels by Guéron, Huet and Peirson and the necessary changes that had been made to the draft agreement. While a few minor modifications were still required on the question of information rights in the various countries, that same evening the OEEC was able to release a press communiqué announcing that "the third jointly-run project . . . of the European Nuclear Energy Agency . . . had been approved". Huet had his English reactor at last.

The formal agreement was signed on 23 March, 1959 at the OEEC Headquarters in Paris in the names of: the UK Atomic Energy Authority, the Republic of Austria represented by the Federal Chancellery, the Danish Atomic Energy Commission, the Commission of the European Atomic Energy Community (Euratom), the Institutt for Atomenergi, Norway, Aktiebolaget Atomenergi of Stockholm and the Government of the Swiss Confederation. It entered into effect from 1 April, 1959 and covered a collaboration lasting five years. Detailed direction of the Project was placed in the hands of a (British) Chief Executive who was responsible to an international Board of Management that was aided in its work by an international General Purposes Committee. The Project was sited at the Winfrith Atomic Energy Establishment of the Authority which provided the legal personality of the Project and also administrative and other services at an agreed price. Staff were to be seconded to the Project by the Signatories.

CHAPTER 10

THE PROJECT TAKES FORM

OF THE two most difficult problems which can face an international project – the choice of site and the appointment of leader (or for Dragon, the Chief Executive) – the first had been resolved before negotiations ever began and the second was void of controversy. Prior to the signature of the agreement, Guéron spent the first Sunday in March 1959 with Cockcroft at his home, quietly going over the main subjects. It had been a stipulation of the UK that the Chief Executive should be British and for some time Cockcroft had been clear that his candidate would be Compton Rennie; Guéron had little hesitation in supporting this choice.

FIRST CHIEF EXECUTIVE

Rennie was already known on the Continent through his work with the Authority's overseas liaison office, in the course of which, he had travelled widely and had come to know personally the technical and administrative leaders in the atomic energy organisations of the participating countries. He had been elected the first chairman of the Halden Technical Committee and had shown himself to be able in conducting meetings, sensitive to the wishes of delegates and a confirmed European in his approach to the matters in hand. To many at Harwell who had not been close to him, the choice was unexpected. Habitually stooping from his height of nearly two metres, head a little to one side, arms folded across his chest with a pipe cradled in his right hand, usually elegantly but casually dressed, he gave the impression of a comfortably situated country gentleman, lacking the thrust and drive of an international project leader. His inveterate courtesy, calm, even hesitant speech, and habit of throwing himself backwards to emit a deep braying laugh added to the illusion of a person who preferred the relaxed atmosphere of the drawing-room to the competitive pressures of international management.

Son of a marine engineer, he had quickly shown his academic brilliance, passing easily into Cambridge University (Sidney Sussex College) where in 1938, he gained at the age of 22, first class honours in mathematics and so the distinction of being called 'wrangler'. He turned to teaching, then with the advent of the second world war he became involved in the development of radar at the Telecommunications Research Establishment of Malvern (the latter providing the source of many of the early recruits to atomic energy). Without knowing on what he would be working, he joined the British atomic energy team in Canada in 1945 and was one of the small nucleus of people who returned in the summer of 1946 to found the Atomic Energy Research Establishment on the airfield at Harwell. Theoretical physics he found boring and quickly determined that this was not where he would make his career. Reactors were much more interesting and he enjoyed the direct contact with engineering, welcoming the opportunity he was given to act as the link between Harwell and Risley.

In this position he was able to gain an insight into the complete range of scientific, engineering and materials problems raised by the new technology. In addition, he became aware of the administrative problems posed by the admixture of fundamental research and tight construction schedules, which brought together strong and often abrasive personalities, all imbued with the driving sense of purpose that was characteristic of the early days in atomic energy. In the international field, he saw a new challenge, and well before Dragon became a reality he began to formulate his own ideas on the essential ingredients of an efficient administrative structure, believing that there were better ways of running a project than many he had seen.

It was not just in the light of a career opportunity that Rennie saw the OEEC initiatives. He was a confirmed believer in the need for European co-operation and he brought to its promotion, as one of his elders described it, a missionary zeal, a zeal that later was translated into a flame of enthusiasm for Dragon and thence for the HTR system.

Although Rennie's experience had given him the technical background for becoming the Head of an international research project, this would not have been sufficient without his other qualities. Belying his self-deprecatory manner and casual air was a well-nourished ambition and a great reserve of persistence and determination. His staff soon learned that he was open to discussion but could be firm; they were never in doubt who was in charge. Those used to exercising their temperaments on their superiors found that there was a definite limit beyond which they must not step or their time with the Project would be summarily cut short, however indispensable they felt. To some, he seemed remote, inclined to share confidences with a select group of long-standing associates only. In fact, he was entirely approachable, but he maintained a certain distance from everyone and knew how to keep his own counsel. When he left the project in 1968, few knew his real reasons. It was generally accepted that he had sacrificed himself to outside influences for the good of the Project, so adding the final touch to a heroic picture that so many of his staff had built up.

His administration was hierarchical (supporting the impression of aloofness) and whilst keeping a firm hand on the general progress of the work, he was able to delegate; he avoided fussing over detail. He also refrained from making promises but would wait until action had been taken and then announce the outcome. For those who had grown up in a system where the management was expected to harass its staff continually, his control seemed too tenuous and the light administrative reins were read as a sign of weakness. The majority responded with enthusiasm and an esprit de corps was developed which old members of the Dragon team now remember with nostalgia.

He put complete confidence in his staff, assuming that when an order had been given, it would be carried out without further intervention. He had complete confidence also in his own ability to see the Project through, never doubting that he was the most competent to direct its course.

Rennie was, however, careful to carry the Signatories with him, providing an unending flow of well-written documentation. Meetings were prepared with care and when the delegates finally assembled, personal discussions outside the conference room had already taken care of potentially conflicting interests. He interpreted his role of Chief Executive as being there to implement the Agreement, while that of the delegates was to assure themselves that the Project was managed efficiently, and to give what help they could. In this sense he was autocratic, but possessed the political sense and skill to convince his mentors of the soundness of his actions.

His appointment as acting Chief Executive, made at the meeting of the Signatories immediately following the signature of the Agreement on 23 March, 1959, was confirmed without discussion at the first Board of Management meeting of 1 June. Soon, a strong bond of trust was developed between the Signatories and the Project and Dragon's reputation in both the diplomatic and scientific world stood high. Clauses in the Convention which had been argued over for months, became of negligible importance, and seldom was the Agreement consulted to settle any difference of view. It must be rare to find the reactions of delegates of so many countries so uniformly positive, or years afterwards, so many people (including those with whom he had been in profound disagreement) speaking of a project's leader with such unqualified approbation.

STAFF

It had been agreed between Guéron and Cockcroft that Shepherd was the most qualified man to look after Research and Development and there should be two other Division Heads who would be from overseas. Euratom was to provide the Chief Engineer, and the other countries the Head of

Administration. Rennie's first choice for Chief Engineer was Gianfranco Franco in charge of reactor construction at the Italian research centre established at Ispra in the north of Italy. Just at the time Dragon was coming into existence, however, the decision was taken to hand over the establishment to Euratom as a joint centre, as a result of which, the Director, Carlo Salvetti resigned. Franco was made Acting Director and could no longer be seconded to Dragon. In his stead, Euratom proposed Anselm von Ritter of Germany, not a nuclear expert (which worried the Authority), but an engineer who had a great deal of experience in power station construction. His appointment was duly confirmed and to the dismay of all, within days of his coming to England, he collapsed and after a short illness, died on 23 January, 1960. A sad beginning that brought the recompense that the Ispra situation had become more settled and Franco agreed to join the Project. Norway provided the Head of Administration in the person of Rear Admiral R. K. Andresen. Of the nine Branch Heads answering to the three Divisions Heads in post at the end of the first year, four came from the UK (including the two in administration), three from Euratom, one from Switzerland and one from Norway. Lockett was rather outside the pyramid, occupying a special position as Chief Designer under Franco.

One of the first tasks of the Chief Executive and his nucleus of British staff that could be seconded immediately, was to build up an international team also at the lower levels. Vacancy notices were published by the Signatories in their own territories, likely candidates were picked out and then accompanied by Alex Prichard, a good linguist who had been transferred to the Project from the Overseas Relations Branch of the Authority and made Branch Head responsible for the general secretariat and administration, Shepherd and Lockett made a tour of the Continent to make a selection. The recruiting campaign elicited a satisfactory response, particularly in Milan where the nuclear company, AGIP Nucleare (formed in 1957 only), was laying off staff. Dragon was able to choose from almost a complete team of mathematicians, physicists and engineers. These and other unsponsored candidates from the Six found that they first had to become Euratom staff in order to be appointed, which incidentally resulted in a number of them later pursuing a career in international administration. By the end of the first year, 55 overseas staff had established themselves at Winfrith – over 25% of the total of 188 that included 48 administrative and junior grades. By the end of the third year when the number of skilled staff had risen to 182 in R & D and Engineering (plus 41 in administration and 26 juniors), the overseas contingent counted 68.

For the Authority, the take-over by the international team of its HTR project meant re-deploying many of its qualified staff, as not all could find posts in Dragon. Few of the physicists joined the Project although a number were transferred to Winfrith to work on the Zenith rector which was commissioned in December 1959. Others continued with research relevant to the HTR at Harwell. Thorn and his team at Risley continued for many months, in collaboration with Lockett, to work on the design of the plant, progressively handing over responsibility as the Winfrith team gained in strength.

Meanwhile Rennie and the Director of the Winfrith site, Donald Fry, set about the task of creating the type of working relationship between the international Project and its host that would permit the necessary freedom of action and of movement to Dragon and its staff, bearing in mind that Dragon procedures had to conform to the discipline of the site as a whole, and the system of administration that was current in the Authority and which Dragon needed to use. With commendable speed, the Authority side of Winfrith made welcome the new arrivals and accepted the special situation that had to be developed.

The ambiance favoured the enterprise. Winfrith is set in the 'Hardy' country (the home of the poet, playwright and novelist Thomas Hardy (1840-1928) celebrated for such works as 'Far from the Madding Crowd'). Situated a few kilometres inland from the coast between Bournemouth and Dorchester, close to such beauty spots as Lulworth Cove, it has developed into one of the most attractive atomic energy sites in Europe. The Atomic Energy Establishment, Winfrith – to give it its full title – comprises a collection of low, functional but elegant buildings widely separated by close cropped lawns and flower beds. Outside the perimeter fence are, on one side the gorse and heather of the heathland, and on the other, the rolling grasslands of the English countryside. Washed intermittently by the showers that drift along the Channel coast, but seldom last, it enjoys a warm and mellow climate. To the overseas staff, conditioned by the traditional tales of weather in England consisting of freezing fogs interspersed with continuous rain, Dorset was a relevation. The people match the countryside, warm and open, happy to receive strangers be they from other parts of England or across the Channel. There are few areas more pleasant to work and live in, and many of the staff who came with reluctance from abroad, found themselves regretting their departure and a number held on to the houses they had acquired, pending a hoped-for return.

On the question of accommodation – a subject that had a tendency to arouse considerable controversy within the national centres in the UK – the Winfrith management made every effort to be hospitable. On the site, staff were first given room in the building that housed Zenith and in the main office block, additional space being provided in some of the site services buildings. Then when it became obvious that a more permanent arrangement was necessary, a new office building (A.10) was put up in one corner of the site. A technical services building near Zenith was put at the disposal of the Project for non-radioactive research and development and when the decision was taken to make, as well as assemble fuel, the necessary buildings were put up inside the reactor enclosure. This last belonged to the Project; otherwise the Project hired the space it used – at rates that were deemed high at the time but look perfectly reasonable in present day terms.

At the personal level, the Authority made available to the Project a number of flats and houses that local authorities had erected for Winfrith, giving priority to overseas staff. It also opened the doors of the hotel it had bought in the middle of Bournemouth – Durley Hall – that was used for short stays and also served at times as a centre for conferences. Many of the overseas staff began their life with the Project in Durley Hall while they looked for their own place in the surrounding villages and towns, encouraged and helped by the site office and in particular, by the first Dragon Personnel Officer, Don Rusden. All but a few were able to find something to suit their taste and pocket in a reasonably short time and merged into the local atmosphere, their children rapidly picking up English and, no doubt, a Dorset accent. Not all the families settled down happily; that would have been too much to expect, but the large majority soon felt at home in a way they had not expected.

THE DRAGON REACTOR EXPERIMENT

Until the new staff had been assembled, the majority of the research and development in the physics and engineering fields was done extramurally and mostly in the UK. At the same time, plans were put in hand for carrying out irradiation work overseas as reactors became available. As a matter of principle as well as utility, atomic energy commissions in the Signatory countries were encouraged to become involved in the Project's R & D programme.

Work on the Dragon Reactor Experiment, DRE* went ahead largely on the basis of the designs prepared by the Authority, although a great deal of detailed work had still to be done and even complete systems developed (such as the gas clean-up plant) before plans could be finalised. Essentially, however, DRE looked very like the reactor that Shepherd had described to the experts in Paris in May 1958 and again at the second Geneva Conference in September. One change was the maximum operating power which was quietly increased to 20MW(th), to ensure that the power density, i.e. the heat output per unit volume, everywhere in the core would exceed that to be expected in a power reactor.

*Outside the Project the name Dragon was used to describe both the Project as a whole and also the reactor. Within the Project the reactor was termed the Dragon Reactor Experiment and referred to by its initials DRE, a contraction that will be used from here on.

DRE, according to the programme annexed to the agreement between the Signatories, was a reactor experiment designed "to demonstrate the principles on which any high temperature gas-cooled reactor must be based. The main features are the use of helium as a coolant, with fuel elements made of impermeable graphite, containing fuel inserts with thorium and enriched uranium in a suitable ceramic form. Provision is made for removing fission products which escape from the fuel inserts by passing a fraction of the total coolant through a fission product trap." Alternative coolants had by now disappeared from the scene; the fuel cycle was uranium 233/thorium (although of necessity fuelled initially by uranium highly enriched in uranium 235); the fuel and fertile material were in the chemical form of carbide intimately mixed with graphite, forming a compact that was porous to fission products; the compacts were contained in an impermeable sleeve; the moderator was graphite.

Design of the fuel elements went through many stages following a constant basic idea that one assembly was of hexagonal shape, 19cm across the flats. Later they were made as a piece from drilled blocks of graphite but to begin with they comprised an array of seven hexagonal rods 2.6m long. Gaps between the rods provided the channels for the main coolant flow. Typically, fuel was in the form of a stack of cylindrical rings loosely fitting inside the rods in the centre of which was, normally, a graphite cylinder that was also a loose fit. Through the spaces flowed the purge stream down to the bottom mountings and from thence via a selector valve to the clean-up plant. A core consisted of 37 elements, with an active volume about 1.1m in diameter by 1.6m high.

Surrounding the core was a removable reflector, an inner removable ring of graphite blocks going out to a diameter of 1.7m and containing the 24 boron carbide control rods suspended on steel cables. Outside this, a permanent external reflector of just under 3m maximum diameter completed the reacting volume. The whole structure was contained in a pressure vessel 18m high (to accommodate the lifted control rods and the internal charge mechanisms) and 3½m diameter, able to operate at a pressure of 20 atm. Helium, circulated by six blowers, entered the core at a temperature of 350°C and left at a mean gas temperature of 750°C. The heat generated was transferred in heat exchangers to a closed water circuit contained within the reactor building and from thence to another water system cooled in an external battery of forced circulation air coolers.

DEDICATION CEREMONY

Design and construction of the reactor building and civil engineering works, and site supervision of the reactor construction, as well as the testing and inspection of certain components, was entrusted under contract to the Authority's Development and Engineering Group at Risley (formed in 1959). Actual site work began early in 1960 and at an informal gathering prior to the meeting of the Board of Management in February, Eklund, the Chairman of the Board, turned the first sod. By April of that year, excavation work for the reactor building had been completed and the concrete floor laid. A small hole was left in the middle in anticipation of the ceremony which would celebrate the launching of the construction programme.

On 27 April, 1960, in the presence of delegates from the Signatories, members of the staff and a small number of invited guests, Eklund, re-elected Chairman of the Board of Management for a second year, laid in the central hole, a sealed steel cylinder containing a copy of the Dragon Agreement together with coins from the various participating countries, symbolising their financial contributions. Concrete was duly poured on top, from out of which, to the general embarrassment of the Project staff, rose the cylinder, obstinately floating instead of remaining submerged. It was perhaps a warning that the question of money would keep popping up throughout the life of the Project, but the more immediate lesson, one that has been learned everywhere in atomic energy often at great cost, was that even the smallest details are important, and it is not just the core of a reactor that presents problems.

END OF THE NUCLEAR EUPHORIA

Dragon was by this time fully into its swing. The international team had come together into a cohesive group which had a clear task to do and the means to do it. All attention could be concentrated on building DRE and developing the components to go into it, without worrying about what was happening in the outside world where the talisman was steadily losing its charm.

The bubble of euphoria that had carried nuclear energy along with such élan, started to collapse soon after the second Geneva Conference in 1958, as countries took stock of the real contribution that nuclear power would be making to total energy supplies. Even generous forecasts had not foreseen this being more than 10% in Europe before 1975; although the British and Euratom programmes had seemed dramatic at the time they were announced, 5000 MW by 1965 plus 15 000 by 1967 still represented only the equivalent production of about 50M tons of coal over a year, compared with an annual consumption of energy in Europe equivalent to around 1000M tons.

Moreover, it was becoming clear that neither of these programmes would be achieved. Everywhere the effort needed to translate prototype or, more accurately, experimental results into big operating plant was proving to be much greater in terms of both manpower and money than foreseen. Scaling up to bigger sizes which, on paper, was the key to lowering unit capital costs to a level where nuclear power could compete with conventional fuels, was exposing new problems that were time consuming and costly to resolve. Construction programmes were in consequence delayed and industry found itself required to put in much more development effort than had been allowed for, while at the same time supporting a cut in the number of new stations ordered.

In Britain the effect was a stretching of the nuclear programme to 1968 and a contraction of the five nuclear consortia to three. In the Euratom countries, the 15 000MW target by 1967 was appearing more and more unrealistic as the months passed. Over Europe as a whole, total energy consumption was growing at a rate less than had been foreseen due to an economic recession and had even been static between 1956 and 1958 **(23)**. Electricity consumption continued to rise but the electricity supply companies were reluctant to invest in unproven systems with escalating costs, when for the first time for 20 years, a buyer's market was developing in fossil fuels. Coal output had gone up and whilst continuity of supply could be considered uncertain due to the vulnerability of the industry to industrial dispute, the discovery of new oil and gas reserves made petroleum products an attractive alternative. With the opening of new trade routes, Suez was forgotten and there was little encouragement to make risk investment in a new technology. Moreover, the design of fossil-fuelled power stations had not stood still; efficiencies were rising and capital costs dropping. In the UK, for example, the quoted cost of new stations fell from £55/kW to £37/kW between 1955 and 1960. The nuclear industry was aiming at an economic target that moved away at a speed that was comparable with its own advance.

A further problem faced by the nuclear industry was the relatively high unit cost of stations compared with conventional stations, even if this was compensated by lower fuel costs. In Europe, money was in short supply and electricity companies tended to prefer systems with lower capital cost and operating costs that were a function of demand. The nuclear industry argued, rightly as it turned out, that fossil fuel costs were bound to rise and inflation favoured the system with the higher initial investment. Continuous battles were waged between rival economists who juggled with amortisation rates, availability factors, interest and inflation figures, deriving their final power costs with a precision far exceeding that inherent in the original data. Throughout the sixties, decisions on power supply systems were taken on the basis of calculated marginal differences, that later could be seen to be of negligible significance in comparison with the grosser effects of unforeseen technical and political difficulties.

For Dragon in its early phase, such considerations were of little moment. In any case, should the HTR system ever be applied commercially, as it was a system capable of high power density and so, in principle, modest capital cost, it was evidently going in the direction into which nuclear power was being pushed. More immediately, Dragon was able to take advantage of the slow development of industrial markets as it went out to tender for reactor components. Companies struggling to find a place in the new industry were prepared to buy their way in and offer prices that were keenly competitive, in many instances with only a nominal charge made for development, and manufacturing profits cut to a minimum. On the research side also, Dragon was able to make use of national facilities that had been built up in anticipation of a rapid growth in demand that was, in practice, maturing only slowly. Particularly important was the coming into operation of a number of big materials testing reactors in the Benelux countries and Scandinavia which had space available for fuels and materials irradiation, and staff pleased to have the work to do.

CHAPTER 11

DRAGON'S LONG-TERM ROLE

WHEN presenting the Dragon Project to the press following the Signature to the Agreement in Paris, Plowden described it as having two main objectives:

- To carry out a programme of research and development work concerning high temperature gas-cooled reactors, and
- To design and construct a reactor experiment which would be the first stage in proving the feasibility of HTGCR systems.

"Thereafter", Plowden continued, "the participants will be free to take whatever steps they desire to exploit the system further individually."

Huet defined a third objective: namely, to show that a Project of this nature could be realised internationally as efficiently as in a national context. So long as this had not been done, the value of common action in the technical areas would not have been clearly shown. He saw Dragon as a demonstration of European co-operation which, if successful, could catalyse further collaborative efforts in the technical field.

The Agreement covered a fixed period of five years and all the Signatories had indicated that they were in favour of a project of limited duration. However, Plowden, recognising that the period was too short to give definitive results commented; "Since much of the value of the reactor experiment will lie in the practical experience of operating the reactor, it is to be expected that some at least of the initial participants will wish to carry on with operating the reactor as a joint undertaking for a further period."

Despite the considerable effort already put into basic research on the HTR system at Harwell and the existence of Lockett's conceptual design for the reactor, even on the most optimistic forecast, the reactor could not operate at power for more than a few weeks before the Agreement was due to end. If, moreover, staff numbers had to be run down in anticipation of this termination, then actual experience of operating the DRE would be small indeed.

According to the Agreement, the Signatories were to consult together six months before its termination, to decide whether and under what arrangements it might be extended. Long before this, however, Dragon's programme would be conditioned by whether an extension was to be concluded or not. Realising this, the Board of Management, when approving the draft budget for 1961/62 at its sixth meeting on 14 February, 1961, instructed Rennie to prepare a paper for its next meeting setting out the technical and financial implications of an extension.

In his paper, Rennie pointed out that for the Signatories to derive full value from their investment, it was necessary to run the reactor for many months at full power and then analyse the experience and in particular study the effect on the fuel. At the minimum, one could negotiate an access agreement with the Authority (which would assume ownership of DRE) or, with rather more advantage, operate the reactor as a joint project. Either course would have the unfortunate effect of divorcing research and development directly connected with the reactor from other development work that presumably would continue elsewhere. There was still a great deal to do in assessing how the experience gained could be applied and there was still much basic work to do; for example on using beryllium oxide as a moderator in place of graphite, on different methods of keeping the reactor coolant clean, on the development of larger gas circulators, and on heat exchangers which would raise steam rather than as in the DRE, transfer the heat to a second circuit from where it was thrown away. There was no shortage of subjects if Dragon was to be more than simply an exercise in building an experiment. Moreover, consideration should be given to the information that would flow in from the Project's collaboration agreement with the USAEC. Concluded in the Spring of 1960, this provided for an exchange between Dragon and the US high temperature gas-cooled project centred on the construction by General Atomic of the reactor at Peach Bottom, designed to generate 40MW(e) for the Philadelphia Electric Company. Owing to delays, the reactor would not be operating before 1964 and it was, therefore, after 1964 that this exchange would be most valuable.

Relations between Dragon and its Signatories were excellent and from the soundings he had taken, Rennie had every reason to believe that an extension based on a broad programme would be well received. Notable was the enthusiastic support coming from Euratom. Earlier hesitations were forgotten and Dragon was seen by Guéron and his colleagues as one of the most significant collaborative reactor programmes in Europe. Euratom's own programmes were only slowly getting off the ground and so finding funds presented no difficulty.

It came then as a considerable shock when at the next meeting of the Board of Management held in Stockholm on 25 May, 1961, its new Chairman Sir William Penney made, in the words of one participant, "his dreadful announcement". Penney as Member of the Authority for Weapons had previously been the architect of the British atomic bomb programme and the principal technical advisor to the Government during its negotiations with the United States on co-operation in nuclear defence. On Cockcroft's ceasing to be a full-time member of the Authority in July 1959, Penney moved over to Research and took Cockcroft's place on the Dragon Board of Management, serving as Vice-Chairman under Eklund until March 1961. Penney had no personal commitment towards European co-operation as had Cockcroft, nor the same personal interest in the well-being of Dragon.

It cannot have been so much the words he used that caused the dismay but the unfeeling, categoric manner in which he presented the UK position from the chair, before any other delegate had spoken. Opening with the statement that "the Authority expected that the Signatories would want some form of extension to the Dragon Agreement", he continued that the Authority "would be glad to continue collaboration on some agreed lines. There was, however, a domestic problem in the UK . . . that concerned the probability that the Authority would start work on its own HTR systems . . . and although the Authority were anxious to help the Project as much as possible, the domestic programme had to be safeguarded. Whatever form of continuation should come about, therefore, the Authority felt the reactor should have two uses, one for Dragon and one for the Authority."

FUEL POLICIES

Within the Authority at this time there was a conflict between its responsibilities towards the international project and its own views on the direction in which HTR work should be going. Independent HTR research had continued in parallel with Dragon and Cockcroft had set up an HTR development committee that began its work under Penney with the object of planning forward strategy. Rennie was a member of this committee and came under pressure to finance a major part of the research from Dragon funds. He was, however, also under pressure from the Continent to limit the dominance of the UK in Dragon, and only part of Harwell's HTR research was covered by Dragon contracts.

The development committee lasted less than a year but before dissolving, it requested Risley to set up a design office to conduct studies on a full-scale (say 500MW(e)) power reactor as a guide to the research programme. Thorn was given the task and, in view of the optimistic claims being made for Dragon, proceeded to base his assessment on DRE design concepts. By the Autumn of 1960 he had already progressed far enough to conclude that it would be impossible to build an economic power reactor that had fuel from which the fission products were continuously purged, even if the technical problems posed by the inevitable escape of some of the fission products into the main coolant stream could be solved. The only way out was to concentrate on the development of a fuel that would retain the fission products *in situ*. By then, it was realised that only a small penalty would in practice, arise from leaving them in the reactor, because fuel bodies could not be made which would allow the crucial elements to escape at a sufficiently high rate. Many of the most highly absorbing fission products have short lives and so would have done all the poisoning (i.e. absorbed all the neutrons) of which they were capable before they could be extracted. Thorn's report sparked off a big debate within the Authority and provoked serious doubts about the value of Dragon as a vehicle for pursuing the HTR.

Even in the early Harwell days it had been argued that a system which might have a primary coolant significantly contaminated with fission products would be unacceptable as a power reactor, and the development of a sealed ceramic fuel had been viewed by many as a long-term necessity. To try and confine the fission products, experiments had been made on reducing the fuel to a finely dispersed powder and then enclosing each particle in its own protective coating. Huddle had made the suggestion in 1957 that the coating material might be pyrolytic carbon and had instituted work in the Chemical Engineering Division aimed at coating the particles in a tumbling bed. A fluidised bed was preferable but the idea had been abandoned as inoperable with the very fine powders (1 micrometre) that the physicists were saying were necessary. Tumbling the particles was not a success; they were far from spherical and the resulting coatings were irregular and unreliable. DRE design had therefore gone ahead on the assumption that the fuel would be an emitting fuel, and Dragon concentrated initially on devising the best form to 'drive' the reactor. Harwell meanwhile continued with research on particle coatings under Henry Lloyd.

Nevertheless it was Huddle who took the first essential steps towards the development of a coated particle that would retain the fission products. On an information exchange visit to the USA in October 1960, he learned that Battelle Memorial Institute, to protect fuel against hydrolysis during its subsequent manufacturing process, had successfully developed a fluidised bed process for coating with pyrolytic carbon, coarse fuel particles that could be made as spheres. Rushing back to Winfrith he obtained Rennie's authorisation to try some experiments on the same lines at RAE Farnborough where carbon work for Dragon had up to that moment concentrated on the development of an impermeable bulk graphite for the sleeve surrounding the fuel inserts. (This development would never have fully succeeded as although much progress was made towards producing a gas-proof graphite by successive impregnations, one of the most tiresome fission products, caesium, was solid and would still diffuse through.) Within a very short time a fluidised bed experiment had been set up and Huddle returned triumphantly to display particles of sand encased in carbon.

Huddle threw himself into the development of the technique, made much more feasible by the relaxation of the demands of the physicists who by then were quite ready to accept particles with a diameter expressed in tenths of a millimetre. RAE Farnborough switched its studies on impregnation to coatings, and contracts were negotiated with Metallwerk Plansee and OSGAE in Austria, and with CEN/SCK in Beligum, for further work on coating processes and the production of spherical fuel particles (kernels).

THE AUTHORITY IN CONFLICT WITH OTHER SIGNATORIES

In the Spring of 1961 however, the Authority believed that Rennie was intent on continuing with the emitting principle and convinced itself that other Signatories would not be interested in the alternative even if they were made aware of the conclusions in Thorn's report. Rennie, as an Authority man, was aware of the report but it was confidential, and he was unwilling to throw early doubts on the existing design of DRE because if the reactor was able to operate with emitting fuel, it would certainly operate with retaining fuel. He was anxious that no ideas be injected at that stage which could provoke a call for changes in DRE with a consequent delay in construction. Only in July 1962 was the GPC approached with the recommendation that coated particle fuel be adopted for the first charge, and not until May 1963 was authorisation given, whereas inside the Project, the changeover had begun in 1961 and all work on emitting fuel had effectively ceased in 1962. The Authority, for its part, considered its views on the future of the HTR to be an industrial secret, and came to the conclusion that once DRE was operating, the Project should have two functions – the international and the private.

Not unnaturally, the other Signatories, who were unaware of the background to the Authority's stand, which was not explained, did not view the situation in the same light. The delegates had come to Stockholm in May 1961 to discuss Rennie's paper and they felt they were being faced with an ultimatum from the Authority that they were not prepared to entertain. It was decided therefore that Huet should assemble an informal working party in September to consider the situation. A rather warmer-toned *aide-mémoire* was attached to the minutes of the meeting but the Authority's stand was reiterated. The Authority was paying 43.4% of the first £10M and all the additional costs up to a ceiling of £13.6M in return for which the reactor and other installations or equipment built or acquired in the UK under the joint programme would become the property of the Authority at the end of the five-year period. The Authority was very willing to negotiate with the Signatories an arrangement which would cover a further period of operation of the reactor, but wished to reserve some part of the facilities for its own domestic programme.

It was already clear that the whole of the £13.6M would be needed for the Dragon programme, as on re-examining the estimates of expenditure that he had inherited, Rennie recognised that the costings had been optimistic. Indeed, on his own responsibility, he had decided that there could be no serious research into the reprocessing of irradiated fuel, for example, in order to be able to stay inside the total financial envelope. There would certainly be no surplus. In effect, therefore, the Authority was being required to pay nearly 60% of the costs of the programme, and Penney especially considered the figure to be excessive.

Even within the Project itself, the Authority attitude was interpreted as a bid to close the door on continental involvement so as to gain a commanding lead in HTR technology. In Euratom the conviction was even stronger. Rumours of a Risley study had been circulating and it was believed that the UK was already designing a prototype on its own, with a view to beginning construction by 1966/67. Euratom was also very conscious that its investment in Dragon was being frittered away because of the absence of any centralised HTR effort in its member countries. It saw the extension of the Agreement as an essential, and at the same time looked for ways of countering the existing advantageous position of the UK.

In the light of comments made at an informal session of the GPC in July, Rennie prepared a programme which envisaged a three year extension costing £12.5M or £11.0M depending on whether the 'desirable' or only the 'essential' research and development was to be included. In both, a figure of £5M had been set against the operating costs of DRE.

The Authority for its part was prepared to agree that a figure of £5M was reasonable but wanted the R & D part cut to one third. If that were done, the Authority was prepared to continue the same scale of payment. A rent for the use of the reactor would be charged but the Authority was prepared to pay for the proportion of its operation reserved for its own use. It then emerged that the Authority was proposing to charge a rent of £1M per annum, supporting this figure with the argument that the installations including the DRE would be worth £5M and a straight line amortisation should then be applied over five years according to the practice current in its own internal accounting.

It can be assumed that the informal meeting of the Signatories called in October 1961 to exchange views on the extension (or, as the Authority preferred, a new agreement in view of the fact that "we shall be providing for collaboration of a different kind and with a different object than that which underlay the original agreement") was not too cordial. Guéron's comment when reading the minutes prepared by Weinstein was "you did polish the stone, but why should I work to put the roughness back in".

Not only Euratom was promoting a substantial R & D programme. Austria and Switzerland both expressed themselves firmly in favour of a programme which would allow the participants to gain a better understanding of the problems inherent in the HTR system. They also could not accept the method used to calculate the rent, as the explanatory note accompanying the original Agreement documents made it clear that the UK's taking possession of the installations at the end, was directly linked to its supplementary contribution of £3.6M. (Rennie, in his draft, had assumed a rent of only £1.75M for the three years.) If the Authority figures were accepted, the erosion of the R & D programme would be very serious and there would be little chance of achieving the minimum goal of developing a fuel element suitable for power reactor applications.

But the political scene had changed. At the end of July, the Government of the UK had applied to join the Common Market and was preparing its application to join also Euratom and the ECSC. The Authority was given to understand that it would be unfortunate if, during the negotiations, the principal technical project in which the UK and the Six were co-operating should turn sour. The conclusion was drawn in the Authority that it would perhaps be better to relinquish some of its private information gained through the Thorn study in order to re-orientate the Dragon programme towards a more productive goal. Dragon's research into coated particles, particularly that undertaken in collaboration with RAE Farnborough, had also, it might be noted, made great progress in the preceding 12 months and the first irradiation results (made at Harwell) were distinctly promising. At the end of November, Penney, accompanied by Cook, met Guéron privately and probably for the first time, a frank exchange of views on where the Project should be going took place. One result was an agreement between them that a strong accent should be put on fuel development even if this meant slowing down construction of DRE.

On this same occasion, Guéron also aired a new approach to the financing of the Project which would suppress the special conditions attached to the Authority's contribution. To his own Commission he had already put forward his view that the Agreement should be extended for at least three years and possibly five; no rent should be paid to the Authority to compensate their £3.6M; instead all should pay an appropriate share; the research programme should be considered more important than the reactor; the request of the Authority to have 25% of the reactor reserved for its own use should be refused; the UK should be made to understand the irrationality of cutting down collaboration because the system could be commercially interesting.

With the two main partners now much closer together, the Board of Management decided to set up a joint working party, first to define Project objectives and then to consider the programme necessary to achieve them. This programme should be regarded as beginning in April 1962, and continuing for five years, i.e. three years beyond the existing Agreement. Discussions about financing could come later when it would be necessary also to agree on the way to treat the additional £1.4M that Rennie believed would be required if the original programme (minus fuel processing) was to be completed.

That such a move should be so welcomed by all the Signatories throws into relief the uncertainties that existed about the real aims of the Project, apart from the construction of DRE. Even if in general terms it was recognised that the central goal was to explore the possibilities of the HTR system as a source of power, it was far from obvious where the Project's mandate began and ended. Much of the research programme agreed at the time of the signature was in direct support of the DRE; the rest was a general exploration of the field without any real end in sight.

PROGRAMME RE-DEFINED

In practice, the working party found little difficulty in arriving at a definition of Dragon's objectives, namely: "to provide the Signatories with information leading to the design of an economic land-based, carbon (graphite)-moderated, gas-cooled, high temperature power reactor". It was also agreed that the continuing research and development programme and the use of the reactor should be directed towards this end. Ship propulsion was out, the substitution of beryllium oxide for graphite was out, whereas it was clearly necessry to have assessment studies made, either by the Project itself, or by outside organisations under contract. If the word 'construction' had been used instead of 'design', the subsequent history of Dragon might have been different, as it might have paved the way for a joint project to build a prototype power reactor. Even so, this clarification of the objectives, accepted by the Board of Management, was a considerable step forward, and the insistence on the preparation of a reference design (Euratom suggested a 300MW(e) station) placed a definite requirement on the Project to direct its research towards a commercial application.

It was appreciated now that the schedules that had been talked about early in the Project's life, when fuel loading was envisaged for 1963, had been over-ambitious and a more likely date was 1964. That meant that given an extension to March 1967, only about 2½ years of operating experience with DRE at full power would be possible before the Agreement ran out. It seemed advisable therefore to provide for the possibility of later extending the Agreement yet again.

Two proposals for the overall programme from April 1959 to March 1967 had been drawn up, one costing £23.5M and the other £25M. Rennie had done this because of the gap that still existed between the scale of effort deemed necessary by the Authority and the other Signatories. By the Spring of 1962, however, with the UK making every effort to join the Communities, the Authority was ready to put its weight behind Dragon. The higher sum was accepted as the basis for negotiation and the UK announced that it was happy to give up its special position, and Euratom's plan for sharing the cost of the total programme amongst the Signatories won general support. The other non-Euratom countries were prepared to pay their current percentage contribution based on their national revenues, and left it to the UK and Euratom to share the remaining 86.8%.

Some of the Signatories had already begun parliamentary proceedings on the basis of a total budget over the 8 years of £25M so could not follow the proposal that emerged from Euratom-Authority discussions to raise the sum to £26M, largely to cover research at Harwell that would be incorporated into the Dragon programme. Instead it was agreed that should Spain, which had again been making overtures, agree to participate, its contribution of £0.72M would be additional. Otherwise, to avoid reducing the programme, the contingency allowance was cut to almost zero and both Euratom and the Authority undertook to make available a supplementary sum of £200 000 each, should this prove necessary. As a result of the restructuring, the Authority effectively abandoned private work on the HTR and announced in its annual report for 1962/63 that all effort on HTR was now directed through Dragon. Spain, in the event, was unable to find the necessary funding.

The integration of the Harwell work into the Dragon programme had a strongly positive effort on coated particle development and techniques advanced so rapidly that it was possible to install at Winfrith during 1962 prototype production equipment for the manufacture of fuel. In addition to the coating of pyrolytic carbon, silicon carbide was introduced and a multilayer process evolved, whereby a porous inner layer of carbon took up the expansion of the fuel particle under irradiation, a silicon carbide layer provided the barrier to fission products and an external dense carbon layer formed the pressure vessel. By the end of 1962, there were no doubts remaining that an excellent coated particle driver fuel could be made for DRE and there was considerable confidence that fuels could be made which would be suitable for use in a power reactor.

An important introduction into the Dragon programme upon which the Board of Management placed increasing emphasis, was the preparation of assessment studies relating to the reference design of a power reactor, the essential objective of these studies being to identify the areas where research and development was most needed. At the same time, by going through the exercise of preparing a design, some feel could be gained for the economics of an HTR power system. The Authority offered to make available the Thorn design (at a price) and industry was invited to tender for a new study that would be based on the most recent information. On the basis of the offers received, the joint proposal of AGIP-Nuclear (Italy) and Indatom (France) was accepted; ASEA's offer from Sweden was not taken up to the disappointment of the Scandinavians while BBC/Krupp of Germany declined to undertake such work on the grounds that it was not prepared to share rights to patents that might conceivably materialise.

The only difficult point in drafting the revised Agreement between the Signatories concerned the disposal of the assets and their valuation at termination. Communal ownership presented a problem as not all the Signatories might wish to participate in a further extension or fuel testing period beyond 1967. Essentially there was no disagreement on the substance. The UK recognised that as the reactor was on its territory it would have an obligation to continue operation should the other Signatories wish it, whereas if the UK were to continue operation alone, after termination of the collaboration, it should pay something to the other Signatories in compensation. Calculation of the residual value of the fixed assets was agreed to be on a straight line 20% per annum rate of amortisation, starting six months after DRE achieved criticality for the first time. Provision was also made for a modification of this profile if the reactor ran into unexpected difficulties coming up to power. By November 1962, following some hard bargaining between the Authority and Euratom, which resulted in the latter's share of the outstanding 86.8% becoming 46% of the total 8 year programme, full agreement had been reached and before Christmas, Weinstein could report to Rennie that everyone had signed.

Euratom could afford to be generous. Negotiation of its second five-year programme to begin in January 1963 had gone remarkably smoothly (as it turned out for the last time), agreement being announced already in June 1962. Of the 480M EUA that formed the Commission's original proposal, 425M EUA had been accepted which was a sizeable increase over the first five year budget of 215M EUA, even taking into account the fact that the rate of expenditure had progressively grown since Euratom's programme was launched. For Euratom at this time, money was not so much a problem as finding viable projects where its participation was welcomed. Dragon was proving to be a sound investment and HTR technology something of a European speciality.

For Dragon, the new agreement was of capital importance, as it could now look forward to more than four years' assured life, supported by Signatories more united than ever before. When Huet left the Agency at the end of January 1964, he could look back with satisfaction on a project going well, internationally applauded, which had gone far to proving the point he had been so keen to demonstrate: that a European technical development could be mounted that was as successful as a national enterprise.

Many have speculated since, that a great opportunity was missed in this period to plan for a joint power station experiment. Coated particles were pushing the HTR into the forward ranks of advanced concepts and it was, moreover, the only system designed to exploit the uranium/thorium cycle. Collaboration between the Signatories had been given a new lease of life during 1962 even if, on the broader front, the rejection by Général de Gaulle on 29 January, 1963 of Britain's application to join the Communities had cast a shadow over relations between Britain and Euratom's member countries.

Nevertheless it is difficult to see how a joint project for a prototype could have been launched. If industry had been invited to join the Signatories, it is most unlikely that they would have been prepared to make any investment or that either the Authority or Euratom would have been willing to give up its dominant rôle. The power companies were not geared to making investment in research into new nuclear systems, except to a small extent in Germany and Switzerland. Those who were prepared to consider the installation of nuclear power stations, looked for proven systems and with increasing frequency were turning to the light water reactors backed by the consolidated industrial strength of the USA. Perhaps a joint initiative by the Authority and the CEA, the main proponents of graphite moderation, might have succeeded, but the CEA was far from convinced that the HTR would find a place in the French programme and the de Gaulle veto was hardly an auspicious background.

In any case it was still too early. DRE was some 18 months from completion, Peach Bottom in the USA was trailing by another 18 months and AVR in Germany by a few months more. No serious evaluation had been made of the engineering problems to be expected in a power reactor fuelled with coated particles and no valid estimates had been made of the economics of a power system comprising HTRs. Until DRE had operated, the HTR was no more than a paper reactor.

CHAPTER 12

EXPLOITATION

RENNIE gave overriding priority to the construction of DRE and this was proceeding with commendable speed, especially when it is remembered that companies from many nations were contributing. By April 1961, erection of the reactor building was more than half complete and the concrete shielding inside, that was to surround the reactor vessel, was taking shape. A number of major contracts had already been placed for the reactor plant and prototype work on certain of the critical components was well advanced. Zenith, the Authority's zero energy high temperature reactor had gone critical at the end of 1959, since when its programme had been devoted to measurements on core assemblies such as would be used in Dragon, so that the core of DRE could be designed with a detailed knowledge of how it would behave from the physics point of view, including the effectiveness of the control system.

During 1961, the complex steel vessel that would house the reactor proper was pressurised and leak-tested at the contractor's works; design of all the internal parts of the reactor was completed and dummies made of the core elements. The prototype circulator was running under test and orders had been placed for the circulators to be used in the reactor; the main gas valve was also under test and standing up to conditions well.

An important policy decision was taken during the year to equip the Project to manufacture fuel elements; orders were placed for the necessary plant and work started on an extension to the fuel storage building in the reactor compound. Dragon's determination to be fully self-sufficient in the manufacture of its fuel components met some opposition, in particular from Germany, where it was considered that the Project was undertaking work that could be done by industry, but any other course would have lost the Project the flexibility in fuel specification it considered necessary, and start-up of DRE would have been delayed.

A second important policy decision taken at this time concerned the future operation of the reactor. The Director of Winfrith would be carrying responsibility for its safety and in order to keep the chains of command clear, it was agreed that he should provide from Authority staff the operating group under contract to Dragon. The Project had full right to attach its own staff to the group and it was understood that Dragon would determine the running programme, and DRE would be operated on its behalf. An incidental advantage was that Dragon did not have to recruit its own team and avoided problems that would have arisen from employing as operators, staff on relatively short term secondments. Members of the group, the nucleus of which was formed in 1961/62 and which grew to about 100 strong, were fully identified with the Project's aims but did not appear on its staff list. Their influence was immediately felt. For example, it was in discussions on reactor safety with the group that it was decided to recommend that the space inside the steel shell that sealed off the reactor compartment from the rest of the concrete containment building, should not be filled with nitrogen but air only, although designs were not changed in case there should be a reversal of policy in the light of experience. As commissioning proceeded, the group exerted an increasing influence on component design to the benefit of the Project.

Three years less a month from the coming into force of the Dragon Agreement saw the reactor pressure vessel being lowered into place after which work on linking up the primary circuit could begin in earnest. Steady progress had been made in the fabrication of the trapping system for fission products carried away in the purge stream and during 1962 the pilot helium purification plant was brought into operation, allowing the contract to be placed for the main plant. Designs still foresaw a significant leakage of fission products into the purge so that the efficiency of the purification plant would determine the rate at which contaminants built up in the primary circuit and this in turn could be the crucial factor in determining the useful life of the reactor.

By the end of 1963, this was the only major plant item not nearing completion and already a great deal of commissioning work had been done. The six primary circulators were installed and the prototype had passed its 1000 hours on the test rig at Winfrith. The intricate charge machine located above the reactor to handle the fuel elements had been assembled and put through its paces, much of the reactor instrumentation was installed, while in the fuel element plant, highly enriched uranium was being used in the fabrication of fuel inserts following process testing with natural uranium.

The first fully completed fuel element was handed over to the operations group in April 1964 and loading of fuel into the reactor finally began in August 1964. A chain reaction developed for the first time on 23 August under conditions satisfyingly close to those that had been predicted.

The timing was opportune, as one week later the Third United Nations International Conference on the Peaceful Uses of Atomic Energy opened in Geneva. Both the Project and the Agency made full use of the occasion to publicise Dragon and on the eve of the opening, presented a film that recorded the construction of the reactor. On 22 October, DRE was officially inaugurated by Thorkil Kristensen, Secretary-General of OECD* before an invited audience of delegates and senior scientists and the Winfrith staff. The ceremony included a formal start-up and shut-down of the reactor by Cockcroft and Perrin and the unveiling by Urs Hochstrasser (Chairman of the Board of Management and of the ENEA Steering Committee) of a commemorative plaque and the Dragon symbol executed in stainless steel. Both of these, this time remained firmly in place, symbolising it might have been hoped, the stability of the enterprise.

Over the rest of 1964, zero energy tests on DRE were made to check the physics behaviour of the core and the integrity of the external circuits. Hot tests began in the Spring of 1965 and the approach to power in June. An extended run at a power of 10MW followed and the full power output of 20MW was attained for the first time on 24 April, 1966, just over seven years from the start of the collaboration.

From the beginning of operation, it was evident that DRE was a highly successful design. The fuel behaved remarkably, retaining fission products and withstanding burn-ups beyond what even the most optimistic would have dared forecast just a few years before. On the engineering side, the circuit was astonishingly leak-tight; the gas circulators running in their gas bearings behaved impeccably; the charge mechanism with its complex manoeuvres inside the pressure vessel gave no trouble; corrosion was negligible in the primary circuit; control was straightforward and the reactor proved not to be temperamental—altogether a reactor experiment that gave every cause for satisfaction.

ASSESSMENT STUDIES

Although in the formulation of the new 'main aim of the joint programme', agreed during 1962, emphasis had been placed on the production of reference designs, the immediate impact on the Project's direction was small. Assessment studies were seen as a secondary activity. Pressure from the two main Signatories was taken care of by the purchase of Thorn's design analysis and the commissioning of the study from AGIP Nucleare/ Indatom **(24)**, which Rennie viewed more as a sop to Euratom than an important design exercise. Out of the million pounds set aside in the budget for assessment studies, about £40 000 was paid for the Thorn study and £100 000 allocated to the Italian/French contract, not a great sum for a group faced with a totally new concept. In addition to the specifically HTR

---

*OEEC was transformed into the Organisation for Economic Co-operation and Development on 14 December, 1960, when the United States and Canada became full members instead of observers. The structure of the Agency remained unchanged, the USA and Canada continuing as Associate Members.

aspects of a ceramic core and helium cooling was added the principle of using a pressure vessel made out of concrete, a technique that had been pioneered by France for its own gas-graphite power programme and taken up in the UK for the most recently ordered Magnox stations. This was the second development after coated particle fuels that made the HTR an interesting concept for central power station applications. A steel vessel was quite satisfactory for DRE, but for a reactor developing 50 times the power or more, the design problems of a similar vessel would have been huge. When the group's report was presented at a Dragon Symposium in April 1964, it was subjected to a degree of criticism that did the contractors less than justice and exposed the naiveties of the critics as much as those of the designers. Nevertheless the study provided a useful starting point for future work and it provided a considerable encouragement to proponents of the HTR as it indicated that a big HTR could be built for a capital cost of £50/kW(e) as compared with, say, £75/kW(e) for a light water reactor and £100/kW(e) for a Magnox reactor.

Within the Project some engineering effort had been diverted to assessment studies from the middle of 1963 onwards, and this was supported by a small number of secondments from British industry. Clear terms of reference were, however, not laid down and the group felt left out of the main stream of work. In part, this was due to Lockett's direction. An inspired engineer when faced with a specific engineering problem to solve, requiring ingenious detailed design, he was unsuited to the task of directing a system analysis and coordinating the efforts of various groups working on the different station components, all of which interacted with each other.

This situation which had begun to worry Saeland, the new Director-General of the Agency, was not allowed to persist. With the completion of construction of DRE, the character of the Project changed and it became necessary and desirable to reorganise. Franco returned to Italy and ceased to work full-time for the Project from the beginning of 1964; others followed. Staff numbers were reduced by almost 40 in the year 1964-1965 and a further 60 during the following year.

Amongst the staff changes that were made, Samuel Hosegood, who had been wrestling with the fission product clean-up plant was put in charge of assessment studies from the beginning of 1965. Hosegood was a very different personality from Lockett. Whereas Lockett was quiet and unassuming, fascinated by the detail, Hosegood was vigorous and ambitious, keen to make a name for himself and to launch the HTR as a system. Rennie was prepared to allocate £180 000 for work contracted out, and Hosegood proceeded to reorganise the Dragon assessment team as the coordinating centre for more detailed studies of plant systems undertaken by a number of experienced industrial companies within the Signatory countries. The two studies he inherited gave valuable background, especially that of AGIP Nucleare/Indatom which pin-pointed two main problems: gaining access to the boilers in the event of a tube failure (common enough even in gas or coal-fired stations) and replacing core components as this became necessary.

THE DIRECT CYCLE

It was already obvious at the DRE inauguration ceremony that, as had been foreseen when the 8-year Agreement was negotiated, the amount of experience of reactor operation which would have been accumulated by March 1967 would still be small and Rennie prepared an informal paper for the Board of Management suggesting two scales of programme which could be the basis for a further extension to the collaboration. This it was argued should be for five years to give time for a satisfactory power reactor fuel to be developed – reliable, and cheap to fabricate and reprocess. Three years was regarded as the minimum that would be useful. If the smaller programme were adopted it was proposed that DRE be operated solely as a fuel test bed, whereas if the Project's ideas on a continuing research and development programme of broader scope were accepted, it would be possible to explore the higher temperatures relevant to the use of gas turbines for generating electricity. Assessment studies on the steam cycle system would have been completed by the end of the present Agreement and futher work along this line could be left to industry.

Euratom was far from convinced that the specification of a fuel for a steam cycle system would be so advanced that attention to this could be relaxed. The general opinion was that the time had not yet come to start thinking about gas turbines, and there were strong doubts about their suitability for large power stations.

Gas turbines had always had a great attraction for the old Harwell group as being the logical conclusion of a system able to operate at very high temperatures. Some even needed persuading that a steam cycle was worth pursuing at all, and there was an inherent attraction in aiming for a goal that no other system could attain. It should also be noted that by this time Rennie and his staff were developing such a confidence in the validity of the HTR system, they believed it inevitable that the Authority would take it up and industry would be called upon to prepare commercial steam cycle designs. Dragon was also discouraged by the UK from taking design work on a steam cycle too far as this was not the role of the international team which should concentrate on the R & D area. In contrast, positive support for gas turbine work came from Germany where Gutehoffnungshütte Sterkrade (GHH) was seeking a place in nuclear development and saw gas turbines as an opening. Close personal contacts were established with the Project, in which enthusiasm and optimism tended to override the more sober scientific judgments.

If the Project thought at this time only in terms of the UK exploiting in Europe the steam cycle prismatic HTR (the term used to distinguish it from the pebble-bed) it was understandable. Germany's main effort was directed towards the pebble-bed approach and in 1964 an agreement had been concluded between the research centre at Jülich (KFA), industry and Euratom, setting up the 'THTR' Association to study the design of a pebble-bed prototype power reactor as a follow on to the AVR reactor. This in no way diminished Germany's interest in Dragon's research programme; rather the reverse, as so much of the work on fuel and primary circuit components was directly relevant to the pebble-bed system. Indeed so complementary were the two programmes, a special agreement for co-operation between Dragon and the Association was made, that continued with KFA when the Association dissolved. Elsewhere in Europe, there was little practical effort being directed to the HTR system of either type.

Not that the UK position was at all certain. Apart from one reactor in Italy and one in Japan, no export orders had flowed from the Magnox programme and the confidence of both the Authority and industry in gas and graphite was weakening in the face of the global trend to water cooling led by US industry. It was felt necessary to explore a water-cooled system and Treasury approval had been given in February 1963 to build at Winfrith a 100MW(e) Steam Generating Heavy Water Reactor (SGHWR) – a system it was argued which would exploit the advantageous characteristics of light water cooling and heavy water moderation. Such a system could not be ready for the next round of power reactors and the Generating Board went out to tender in April 1964 for either a light water reactor series based on American technology or a series based on the Authority's Windscale Advanced Gas-cooled Reactor that had been operating successfully at full power for more than a year. This employed enriched fuel clad in stainless steel, graphite moderation and carbon dioxide cooling.

Within the Dragon Board of Management there was a general consensus that some extension to the Dragon agreement was desirable and Rennie was instructed to prepare concrete proposals. His paper, first presented to the GPC, extolled the virtues of gas turbines and won support from Schulten at least, present as a Euratom delegate. He too looked to the long term promise of the very high efficiencies theoretically to be obtained from a gas turbine system. The Board of Management was a little less sanguine and it was left to Richard Polaczek from Austria to force home the point that for delegates to be able to convince their governments of the value of continuing the Project, they needed a clear statement on the imminent commercial appeal of the HTR system. The first aim must be to have a reliable reactor system before embarking on a gas turbine development.

At this meeting of the Board in March 1965, effectively for the first time, acknowledgement was made that there had to be commercial justification for an extension. The HTR was in competition with other systems and must prove itself better if it were to justify an on-going development. It had also become appreciated just how much effort was needed to commercialise a system. Sweden, for example, in spite of all the work that AB Atomenergi had put into the development of a boiling heavy water system which led to the start of construction of a 200MW(e) reactor, was witnessing, on the part of industry, a steady trend away from the national line towards the American light water reactors. Sweden saw little chance of engaging in a third system for a long time to come and in view of recent budget cuts gave a warning that a further participation in Dragon might not be endorsed by the Government. Persuasion came in the form of a promise of substantial extra-mural research contracts, which proved effective in keeping the country a member.

Doubts were again expressed at the Board's meeting in June of the value of the gas turbine exercise, in spite of the Project's efforts to generate enthusiasm by organising in May, in collaboration with the Agency, a colloquium on nuclear gas turbine plant. Wisely, from then on, although the Project continued to keep in close contact with GHH, Rennie concentrated his attention on obtaining acceptance for the first part of his original proposal, limited to a period of three years. In this, out of a total cost of £5.71M + £0.59M for contingencies, the main expenditure was on DRE operation and improvement. Of considerable significance had been Peirson's statement that this "second extension should be the last and [he] considered it inadvisable without serious thought, to embark on an extension which contained within itself, the seeds of a further extension". Such a comment could not be taken lightly following the announcement in May 1965 that the CEGB had selected the AGR for its second nuclear programme, that foresaw about 5000MW of nuclear capacity by 1975. This did, however, suggest that the British were reaffirming their faith in gas cooling.

AGREEMENT FOR LIMITED EXTENSION

Clouding the horizon was Euratom's inability to make any definite pronouncements about its future participation **(25)**. Since 8 April, 1965, the three European Communities had been grouped under a single Council which took over the responsibilities of the three Special Councils of Ministers that had previously dealt separately with Euratom, the Common

Market and the Coal and Steel Community. As a result Euratom policy was directly enmeshed with that of the Common Market and when France in its determination to force a change in Community policy, refused to take part in the work of the Council or its Committees, the decision-taking machinery was blocked. Even though in June, the Council agreed to the addition to Euratom's budget for the second five year plan (due to end on 31 December, 1967), of some 3M EUA, earmarked, amongst other things, for a possible extension of the Dragon Project, no clear instructions were given to the Commission on how to proceed. For the Commission the immediate problem was its budget for 1966 and while Guéron and his collaborators were strongly in favour of a continuance of Dragon, they could not hope to get approval for an extension, stretching over two years into the third five year plan, when the whole future of their organisation was being called into question. They also could be forgiven if they themselves were reticent. If Euratom's own budgets were to be seriously cut in the near future, they would have to reconsider the whole of their programme including the funding of the Joint Research Centre which, being on their own territory, would take precedence over external operations. There seemed to be some hope that some form of extension up to the end of the second five year plan would be accepted, i.e. to the end of 1967, but beyond that it was impossible to make any commitments.

Yet most of the Signatories were agreed that an extension of three years was highly desirable and that anything less was of greatly reduced value. Rennie was, therefore, instructed to devise a programme for the period up to 31 December, 1967 which could accommodate a prolongation, without interruption or financial loss. This was extremely difficult to do considering the long lead times associated with new fuel manufacture, irradiation in the reactor and then analysis of the results, quite apart from the impact on staff and the need to run the Project down in a logical way, if it was to close. Some six to seven years are required for the design, development and testing of a new fuel form, including three to four years irradiation in DRE. An extension of even three years was of limited value if the HTR was to be taken seriously. Moreover, fuel loading and reactor operation were not simultaneous processes. They were conducted successively and the end of 1967 was not a logical stopping point, being some time after the discharge of the second core, that was due to be loaded into DRE in the Summer of 1966, and before the third charge would have been operated for a useful length of time.

If the collaboration was to be terminated at the end of 1967, there was no point in going ahead with the third charge and if other programmes of research and development were to be run down smoothly, there would be no point in committing more than £0.66M for the extension as, against £1.69M if the Project were to go on. This was the dilemma facing the Signatories;

whether to choose the first and have any prolongation seriously compromised, or choose the second, recognising that if the Project was not prolonged, more than £1M had been committed unnecessarily. No doubt encouraged by the announcement by Switzerland that Brown Boveri of Baden was proposing to begin a serious industrial design study of a prismatic HTR and would be expecting to work in close collaboration with the Project, the Board, at an informal meeting in February 1966, agreed to adopt the bolder solution, modified to limit the expenditure of Euratom to 2M EUA. The other countries agreed to contribute at the same proportional rate, bringing the total for the restoration of the 1966/67 budget to a constant level plus the nine months to the end of 1967, to £1.555M.

The Agreement was signed in Paris on 11 May, 1966 (three weeks after DRE attained full power) although Norway was able to give only conditional acceptance because of a reappraisal that was being conducted on its future nuclear policy. Norway's ambitious programme of nuclear research that had been carried forward by the personal dynamism of Gunnar Randers, Managing Director of the Norwegian Institutt for Atomenergi (incidentally Chairman of the Board of Management for 1965/66) was being questioned, particularly by industry. The debate was prolonged and it was not until March 1967 that a decision on Dragon was eventually taken – a decision that was positive but constituted a further warning that research on nuclear energy had to be justified by application. Throughout the 1960s coal and oil were in plentiful supply and, in the nuclear field, the HTR was in competition with systems that had had enormous sums expended on their development.

LOOSENING OF TIES WITH AGENCY

The start of 1966 was marked by the tragic death of Jerry Weinstein. When Huet had left the Agency early in 1964 and Saeland had been made Director General in his place, Weinstein had been appointed Deputy Director General. From that time onwards he had assumed much of the load of smoothing contacts between the Signatories and Dragon, keeping the delegates informed of developments and helping Rennie to interpret the changing moods. Weinstein had always been much more than the legal expert in Paris and Rennie felt the loss keenly. He was a close friend, an able judge of the international situation and someone with whom Rennie could discuss his problems in confidence, knowing that he was dealing with a man dedicated to the principle of European co-operation and profoundly interested in the success of Dragon. Weinstein had been concerned with the Project from the beginning; he understood it in all its essentials, yet was content to leave technical decisions to the technically qualified. He also had a quite special relationship with the delegates (in particular with Peirson), all of whom regarded him as a personal friend and one of outstanding ability. With his passing, the bond between Winfrith and Paris inevitably diminished as there was no longer at the Agency anyone with the same personal involvement in the fortunes of Dragon. Saeland was more identified with Eurochemic and with Halden. Weinstein was the continuous thread that had stretched from the original Nicolaidis Committee and which now was broken. Rennie seemed to prefer to leave it that way, repulsing the first overtures of Weinstein's successor, Ian Williams (formerly with the Authority) and even discouraging his participation in Board of Management meetings. On its side, the Agency showed no lessened responsibility towards Dragon and soon had to make a great effort to save the Project. Nevertheless, something had been lost that was not to be replaced.

Rennie's reaction was in keeping with his general attitude towards the Agency. The one attempt that it made to have its own man in the Project was not a success. Klaus Stadie, having worked for General Atomic in the USA, returned to Europe in the Summer of 1964 to join the Agency, and was seconded to Dragon. He had understood from Weinstein that his role was to be that of industrial expert acting on behalf of the Agency, yet found himself occupying a minor position at Winfrith in a disorganised assessment team, with no special access to Rennie. This was a position which he strongly resented, and he made no secret of the fact to Saeland. When Hosegood took over the assessment group, Stadie was able to make a technical contribution to the work but the status of Agency delegate was never acknowledged. After two years, he returned to Paris and the experiment was not repeated. The few people subsequently seconded to Dragon by the Agency were essentially placed for their individual capacity and were not expected to report back.

POWER STATION DESIGN

Two important steps were taken in the 1965/66 period which marked the progress of the HTR along the road to commercial acceptance. In the USA, the Public Service Company of Colorado, convinced of the economic potential of the system, made an agreement with General Atomic covering the construction of a 330MW(e) HTR power station at Fort St. Vrain near Denver, even before Peach Bottom had gone into operation. In Europe, Hosegood's assessment study reached an advanced state and commercial interest was awakened. Accordingly it was arranged that the British Nuclear Energy Society – the joint organisation of Britain's scientific learned societies – should hold a symposium devoted to the HTR. This took place on 23, 24 May, 1966 and for the first time, all aspects of Dragon's work were presented in an open meeting **(26)**. Seventeen papers were read, with contributions coming also from Germany and the USA. General Atomic was able by then to report on operating experience at Peach Bottom, the reactor having gone critical early in March, and in addition, talk of its plans for Fort St. Vrain.

Detailed descriptions were given by the Project of DRE and the research and development that had made it possible, and a presentation was made by Lockett and Hosegood of their 'Design Concepts for Power Reactors'. New features that were introduced were the use of 'podded boilers', i.e. steam raising units enclosed in cavities in the concrete pressure vessel, and a 'feed-breed' system for the core, whereby the feed elements, i.e. those producing most of the power and so most subject to stress, could be replaced at regular intervals with the reactor running, leaving to a major shut down, say every five years, the replacement of the breeding (fertile) elements. Coated particles made purging of the elements unnecessary and all the prismatic designs were based on a core built out of graphite blocks into which the fuel was inserted. The main differences in fuel element design related to the form of the fuel inserts and the method of cooling.

Rennie, however, was conscious that with the UK having selected the AGR for its next round of power stations, the HTR requiring highly enriched fuel, a different fuel cycle (with thorium as the fertile material) and an expensive coolant, was under a strong disadvantage. Studies within Dragon had ranged over a wide spread of variables including the uranium/plutonium cycle and Rennie became convinced that there was a real possibility of designing an HTR with a low enriched uranium feed, the fertile material then being uranium 238 (rather than thorium) converting into plutonium. More detailed physics calculations were made and he was able to announce at the BNES Symposium that, further to the written papers, Dragon was also working on core designs that would require an enrichment of under 5%. Although still somewhat higher than that required for an AGR, the exact level was not too important, particularly with the prospect of centrifuge processes for enrichment becoming practicable.

The news had a profound effect upon the Authority's attitude towards the HTR. Under Penney, Dragon and the highly enriched uranium/thorium HTR had been relegated to the rank of a subsidiary insurance should its overall strategy of Magnox, AGR, fast reactors, not proceed as expected. Penney, in particular, felt that the UK was trying to do too much and since becoming chairman of the Authority in 1964 had been urging his colleagues to consider dropping either the SGHWR or the HTR, neither of which seemed to fit into a coherent long-term programme. Both, however, had their supporters and no decision was taken. A low enriched HTR seemed to slip more into line and a committee was set up in October 1966, chaired jointly by Rennie and Thorn and in which the CEGB participated, to analyse Dragon's power reactor studies and to determine the UK's attitude towards further development.

Very quickly this committee concluded that the UK had little interest in thorium, but a great deal in a low enriched uranium/plutonium system and from then on, Dragon's attention swung more and more to the low enriched cycle, as the outline engineering designs were given body and a more solid basis for economic analysis was established. It should perhaps be emphasised just how much remained unknown at this time. If DRE was proving that coated particle fuels were very good, there were still an enormous number of variables to be considered before a decision could be taken on a fuel element design; graphite specifications were still fluid and the whole of the engineering had to be worked out. A power HTR would be quite different from DRE, and from the current designs of gas-cooled reactors in the UK and France.

In spite of all these unknowns, by early 1967, Rennie and Thorn had together assembled a persuasive argument in favour of the HTR (their final report was dated July) and it was significant that the new Authority delegate to the Board of Management meeting in March 1967, was James C. C. Stewart, Member for Reactors, replacing (Sir F.) Arthur Vick, the Member for Research. In Authority terms, the HTR had become a serious power reactor system – by no means generally accepted, notably by Hans Kronberger, Scientist in Chief of the Reactor Group, yet sufficiently relevant to merit close attention. Kronberger, as the proponent of a developed AGR with silicon carbide canned fuel elements and defender of the SGHWR against Penney's efforts to reduce the Authority's breadth of operation, was the HTR's most powerful opponent. He above all, had to be convinced that helium cooling was necessary or desirable and that core designs for an HTR could be made superior to a stretched AGR.

Meanwhile Dragon's assessment studies had been pursued and the outline concepts described at the BNES Symposium were presented which much more substance at a jointly sponsored Dragon/THTR meeting in May 1967 **(27)**. Held in Brussels, the meeting was attended by 350 representatives of European industry, the electricity companies and other organisations. Dragon came forward with two distinct systems – a highly enriched uranium/thorium system with the fuel contained in hollow pins supported in the graphite core blocks, and a low enriched uranium system with the fuel contained in a series of annular rings mounted in channels in the blocks. (So delicate was the model exhibited in Brussels, it returned to Winfrith in pieces). The general core configuration of this second design, including the size of channels, looked very like an AGR.

Many reactor companies and the Reactor Group of the Authority had been involved in the studies and so were conversant with the main lines of thinking and the promise they held. Moreover, Rennie had taken the precaution, once the essential aspects of the design had been fixed, to have a separate costing conducted by two of the British consortia. Their results came out a little higher than those of the Project but the difference was not significant and could be explained by the difference between national prices and those obtainable from international competition.

Certain features of Dragon's engineering design could still be criticised as reflecting too much a research rather than an industrial approach, notably the adoption of 18 separate cooling circuits, each with its own heat exchanger and gas circulator based on the use of bearings lubricated by the circuit gas. Gas bearings had always had a fascination for the old Harwell team and their development owed much to the basic research put in from Fortescue's days on. They could not, however, be scaled up to very large sizes without a great deal of development work and the number of coolant circuits was dictated more by the desire to continue their use from the DRE, where they had been highly successful, than from a true analysis of system costs. This apart, the reference designs were regarded as being of solid worth and the reactor industry was prepared to be impressed.

BRITAIN'S OWN ASSESSMENT

The Authority's reaction to the report produced by Rennie and Thorn, was to set up in July 1967, a sub-group of its Reactor Policy Committee to examine further developments of gas-cooled reactors. Taking part in the study were representatives of the Central Electricity Generating Board, the South of Scotland Electricity Board, the three industrial consortia – Atomic Power Constructions (APC), English Electric consortium and The Nuclear Power Group (TNPG) – and Dragon.

Three basic lines of possible development were identified; the two Dragon helium-cooled systems presented at the Brussels symposium in May and the Authority's own AGR development of uranium carbide fuel in silicon carbide cans, with carbon dioxide cooling. As variants on these themes were a low enriched HTR, cooled with carbon dioxide, and an AGR cooled with helium. Rennie was clearly in favour of omitting the highly enriched system (for reasons already explained) and dropping any consideration of gas turbines. APC was assigned the task of producing a paper on the Dragon low enriched concepts, referred to as a Mark III G (for graphite), and TNPG on the developed AGR, referred to as Mark III S (for silicon). When these were presented, the CEGB expressed a strong preference for a design with a replaceable core as it had growing fears that the graphite moderator, common to both systems, would not last the life of the plant. Both papers expressed reservations on the fuel element; the first because although there were grounds for confidence in coated particles there was no experience of the multi-annular element that had been proposed; the second because of the little data available on the integrity of silicon carbide tubes.

A series of visits was then arranged to the establishments concerned with different aspects of the development of gas-cooled systems, beginning with a visit to Winfrith in September. In the course of this, Hosegood presented the latest ideas on which the Project was now working which included an engineering concept based on a downward flow of coolant through the core (an idea initially dismissed as impracticable but later adopted) and a new core concept. Already, at Brussels, the Dragon group with the responsibility for making calculations on the DRE core had realized that the enrichment that was being quoted was too high. Coated particles should not be treated as a single continuous fuel and elements like the annular form (which in any case implied uranium densities beyond what could probably be achieved) concentrated the fuel more than was necessary. A more homogeneous distribution could be devised, still with a low enriched feed. Even Rennie took some persuading that this was in fact so, but the latest calculations were confirming the point. The Project had begun, therefore, to concentrate on what became known as the homogeneous system and on the development of a fuel element that would give maximum performance.

After examining this new approach being made by Dragon, the subgroup made visits to the Authority's laboratory at Culcheth near Risley and the production factory at Springfields. Two further reports were prepared: one by the Authority on the III S which was submitted then to TNPG for comment, and the other by Dragon on a homogeneous low enriched reactor, which was vetted by APC. In view of the CEGB's views on the risks of leaving graphite too long in the reactor core, reconsideration was also given to the priority that had previously been attached to the ability to refuel the reactor while it was operating and the conclusion drawn that the disadvantages of refuelling off-load were over-ridden by having a fully replaceable core. By October, the English Electric consortium had prepared its analysis of fuel cycle costs and had concluded that whilst a III S might be able to show a gain over the existing AGR of a discounted capital cost of £4-5/kW installed capacity, the III G showed a gain of double this figure. In November it was concluded that it was no longer certain that a carbon dioxide-cooled III S was indeed feasible and if decisions had to be taken on the basis of present evidence, the choice had to be a helium-cooled III G.

The CEGB was not intending to put out an enquiry for a Mark III before 1969, so that studies on both systems could continue, but in the eyes of the generating authorities, the industry and the official body of the Authority, the Dragon, low enriched, homogeneous prismatic HTR was now the front runner for thermal systems beyond the basic AGR. TNPG was so convinced of the promise of the system that it went ahead with its own commercial design which was submitted to the CEGB in the Spring of 1968, as an alternative to the AGR that was programmed to be built at Hartlepool. Although the proposal was not fully worked through, and was not accepted, this action made by a company with a very considerable experience in the design of gas-cooled graphite-moderated reactors gave a strong boost to the HTR in both the CEGB and government circles.

It may seem curious in the light of this that Stewart, at the Dragon Board of Management meeting in November, should seek assurance on behalf of the Authority that Dragon would not be putting too much emphasis on low enriched fuel cycles to the detriment of highly enriched uranium/thorium cycles that were of such importance to all the Signatories. An explanation may be that the fuel section of the Authority, which was interested in world-wide markets, did not want to lose contact with highly enriched uranium/thorium development in view of its adoption for the THTR and the USA reactors and the possible emergence of other European projects. The section had undertaken a costing study for Dragon on reprocessing and had concluded that the thorium cycle would present no great problems should a market become established.

CHAPTER 13

DRAGON'S FUTURE IN JEOPARDY

IN THE Autumn of 1966 when it had become necessary once more to begin negotiations on an extension to the Dragon Project, this time from January 1968 onwards, the image projected by the UK was that of a country that had lost all interest in the HTR. Since the choice of the AGR for the CEGB's next nuclear power programme had been announced, both the Authority and industry had been making great publicity for the system and for its development potential, centred on the replacement of the stainless steel fuel cans by silicon-carbide. The developed AGR was to be the last in the line of thermal reactors before fast reactors were introduced. Where then did the UK stand in regard to HTRs? For the other Signatories to the Dragon Agreement, this was a crucial matter as it had always been assumed that the UK would take the lead in any commercial exploitation.

Peirson's reaction when he understood the uncertainty that reigned, was to write to all the Signatories in October 1966, stating that the Authority was still strongly of the opinion that it would be wasteful of the effort already put into the Project to bring it to an end in December 1967. He continued "we have been very much impressed by the favourable forecasts of power costs from the high temperature reactor system put forward by the Dragon Project. Even so the Authority consider that sufficient data are not yet available to determine its merits in comparison with other reactor systems. This is the reason why the Authority had not yet decided whether the high temperature reactor should have a place in the reactor development programme, but it is exactly for this reason that the Authority believe that operation of the reactor should be continued until March 1970." This was effective in dispelling uncertainty of the British view, but the main hurdle had still to be confronted.

Euratom's problems had been steadily becoming more intransigent and although the Commission was keen to continue its support of the Project, it was already evident in September that it would not be in a position to agree to a further participation for a long time to come. Preparations of its third five year plan, to begin in 1968, which would certainly include provisions for contributions to Dragon, were only just beginning and even the most optimistic foresaw a long drawn out negotiation ahead. Consideration had been given to Euratom's participation in a Dragon extension ahead of their discussion of the five-year plan, but it was concluded that this would not be successful. Instead it was proposed that the Agency invite individual members of Euratom to join in further talks, in addition to the Commission's representatives.

Of crucial importance was the attitude that would be adopted by France. France, if it wished, could create the conditions necessary for an interim solution but unofficially Goldschmidt had already intimated to Peirson that Dragon could be treated only in the general context of Euratom and there was a risk that any moves to circumvent the impasse would be opposed. With great care, therefore, a special meeting of the Signatory states was arranged with invitations sent via Euratom to its own members.

In October 1966, representatives from all countries taking part in Dragon with the exception of Luxembourg and the Netherlands, assembled at the Agency to consider Rennie's up-dated programme and his plea that a decision on the future of the Project be taken early in the New Year. On the technical side, it was agreed that the programme should be revised to include assessment and engineering studies beyond the Spring of 1967 but on the Agreement itself, little progress could be made. Austria, Denmark, Switzerland and the UK all signified their intention of supporting the extension (the Authority reiterating its strong interest in a continuation); Norway and Sweden could only declare that their attitude was currently under study. From the Euratom countries, France continued to insist that Dragon could only be considered in relation to other reactor development projects.

Germany was the most positive, fully appreciating the relevance of Dragon research to the THTR programme, and the lasting value of DRE as a fuel test bed even though AVR had just begun to operate. It also now had its own prismatic project, GHH having contracted to build a 25MW(e) gas turbine HTR for KSH at Geesthacht. Belgium and Italy were expressing support, with still the proviso that this would be through Euratom and, it was believed that the Netherlands was in favour too. So, whilst there was clearly a majority for a continuation, there was no machinery for bringing it about. No great enthusiasm was expressed for the UK proposal that the separate countries should underwrite the Euratom contribution pending a decision on the five year plan.

Whilst diplomatic initiatives taken by the UK within the Euratom countries confirmed the view that there was a large measure of individual support, there seemed no way of separating the Dragon question from the infighting over all the activities of the Communities and in particular those relating to the Common Market. Peirson therefore urged Saeland to raise the question at the OECD Council, resisting the proposal that this should await the next meeting of the Steering Committee for Nuclear Energy not due to take place until June. The drive came from the Authority, the UK

Government being reluctant to take too strong a position because of its renewed efforts to join the Communities. Peirson was strongly positive, conscious of what an impossible position Rennie was being pushed into, trying to manage a programme with a staff that might have to disband in a few months' time. Penney, influenced by Kronberger's doubts on the practicability of the HTR, adopted a more detached attitude, but even he expressed the opinion when testifying before the Parliamentary Select Committee on Science and Technology early in March **(28)**, that if international agreement could be obtained, the Dragon experiment should be continued for a further two years. The Authority was also thinking of the problem it would face if the talks should break down. It would then have to decide whether to take on its shoulders the whole load and pay compensation for the privilege, or shut the Project down before it had achieved its objectives, and redistribute its staff – at least those that did not go to the USA.

OECD INTERVENTION

Considering that one of OECD's main preoccupations was the current brain drain from Europe across the Atlantic, the Secretary-General could scarcely refuse to put Dragon on the Agenda of the Council meeting of 24 April, 1967. At this meeting, the convocation of a restricted group of the Steering Committee for Nuclear Energy, composed of representatives from the 12 countries participating in Dragon, and the Euratom commission was approved. A few days before, Dragon's problems had also been broached by Penney at a meeting of the European Atomic Energy Society in Portugal, when papers had been presented on the HTR. Times had changed, however, since the middle fifties, and heads of atomic energy commissions were no longer laws unto themselves. As far as can be gathered, no solutions were offered but some modification to the rigidity of France's position may have resulted.

When the restricted group of the Steering Committee met on 16 May, the suggestion from Achille Albonetti from Italy that the time had perhaps come to involve the industry in the continuation of Dragon was highly pertinent. Even though in Italy the HTR did not figure in the national plan, the component industry was keen to be involved. He received no support, even from Switzerland, where it was considered that Dragon should remain a government initiative. It was also much too late. The Project was due to terminate at the end of the year and unless something was decided within the next two months, Rennie would have to start the rundown of staff. Whereas all the non-Euratom countries, with the exception of Norway, were ready to support a continuation until March 1970, the majority of Euratom's members were unwilling to consider a solution outside the context of a general Euratom settlement made even more complicated by the formal application of the UK, Ireland and Denmark to join the Communities. Dragon was one of the pawns in a multi-sided chess game that was to determine the future structure of European co-operation. Only the Netherlands with the suggestion of a compromise extension of 12 months and France which stated that it was prepared to envisage a direct national contribution on a temporary basis, gave any hope that the deadlock could be broken in time.

UK ASSUMES RESPONSIBILITY FOR EURATOM'S SHARE

It was left to the UK to make the effort. Peirson was strongly opposed to another temporary extension but undertook, if he could obtain the necessary assurances from the Euratom Commission that Dragon would feature in Euratom's future programme, to try and persuade the UK authorities to assume the temporary financing of the Project to ensure its continuity. This offer, the Six were not too proud to accept.

Peirson's idea was that whilst the five non-Euratom countries continued to support the Project at the current percentage level, the Authority should carry the rest of the charges for 1968. If Euratom was able to confirm by the middle of that year its continued participation, the Authority would be reimbursed for Euratom's share. Should, at the date foreseen, Euratom not be in a position to confirm its participation, the Project would close at the end of 1968 and the Signatories would waive their rights to the written down assets. For DRE these would constitute a little over 10% of the reactor value.

Williams, who was by then taking over responsibility for Dragon affairs within the Agency, pleaded the cause of the Five, pointing out that the provision for waiving rights to the assets was somewhat unjust to them. No modification was, however, made to the document prepared for the second meeting of the restricted group convened for the end of July, as Peirson justified the clause on the grounds that the Five were not being asked to share the risk of underwriting Euratom.

There was every reason to hope that a firm decision could be taken then, as Peirson had obtained the necessary authority to put his proposal formally and the Council of the Communities at its meeting on 12 July had given its consent for negotiation to proceed on the lines set out by the UK. Nevertheless, Euratom was not able to commit itself to the same proportional funding beyond 1969 and it was intimated that the Community might in the future be replaced by some of its member countries. A precedent for this had already been established when from the beginning of 1964, Euratom ceased to be a signatory to the Halden Agreement. The Netherlands (through the Reactor Centrum Nederland) became a Signatory instead, and was joined in 1967 by CNEN of Italy and a German industrial group. With some reluctance, therefore, the UK accepted that the draft agreement should be amended to provide for other funding arrangements beyond 1968, provided they were mutually agreed before 31 July of that year.

Even then, Euratom was obliged to take a further step back and on the evening before the special meeting of the Signatories, arranged to take place in Vienna at the end of September 1967 to coincide with the IAEA General Conference, let it be known that it would have to reject the draft, as the Commission would go no further than waiving its rights over the fixed assets to permit the Project to continue. Accordingly in a hasty discussion at which the Euratom representatives, Peirson and Hochstrasser, Chairman of the Board of Management, were present, it was decided, on Peirson's initiative, that no agreement should be signed. Instead, the situation should be covered by an exchange of letters between the Authority and the other Signatories. No reference was to be made to future Euratom contributions other than to note that the Agreement would be retroactive from the beginning of the year and would take account of all contributions that had already been paid if it proved possible to agree terms for an extension by 31 July, 1968. Polaczek successfully took up the rights of the Five and it was conceded that only Euratom should waive its rights to a share of the fixed assets in the event of no accord being reached. Guéron was able to signal Euratom's consent on 9 November, 1967, less than two months before the current Agreement was due to expire. That same month, de Gaulle again vetoed any enlargement of the Communities.

CHANGES IN EURATOM

For the Euratom staff, this had been a traumatic period. On 13 July, 1967, the three separate Commissions that had served the ECSC, the Common Market and Euratom were merged into one, and only one member of the former 5-man Euratom Commission was appointed to serve on the new 14-man joint Commission. All support services such as external relations, the legal division, budgets, etc., were merged at the same time, and their previous identification with particular activities was lost. Although the separate Treaties remained in force, Euratom as a coherent entity ceased to exist, its various projects coming under the control of a Commission more concerned with resolving political dissention amongst its member states than with the well-being of a single external unit such as Dragon.

In the reorganisation consequent upon the merging of the Commissions, new Directorates-General were created, the Joint Research Centre was separated from Research which was then joined to Industry. (The Joint Research Centre that had been explicitly foreseen in the Euratom Treaty, it should be explained, had finished up as four separate laboratories, with facilities at Karlsruhe in Germany, Ispra in Italy and Petten in the Netherlands, plus a measurements centre at Geel in Belgium that was rather outside the main stream of nuclear power development. Only Ispra was totally funded and staffed by Euratom. The others remained partially operated by the host country). Guéron, an unrepentant promoter of a Community with supra-national powers, was relegated to the side-lines. Federalism was finished and nothing was to be allowed to go on without the unanimous approval of the Ministers of the member states.

In the circumstances, Dragon could be grateful for the continuity of support it retained in the Commission services. From the moment the Agreement had been signed, Dragon had a consistent ally in Guéron, and he was amply backed by Pietro Caprioglio who was made Head of Project on joining Euratom in July 1959, a position he continued to hold even after being appointed Head of Euratom's part of Petten. Caprioglio saw in Dragon a most important vehicle for European co-operation and served the Project's cause with great perseverance. Tall and of commanding presence, skilled in handling international affairs and well versed in nuclear technology, he was diligent in representing the interests of Euratom's member states at GPC meetings and in discussions with the Project. He was equally assiduous at promoting Dragon in the Community countries – in Germany when on business for the THTR Association, at Petten and also in France, regularly lecturing at Saclay in an effort to stir up some enthusiasm there for the HTR. In the middle of 1967, however, he gave up his positions in Euratom and returned to Italy.

At that time the man in the Commission service in Brussels with special responsibility for Dragon was Mario de Bacci, who had himself been a member of the Project team at Winfrith. Recruited through Euratom from AGIP-Nucleare he joined the Project in October 1959 as a young engineer and worked on reactor plant instrumentation and control. Impressed by the progress he had seen at Winfrith and deeply convinced of the need for technical co-operation between the nations of Europe, he joined the headquarters staff of Euratom in Brussels in August 1963, continuing his work for Dragon in a new capacity.

The vacuum left by Caprioglio was not allowed to persist, and in the Summer of 1967, Pierre Marien was appointed Project Head, reporting to the new Director-General responsible for research, who was not a technical man. Marien certainly was, and a Dragon man too from its earliest days. As Head of the Technical Department at Mol he had even been present when Shepherd gave his talk on the Harwell HTR studies to the EAES meeting in Rome in 1957. Marien joined Euratom in September 1958, and in September 1960 moved to Winfrith as Shepherd's assistant, becoming also joint secretary of the Board of Management. With the successive reorganisations of the Project in April 1964 and October 1965, he became Branch Head, first of Engineering Research and then Physics and Engineering. He returned to Brussels in August 1966 to head the office of the Commissioner for Industrial Affairs, but through the mergers found himself back with Dragon, being appointed Euratom's principal delegate to the GPC from October 1967 and to the Board of Management the following year.

Marien was a nuclear engineer and heat transfer specialist, and whilst at Winfrith, went through all the possible designs of fuel element to analyse their heat transfer characteristics. In so doing, he became an expert on HTR core characteristics and on the merits and demerits of each type. He was strongly in favour of the prismatic form of HTR as against the pebble-bed and was soon to come near to provoking an international incident by putting his thoughts down on paper. Ambitious and forceful, he could be relied upon to uphold the interests of Dragon in Brussels to the limits of his considerable ability, albeit in an independent and personal way. He was to exert a great influence on the Project's future development.

CHAPTER 14

TRANSITION

WHATEVER the uncertainties on the political front might bring, Rennie's task was to keep the Project functioning at full pressure even though the staff complement had declined to the lower eighties. Those returning to the Continent were not replaced and Dragon came to rely more and more on the Authority's permanently seconded staff for its leaders. First priority in the technical programme was the operation of DRE, its improvement and the preparation of fuels needed to drive it. DRE was the fulcrum around which the Project revolved, and the prime tool for fuel development and the study considerable effort in coordinating the extra-mural research conducted in laboratories in the Signatory countries upon which much of the development depended. In addition to these traditional tasks, two new roles were emerging, namely: the management and servicing of experiments in DRE that had been devised outside the Project and which formed part of the "private" research programmes of the Signatories; the provision of advice to industrial organisations wishing to exploit the HTR. While companies in the Signatory countries had access to Dragon information through their national organisations, and more directly through staff seconded to the Project and informal visits to Winfrith, this was not considered to be sufficient by some. In particular, Brown Boveri, Baden, had sought a more direct and private link and GHH was also discussing a closer collaboration.

Industrial interest in the HTR was growing steadily and for a time it seemed that Dragon might have a counterpart in a European organisation capable of commercialising the system. The foundation for this had been laid by The Nuclear Power Group of the UK seeking associates to exploit its designs of AGR on the Continent. Links already existed with SNAM Projetti of Italy through the collaboration between the precursors of the two companies on the construction of a Magnox reactor at Latina, and a partnership was established with BelgoNucléaire to present a bid for an AGR in Belgium. Contacts were also made with GHH in Germany and as a result of these, attention was turned to the HTR. BelgoNucléaire and SNAM were already working together on coated particle development and were keen to establish a front position in the manufacture of HTR fuels. GAAA of

France expressed interest and on 18 January, 1968, agreement in principle was reached between the five companies to co-operate in the development of HTR technology. Behind the scenes, Marien was vigorously promoting the move as he believed that continued Euratom participation in Dragon would be conditional upon positive action being taken by industry.

On 1 August, the Inter-Nuclear company was founded with TNPG, BelgoNucléaire, GHH and SNAM as shareholders in the proportions 3:1: 3:3, and with Paul-Henri Spaak as Chairman of the Board, thus giving the company a prestigious standing in European eyes. GAAA had an option to join but never took it up. The object of the company which was registered in Belgium, was stated as the construction and the exploitation of high temperature gas reactors or complete nuclear power stations utilising this type of reactor, and the coordination of the research, design and development work of the shareholders in the field of high temperature gas reactor technology. Although not explicitly spelled out, the system in view was the prismatic HTR and a side agreement made it understood that TNPG would be responsible for overall design under contract to the company.

Despite the modest initial capital invested (10M BF) Inter-Nuclear was in a position to draw on the financial resources of its shareholders and so theoretically could have evolved into a powerful organisation. Unfortunately, although several studies were undertaken in its name, Inter-Nuclear never developed to the point where it was in a position to become a serious reactor company in its own right, remaining essentially a loose association between companies far from fully committed to the HTR. When GHH disappeared from the nuclear scene in 1971, Inter-Nuclear ceased to have any real meaning as neither BelgoNucléaire nor SNAM had major engineering development facilities, but in any case, TNPG was too preoccupied with its position in the UK to put in the effort necessary to make such an enterprise a success. In July 1968, just a few months after TNPG had put in a bid for an HTR at Hartlepool, the UK Government announced that the industry was to be reorganised into two design and construction companies and the problem became one of survival rather than making a new offensive. Inter-Nuclear was finally wound up in October 1977, having made little impact upon the commercialisation of the HTR in Europe.

In the USA, at the beginning of 1968, it seemed that the HTR had already made the transition to being a commercial reactor. Since Peach Bottom had gone critical in March 1966, its run up to power had been rapid and smooth and in the Summer of 1967 it had been integrated into the supply system of the Philadelphia Electric Company. Its performance had been satisfactory and Dragon's fears that the rather primitive coated particle used as fuel would bring the technology into disrepute had not been realised. In the Autumn of 1967 Gulf Oil had taken over the General Atomic Division of

General Dynamics, and a new company formed, called Gulf General Atomic (GGA). It was expected that under the new management, construction of the Fort St. Vrain power station would get under way quickly and that early in the 1970s, the USA would have a major HTR plant in operation.

Bright though the industrial prospects now looked for the HTR, there was in Europe, still no single fuel cycle or reactor design which had won general acceptance, even though fuel research in Dragon had become concentrated on the low enriched uranium/plutonium cycle. The 2¼ years that remained until the Project was due to finish, even if agreement could be reached by the end of July for a continuation until March 1970, seemed all too short for the work that remained to be done. Accordingly Rennie, at Euratom's instigation, tabled a draft paper for the January 1968 meeting of the GPC which examined a programme for operating the DRE until March 1973, to permit the full testing of power reactor fuel elements. For the Authority, as Peirson made clear at the subsequent Board of Management meeting, this caused acute embarrassment as it came at a time when the Authority was being required to make serious economies, yet carry the load of Euratom's contribution. Euratom believed, on the other hand, that its position in its own internal negotiations would be strengthened by an expression of intention to prolong the Project for a further five years. It was still far from obtaining agreement for its own programme, but it was understood that Belgium, Germany, Italy and the Netherlands would all be in favour of continued participation in Dragon. Not surprisingly, the non-Euratom Signatories were of the opinion that only when the extension to 1970 had been obtained, could a further extension be considered.

At this same meeting of the Board of Management in March 1968, Rennie announced his intention of retiring at the end of June from the Authority and from his position as Dragon Chief Executive.

Rennie's reasons for leaving were the subject of a great deal of speculation. Within the Project, it became established lore that so bitter had his battles become with the Authority, notably with Penney and with Kronberger, that only after his departure could the Authority recognise the error of its ways in pursuing the AGR and its carbon dioxide cooling, without losing face. Yet by the end of 1967, even though the fights with Kronberger on silicon carbide had been bitter, the main battle had been won, and he had every reason to believe that the choice for the next round of nuclear power stations in the UK would fall on the HTR.

It is clear that by then Rennie had become exasperated by the Authority's leadership, interpreting the slow progress towards appreciating the superior qualities of the HTR over the AGR as being due in great measure, to the 'Not Invented Here' syndrome. He considered the Authority insular and arrogant. He had every faith in the Authority's ability to build things well, to a time-table and to a price, but none in its ability to make essential policy decisions. Peirson always apart, he saw the Authority as a body moved by changing fashions, unable to grasp the essential elements of a problem, unwilling to face all the facts of a case at one time and draw logical conclusions. Rennie was not, however, so disillusioned that he did not discuss with Penney and then his successor as Chairman of the Authority from 16 October, 1967, Sir John Hill, the prospects of moving from Dragon into another Authority post. No vacancies materialised in top positions for which there were not already strong contenders more directly in the line of command, and no post was offered that Rennie considered took due account of his success as Dragon's Chief Executive.

Rennie had also become devoted to the HTR concept and was intent on pushing the system as hard as he could. Rising 52, he was still young enough to take up a new challenge and ambitious enough to want to do so. He had served Dragon for nine years, had piloted it through its formative years, established a highly satisfactory structure, developed and led a dedicated team, built a thoroughly successful reactor, and had seen the HTR system demonstrated to be a promising commercial proposition. Although DRE could still operate for many years as a test bed for reactor components, the Project's essential objectives had been achieved. From then on, the centre of action would be on the industrial front. If he was wearied by the endless negotiations over the past three years on the extension agreements, he did not show it; indeed if the negotiations on the 1967-1970 extension had gone smoothly it is possible he would have left earlier. Nine years he believed was quite enough in the same post. After that, one was likely to become stale; he had contributed as much as he could and it was up to others to take up the reins while he led the crusade elsewhere.

SHEPHERD APPOINTED AS SUCCESSOR

It was for the Authority to propose his successor and the natural choice was his Deputy, Leslie Robert Shepherd, a choice that received the unanimous support of the Signatories. No one knew more about HTR technology or had a better overall grasp of the HTR and its potentialities. As Fortescue's first associate in the original Harwell group, he had participated in the evolution of the system from the earliest days, developing an unswerving faith in its role as the logical successor to the earlier gas-cooled graphite-moderated reactor concepts.

Born in November 1918, his early years were spent in the mining community of South Wales. In 1937 he entered University College, London, gaining a first class honours degree in physics in 1940 and he then moved to Cambridge where he worked on electronic anti-aircraft fuses.

After the war, he stayed on at Cambridge to do his PhD under Sir Laurence Bragg at the Cavendish Laboratory, and made his first contact with the phenomenon of radioactivity. In 1948 he joined Harwell, where at first he was concerned with the physics of natural uranium/graphite reactors, subsequently moving over to fast reactors and leading the group that operated Zephyr, a zero energy fast reactor. He was converted to the HTR when he became Fortescue's deputy in 1956 and was made responsible for research and development. When Fortescue left for the USA, he took over the group. At the meeting between Cockcroft and Guéron prior to the Dragon signature, they agreed that Shepherd should head the Division responsible for Research & Development provided no one else was put forward who could claim to be his superior from the technical and personal points of view. There was no such candidate. With the reorganisation of the senior staff structure as the DRE construction came to an end, he was appointed Deputy Chief Executive as well as Head of the Materials & Chemistry Division (commuted the following year back into the Research & Development Divison with the dissolution of the special Physics Division).

In so many ways, Shepherd was the ideal complement to Rennie. While Rennie was at his best handling matters of policy, Shepherd was a master of technical analysis, enjoying the cut and thrust of scientific debate, and reeling off facts and figures from memory with hardly any reference to notes. While Rennie was an autocratic leader, Shepherd was always prepared to give weight to the other side of an argument and try to see an opponent's point of view. While Rennie would not tolerate exhibitions of temperament, Shephered would listen patiently, trying to sieve out the substance of what was being said. If the softness and lilt to his voice betrayed his Celtic origins, he showed none of the fierceness sometimes associated with his forebears. Nevertheless, he was a dogged fighter and in Rennie's time was always deputed to represent the staff at Authority promotional meetings, as he could be relied upon to defend them to the limit. In appearance, while Rennie was casual in his dress, Shepherd was always fastidious, the antithesis of the popular image of a research physicist. Of the two, Rennie was the bolder, trying to circumvent problems before they solidified, yet prepared to take decisions as soon as events dictated them. Shepherd was quicker to appreciate the technical implications of a situation yet more inclined to wait and see how matters developed, hoping that further analysis would show the right way to proceed.

His scientific competence was immense and in his exposition of technical analysis he demonstrated an exceptionally ordered mind. He would write reports or original papers, not rapidly, but with such fluency that his manuscripts contained hardly an alteration. Yet he was a romanticist, as shown by his life-long preoccupation with interplanetary travel, and an idealist when it came to questions such as European co-operation for which he was a devout enthusiast. As the head of Dragon's research and development activities he had shown himself to be clear in his scientific concepts, never content to accept any advance as the last step, always looking beyond to the next level of improvement and always ready to listen to new ideas and alternative approaches. Such qualities commended him to the delegates of the Signatories, with whom he was on excellent terms, even if his intellectual flexibility, coupled to his considerate attitude to all his colleagues, was a handicap when it came to imposing a coherent policy on his individualistic senior staff. When Rennie left there was no one to provide the complementary qualities to Shepherd that he had given to Rennie. Together they had made a formidable team, even a little too formidable for some of the Signatories who, confronted with such powerful competence, found it difficult to change any course of action once the Project had made up its mind.

BRITAIN INCREASES CONTRIBUTION TO ASSURE CONTINUATION

At the meeting of the Board of Management at which Rennie's departure was formally announced, the decision was taken to set up a sub-committee of the GPC to consider future programmes and the staffing necessary to implement these. This met under the chairmanship of the Chairman of the GPC, Michael Higatsberger of Austria, and consisted of Fry, Director of Winfrith representing the Authority, Marien from Euratom and Shepherd from the Project, with the Head of Administration Branch acting as secretary. They set in motion the re-appraisal while the Signatories waited to see whether Euratom could determine its forward position by the deadline of 31 July, 1968. Progress at first was slow and while the Agency was putting in train some contingency planning in the event of a delay, Peirson was obliged to warn the Board of Management at the end of June that if no decision had been taken by the date laid down, and only a consent in principle given subject to major negotiations on financial terms, the Authority would no longer be able to sustain the Project. Remaining funds would have to be used to resettle the seconded staff. Saeland also pointed out that the other non-Euratom Signatories were in no position to accept any radical changes.

For Euratom, the crucial meeting of the Council of Ministers was to take place late in July and in preparation for this, the Agency, the Authority and representatives of the Commission of the Communities met in Brussels to hammer out a compromise. With much difficulty, Guéron had succeeded in persuading the Commission to make an approach to the Council of Ministers requesting that Dragon be treated as a special case, despite opposition from Germany which was anxious to see continued Community support for THTR and was conscious that the other members of Euratom had a definite preference for the Dragon (and US) prismatic system. By separating the two projects, Germany feared (with reason) that support for THTR would no longer be forthcoming. Continuation of Dragon was, nevertheless, of importance to the German pebble-bed and gas turbine HTR programmes and reluctantly Germany withdrew its objection. At the eleventh hour however, France raised the question of Euratom's proportional contribution and tried to insist on a reduction to 30%, agreeing eventually to a Belgian compromise being put before the Council. The Council's response was to empower the Commission to negotiate an extension on "more equitable terms" with a view to reducing the Euratom participation to 36%. This was the position confronting Peirson; if the Project was to be saved, the Authority would have to assume a larger fraction of the budget.

From the negotiating point of view, the position of the UK was weak as it had such explicit interest in the results that would emerge from the coming period of operation of DRE. While Guéron and the Euratom staff were keen for Dragon to continue, as one positive example of European co-operation in the technical field, only Germany amongst the Euratom members had definite HTR projects. Euratom was in the position of seeking help to persuade its own Council. If the UK refused to move, there was a strong probability that the Council would reject the proposals and the Authority would then have to bear the full burden. Guéron was, in consequence, able to persuade Peirson to accept responsibility for 46.8% of the budget instead of 40.8% diminishing the Euratom contribution from 46% to 40%. In return it was unofficially agreed that the programme of irradiation in DRE would be largely devoted to the proving of the fuel for the HTR that the CEGB was likely to be ordering for its next nuclear station.

With this accepted, the only other difficulty that remained in concluding the terms of the new agreement that was to be back-dated to January 1968, was the exact form of words, notably in reference to what should happen after 1970. Should no further extension be agreed by September 1969 at the latest, so that the Project as such ceased to exist, yet the Authority continued to operate the reactor, the Signatories were to have access to operating experience and given preferential treatment in regard to requests for irradiation space made by companies or organisations sponsored by the previous Signatories.

The day before the deadline, Euratom was able to announce that the new terms had been accepted by the Council. This marked effectively the last connection between Guéron and the Project. After openly fighting the Commission and the Member States of the Communities for more than ten years, he retired at the end of 1968. His last appearanc at a Board of Management meeting coincided with that of Rennie's. It marked the end of an era.

CHAPTER 15

INDUSTRIALISATION VERSUS RESEARCH AND DEVELOPMENT

RENNIE'S departure and the assumption by Shepherd of the responsibilities of Chief Executive from 1 July, 1968 marked a new beginning in the life of the Project. Assessment studies were coming to an end and in the technical programme prepared by the sub-committee and approved by the Board of Management for the coming period, it was concluded that the HTR had reached a stage of development where its adoption as an alternative source of power in central electricity generation was imminent. "The detailed design and assessment of such reactor power plant is now a matter for the industrial organisations concerned with their construction and the utilities ordering and operating them." The optimism seemed justified.

Electricity consumption was rising steadily, and even in the UK where the economy had been in recession and 1967 had seen the pound devalued, growth in demand for electricity was rising again to a rate of about 6% per year. In the Dragon countries as a whole, the average increase was around 5% per year. Moreover, despite the keen competition from fossil fuels, nuclear power was winning on price and markets were expanding, particularly for the American designed light water reactors of Westinghouse and General Electric. In the USA alone in 1968, construction started on 20 new nuclear stations a number of which were over 1000MW(e) in capacity **(29)**. Canada was continuing with its line of heavy water reactors and the UK and France (for a short time longer) with gas cooling and graphite moderation. Even though elsewhere, the light water reactors had largely supplanted national designs, there was still room for the introduction of a thermal reactor of high efficiency and compact form if the price was right, as Fort St. Vrain seemed to prove.

The HTR appeared to offer a real alternative and Europe was well placed to exploit it. In Germany, there were the THTR and GHH projects, in Switzerland the Brown Boveri design, and in the UK the sub-group in the Authority studying the Mark III gas-cooled systems had just recommended that future design and development work should concentrate on the homogeneous low enriched HTR with replaceable cores. With construction started at Fort St. Vrain in September, the Reactor Policy Committee of the Authority in November endorsed its sub-group's recommendation, at the same time making it clear that it was proposing construction of an HTR prototype. Contracts were placed soon after with the two consortia that emerged from the reorganisation of the nuclear industry, for the preparation of competitive designs of this prototype. The site chosen was Oldbury.

OPERATIONAL RECORD OF DRE

The confidence thus expressed in the potentialities of the HTR and which was shared by the Authority, the Central Electricity Generating Board and industry alike, was based largely upon the experience of operating DRE. Some difficulties had been encountered in the secondary water cooling circuit starting early in 1967 which necessitated shutting off one heat exchanger and blocking a number of tubes but this was not related to the performance of the reactor core. In all the important areas, DRE had given excellent service. The fuel, which was being continuously improved, was highly retentive of fission products and, as was clearly demonstrated when the heat exchangers were replaced in the Spring of 1968, contamination of the primary circuit was remarkably low. DRE had shown that gas loss rates could be kept well within acceptable limits, there were no unexpected corrosion effects with helium cooling and wear of moving parts in the water-free atmosphere could be made small. The problems on the heat exchangers were entirely on the water side and even then it was possible to operate the reactor with a steady power output of 18MW.

Moreover as an experimental tool, DRE was proving to be highly versatile as the Project staff, assisted by the Operations Group, learned how to use the available core space to maximum advantage. During the running in period from June 1965 - August 1966, DRE was operated for nearly 230 days. For half that period, the power level was over 10MW, and for the last three months was continuously at 20MW. In the first fuel charge the majority of the elements were devoted to driving the reactor and just a few holes were available for experimental samples, but in the light of the experience gained, it was decided to make all the elements contribute to the irradiation programme, by using the outer six rods as drivers and leaving the inner rods for experiments. In the second charge loaded into DRE in January 1967, in addition to the experimental fuels manufactured by the Project, were compacts of coated particle fuel manufactured by the Oak Ridge National Laboratory, and samples made by German and Austrian companies for the THTR Association.

When the reactor was shut down in March 1968, further modifications were made to the core and, for the first time, elements were introduced which, although they had the same outside dimensions as before, were made from blocks of graphite into which could be inserted fuel pins such as might be used in a power reactor. For the first time also, irradiation of low enriched fuels was begun. Elements were adapted to allow fuel balls for THTR to be tested and compacts of fuel were included made from kernels of SNAM Projetti coated by BelgoNucléaire. There was even some silicon-canned fuel of the Authority in the reactor. On start-up in the Summer, a new operational routine was introduced, whereby only 20% of the driver fuel was changed at a time to give more regular conditions in the core throughout a complete irradiation period.

PROBLEMS OF FUEL DEVELOPMENT

Fuel development is a lengthy process, complicated by the fact that the fuel under test is, as already noted, an integral part of the core. From the definition of the particular make-up of samples to be irradiated, a year would pass before they could be inserted into DRE. They had to be fabricated and assembled into elements while the overall core characteristics were being worked out and the safety of the whole system assessed. Irradiation would vary from a single reactor run of about three months, up to several years for those samples taken to maximum burn-up. After that, the elements would be discharged from the reactor and allowed to 'cool' for a period to allow radioactivity to decay, following which they could be dismantled and the samples extracted. For highly active elements, this might take six months. Only then could a detailed examination begin and a full analysis would take at least a year. From first conception to the verification of fully burned-up samples, a period of about 6-7 years could thus elapse.

It must also be remembered how many parameters have to be defined in the specification of a fuel. There is the initial material – its enrichment and composition, the size of particle into which it is transformed, the density and thickness of each coating applied and then the manner in which the coated particles are bonded together. There is also the question of the shape of the fuel inserts, how they are contained in the graphite blocks that make up the core, how the heat is carried away – all these different aspects must be considered in the knowledge that each separate decision will interact upon the others and, of course, upon the design of the reactor for which the fuel is made. A basic problem, therefore, in the evaluation of a system is to know where to start, where to place priorities and where to accept compromises.

In the design of Fort St. Vrain, GGA had chosen as its basic structural element for the core an integral block; i.e. a graphite block that had separate holes drilled in it for fuel inserts and for the passage of the cooling gas. The fuel heated the block and the block was cooled by the helium. Dragon did not like this design because of the high thermal gradients that would be set up in the block which could result in distortion and possibly failure. Dragon preferred a block structure where the fuel pins were inserted in channels down which the coolant flowed – the so-called 'pin-in-block' design. The UK shared this view and the Mark III sub-group had asked the consortia to evaluate two basic pin forms and two sorts of fuel bonding. Each of the (then three) consortia came up with a different choice, depending upon their particular optimisation procedure and also upon the advice they received from Dragon where opinions on priorities differed from one person to another. When the Reactor Policy Committee of the Authority took its decision in November 1968 to concentrate its advanced work in gas-cooling on the HTR, the fuel element had still to be defined.

PROGRAMME FOR DRAGON

Shepherd by this time was already thinking ahead to the period beyond March 1970 when the present agreement was due to terminate and had been sounding out the Signatories to obtain their views on a future programme. It was evident that there was a large measure of support for a continuation at about the current level of expenditure, with the Project fulfilling a variety of roles. As always its main task would be running DRE, preparing fuels, experimental rigs and improving and maintaining the reactor. It would be testing materials and fuels of its own and also those of industry. It would be pursuing the programme of extra-mural research in the Signatory countries and providing advice and data for organisations within them. It wanted also to contribute in an original manner to the physics and engineering of HTRs.

The staff of the Project was not prepared to let it become a service station, simply providing irradiation facilities for outside organisations. It was determined to continue as the lead organisation in Europe in the development of HTRs. Shepherd stressed the view that if the Project was to attract the quality of staff fitting to an international organisation, it must continue to work at the frontiers of the technology. Although for steam cycle systems, industry would be taking over the design of power stations, the HTR had still a great deal of development potential and it was time, for example, to think again of higher temperatures and gas turbine applications where GHH was already paving the way. Furthermore, the Agency, in collaboration with the Authority, had been giving thought to a joint programme on a helium-cooled fast reactor to which Dragon could clearly contribute.

Shepherd put forward his proposals for the continuing programme in a massive review of the present state of HTR technology, leading on to an analysis of the detailed research that should be undertaken. He estimated that for a three-year extension, plus the additional expenditure needed to restore the 1969 programme to a constant level, a budget of £7.2M would be required and he requested the Signatories to agree this figure. With very little modification (once it had been trimmed to a more manageable size) Shepherd's thesis was accepted by the Board of Management in April 1969. The budget was shaved to £7M to bring the round total for the Project from its beginning to £38M and work on fast reactors was made dependent upon the outcome of the Agency's negotiations (which though leading to the establishment of a Coordinating Group on Gas-cooled Fast Reactors, and indirectly to the Gas Breeder Reactor Association of interested companies, involved Dragon hardly at all).

Dragon was ready to embark on a new period of activity with a wide area of responsibilities, stretching from routine irradiation work to research on advanced HTR systems. Commercialisation could be left to the Signatories and indeed, the German member of the Euratom delegation at the April Board meeting made clear that he "would welcome a general tendency to leave leadership more and more to industrial enterprises in Signatory countries", a view that the Project was happy to accept if, but only if, this did not extend to research on improved systems.

A problem, however, the Project would have to take more seriously, was the allocation of resources, and it was recognised in the Board that there was a need for "greater rationalisation of the Dragon programme and associated HTR programmes in the Signatory countries" reflecting "the new phase upon which the Project had entered, in support of the commercial exploitation of the HTR system". Ad hoc procedures depending upon the willing co-operation of individuals in the Dragon senior staff were no longer adequate. It was agreed, therefore, to convene a special meeting of the Board and the GPC members at which new co-ordinating machinery could be discussed. Such a meeting was held in June 1969 and the decision taken to set up a Programme Sub-Committee of the GPC – the PSC – to advise the Chief Executive on:

1. the use of DRE as a fuel and materials testing facility within the framework of the entire Dragon irradiation programme, and
2. the coordination of the Project's programme of work with HTR work undertaken elsewhere by research establishments, industries and utilities both within the Signatory territories and those of other countries with which the Project had formal collaboration agreements (i.e. the USA).

Later, the Board modified the final phrase to "having due regard to work in other countries". The PSC which was to report to the GPC was to be comprised of one member per Signatory, freedom being given to co-opt additional persons according to need. The Chief Executive was to participate (and could also invite staff members) as well as a representative of the Agency.

In confident mood, the Board of Management met in June to review progress on the negotiations over the new extension and to take part in the somewhat tardy celebrations held to mark Dragon's 10th Anniversary. Spread over three days, they included a presentation of the Project to prominent people from government, science and industry and ended with an open day for the press, staff and public who were able to visit the exhibitions designed to make publicity for Dragon and the HTR system.

Euratom had still formally to come to a decision on the new extension, at the meeting of the Council of the Communities to be held a few days later, but clear indications were given that this time the affairs of Dragon had been de-coupled from other projects under discussion, and a positive outcome was practically assured. Germany, Belgium and Italy could be certain of giving their support in the light of their own programmes and the existence of Inter-Nuclear. The Netherlands, which through Petten was actively involved in HTR work, and Luxembourg could be relied upon to co-operate and France with de Gaulle retired from the political scene since April was unlikely to be difficult, particularly as its attitude towards the HTR was changing. A major review had been under way in France of its nuclear power policy and, although not formally disclosed until November, it was already clear that no more reactors analogous to Britain's Magnox reactors would be built, and the country's programme for the VIth national plan, to run from 1971-1975, would be based on light water reactors **(30)**. Although this meant turning away from gas cooling for the present, it would release a large number of staff who were proficient in the technology and who were keen to pursue the development of a related system. Moreover, with attention being directed more and more in Dragon to the low enriched uranium/plutonium cycle, as against the uranium/thorium cycle, France was prepared to take a new look at the HTR. Euratom's agreement to the new extension to March 1973 was duly confirmed.

Only Norway (the smallest contributor) was reticent and made it known that it was unlikely to be able to contribute more than £50,000 of the £115,000 that was calculated as its due amount. Reluctant to pull out altogether, particularly as this could be interpreted as a riposte to the UK's departure from Halden, recently announced, it nevertheless felt compelled to limit its involvement. The other Signatories were anxious that for "this last extension" as Peirson put it, there should be no fundamental change in participation, and Saeland proposed that the difference be made up in kind. Acting on this suggestion, Shepherd went to Kjeller with the proposition that contracts placed in Norway in excess of £50,000 should be paid for by the national institute up to a limit of £115 000. Such a deal was not possible and in any case, could not affect the formal position. Either way the other Signatories had gone too far with their requests for funding for adjustments to be made to their payments. The solution adopted, was to cut the contingency sum in the budget by £65 000 and change the percentage contributions of the other Signatories slightly. Austria gave notice that this was not to be taken as a precedent.

Of all the extensions, this was the smoothest to be negotiated and the delegates and the Chief Executive could congratulate each other on arriving at a new understanding. No problems were posed in drafting the few clauses necessary in the Agreement (once Shepherd has persuaded the Agency to include a mention of total expenditure from the beginning of the Project to avoid the necessity of closing all the accounts on 31 March, 1970) and a deadline of 30 September, 1972 was set for the determination of any further extension. Of the changes in emphasis which appeared in the annexed technical programme, the most significant were the specific references to work in support of low enriched uranium cycles, limited assessment studies in close collaboration with industry of the direct helium cycle application of the HTR, i.e. gas turbines, and the provision of consultancy services in the design of high temperature power reactors on terms to be agreed.

PROGRAMME IMPLEMENTATION

Although the broad programme had been unanimously approved and the mechanism for defining the detail established, a serious rift developed between the Project and its major Signatory over its implementation. This concerned essentially the choice of fuel element for Oldbury and the irradiations in DRE which were to prove it.

In the discussions which had taken place at the end of 1968 and the beginning of 1969 between the Authority, the CEGB and the industrial consortia, in which Shepherd also participated, the decision had been taken to adopt for the first charge in the Oldbury prototype, a fuel pin in the form of a hollow rod, cooled on the outside only. For the CEGB, increasingly uncomfortable over the difficulties that were being experienced in building the AGRs it had on order, the prime consideration with this first HTR was certainty of operation, and the hollow rod most closely resembled the driver fuel rods that had been used in Dragon for $4\frac{1}{2}$ years and which had given such good service.

Senior staff at Dragon believed this decision to be short-sighted and to be the result of a lack of that experience and insight which only the Dragon team possessed. They were working on different forms and they wished to devote the maximum research and irradiation effort to their development.

In a reactor of high power density, using the hollow rod design, shrinkage of the fuel would give rise to excessive temperature differences across the gap between sheath and fuel and this could damage the coated particles. Dragon was studying a pin that was cooled on both the inside and outside and the fuel was allowed either to press in on the inner liner (tubular interacting, as it was termed) or remain free (tubular non-interacting). In addition, the Project was seeing if it could not get the best of both worlds, by adapting an original design of Marien called the teledial, where thin rods of fuel were embedded in a ring of holes drilled in a graphite tube, that was then cooled externally and internally.

The Authority, which would be responsible for the Oldbury fuel, took the engineering view that even though Dragon was already planning to change the fuel rod clusters of DRE by a pin-in-block assembly, the new pin designs were still largely speculative, whereas several years' experience had been amassed on the old hollow rods. The authority urged therefore an intensive programme of irradiation of hollow rod pins of the form that the British had chosen, to give a statistical basis for the performance that could be expected. If, in the future it could be shown that other designs were better then changes could be made in later versions. For the present, what was important was reliable data on the simplest pin that could be devised. The Project countered with the argument that the experience with the driver fuel was irrelevant and what mattered most was obtaining a thorough understanding of the behaviour of fuel samples of different compositions in the new block elements, following which the final design could be defined. The most meaningful statistical proof lay in the testing of the particles.

Apart from this basic disagreement, which was characteristic of the engineering versus R & D approach, other forces were at work. Dragon staff believed they knew best and they were not prepared to have outsiders decide what the right fuel programme should be. They were jealous of their knowledge, and resistant to other people imposing not just a choice of fuel for their own reactor but, as a corollary, a development and irradiation programme at Winfrith. Moreover, Dragon's Chief Executive and two Division Leaders (Hosegood, Reactor Technology and Huddle, Materials) as secondments from the Authority, felt personally involved in UK programme decisions and constrained to dispute them when they disagreed. Had Oldbury been, say, a German project they would have accepted much more readily the essentially internal policy decisions taken by a Signatory country. It must also be noted that at the personal level, Thorn as the Director responsible for HTR development in the Authority, was continually at cross purposes on professional matters with Dragon's senior British staff – perhaps another facet of the traditional division between engineering and R & D.

Dragon's sensitivity to the threat of outside domination and its conflict with the Authority, were laid open at the first meeting of the Programme Sub-Committee in October 1969, attended by eight people from Euratom plus Marien in the chair (acting in place of Andreas Fritzsche of Switzerland, the current Chairman of the GPC). The main point of discussion was the irradiation programme for DRE, and Shepherd made the Project's attitude clear when he invoked a ruling made by the Board of Management in 1965, that the Project's own programme had priority, applications from Signatory organisations second and non-Signatory organisations last. That this was no longer the will of the Board was shown at its meeting the following month, when it removed the distinction between Project- and Signatory-inspired irradiations and instructed the PSC to work out together the optimum use of space in the future.

Thorn had come to the PSC meeting expecting to have as the main item on the agenda, the Authority's irradiation programme for Oldbury. His approach was conditioned by the understanding which had been reached with Euratom, when the Authority had agreed to pay 46.8% of the total budget for the 69/70 extension (now prolonged at a slightly higher rate to 1973) that full support would be given to the UK prototype. The Authority was thinking in terms of 60% of the available space being devoted to Oldbury irradiations, whereas Thorn found the Project demanding priority, and a battery of representatives from Euratom countries also staking a claim. Only Tor Midtbø (former Branch Head of Engineering II in Dragon and now representing Swiss interests) and, to some extent, Marien, seemed to appreciate the need for a concentrated effort on the fuel the UK had selected for Europe's first prismatic HTR power station based on the steam cycle.

The Project insisted that DRE was unsuited to the irradiation of more than a few pins of low enriched uranium, because of the high absorption of the heavier isotope and it was, in any case, irrelevant to do statistical tests on anything but production fuel which would not be made for some time yet. Dragon's own programme was far better adapted to defining the fuel element and the essential components of a tubular or teledial element could be proved in good time for fabrication of the core for Oldbury. Moreover, it contended that if work were to be concentrated on one design, this design had to be "jointly agreed" (a phrase that was underlined twice in the minutes of the meeting).

Thorn returned to the CEGB and the consortia to rediscuss UK policy in the light of the situation as it was now presented, and the second meeting of the PSC was delayed until January 1970 pending a re-appraisal. As only 30% of the Dragon space would be available to the Authority, the idea of a concentrated effort on the hollow rod (which incidentally had still to be fabricated) could not be put into effect and in these circumstances, it was acknowledged that Dragon's technical arguments should prevail. In January therefore the UK agreed to accept a broad programme of fuel element research and the consortia's designs were held up until more concrete results could be obtained. Subsequently both the tubular interacting and the teledial were shown to give excellent performance – after events had passed them by – but then a hollow rod would have been equally satisfactory at a lower power density.

A curious aspects of the Oldbury story, is the fixation on having a high power density (of about $8MW/m^3$). CEGB was not pressing for such an extreme design, and the English Electric consortium (that was then called BNDC) did indeed suggest that the density should be dropped to half. At this level Dragon's objections to the hollow rod could have been withdrawn. One reason now given is that the Continentals were pushing for maximum performance and the Project too. The Authority, however, is not normally susceptible to outside pressure of this type and although TNPG could have been influenced by its continental contacts through Inter-Nuclear, the more probable explanation is that the anxiety to show immediately what the HTR was capable of, was general. Also, everybody wanted to do better than Fort St. Vrain which had a power density of $6.3MW/m^3$.

PRISMATIC HTR POWER PROJECTS IN EUROPE LOSE SUPPORT EXCEPT IN FRANCE

The arguments over fuel and their acrimonious nature, the delays and the previous doubts expressed by Kronberger on whether a satisfactory fuel could be mass produced, shook the confidence of the CEGB, and the question was raised whether the HTR was, after all, yet ready for commercial exploitation. Huddle did not help, by stressing all the unknowns surrounding the behaviour of primary circuit materials at high temperature. Intent on starting a new metals research programme, he was promoting it by parading all the misfortunes that might be in store.

In its broad objectives, the programme he proposed was entirely justified, and proved to be very popular with the Signatories in the long term. Up to that time, the main concern with metals in the primary circuit had been related to bearing surfaces and the corrosive effects of trace elements, the latter being studied in Norway under contract by exposing samples to possible contaminants at low pressure. Now that operating experience with DRE had shown what an HTR helium atmosphere was really like, Huddle saw the need to make tests under conditions simulating those to be expected in a reactor, and also to measure the influence of stress. At gas turbine temperatures, the field was largely unexplored. His enthusiasm for this new area of research tended to cloud the fact that DRE had been operating for five years without trouble and, for a steam cycle power plant, while further confirmation was desirable, no serious problems with the primary circuit were to be expected.

When the designs of the British consortia for a 750MW(e) HTR power station at Oldbury finally came to the CEGB for assessment early in 1971, it was against a background of economic recession and renewed indecision on the part of the Government as to what should be the basis of Britain's future nuclear programe, in view of the poor progress made on the AGRs and the major problems that were still outstanding. The recession was reflected in the demand for power, and the CEGB, having planned on an annual increase in consumption of over 6%, had seen the figure drop to about 4% in 1969, 3% in 1970 and there were no signs of recovery. Moreover, because of investments made many years before it would be commissioning over 9000MW new capacity in 1971 and '72, thereby adding nearly 20% to its total network. On top of this, as part of the Government's drive to curb the growth in unemployment, it had been obliged to order yet another oil-fired station **(31)**. In the meantime, nuclear policy was being reviewed by the Vinter committee that had been set up in 1970. This was trying to resolve the competing claims of the ailing AGRs, the SGHWR, the HTR and the light water reactors.

When the UK had selected, at the end of 1968, the HTR as its next thermal reactor, it had been decided to phase out the SGHWR, even though the reactor of Winfrith had reached its full power of 100MW(e) the previous January. However, Kronberger, who was made the Authority's Member for Reactors from 1 January, 1969, in succession to Stewart (who joined BNDC), kept it going, and an energetic sales campaign was instituted by the Authority to try and find customers for the system overseas. Support for the SGHWR came also from TNPG which joined in this marketing campaign. Furthermore, when the company, until then the prime mover of the HTR in the industrial field, found the tender of BNDC for Oldbury being preferred to its own, it put its full weight behind the only other British thermal system that seemed still extant. Quite apart from this struggle over the indigenous reactor types, the Government was under continuous pressure, notably from GEC, to go over to light water reactors, so that British industry could participate in the growing world market for components and complete light water systems. Conditions in 1971 were thus hardly opportune for the introduction of an untried gas-cooled reactor into the British economy and the prospects for Oldbury steadily dimmed.

Reorganisation and reappraisal were in the air no less in Germany. Following the exclusion of Euratom from the THTR Association at the end of 1967, a new company (HKG) had been formed, with a view to ordering a prototype pebble-bed power station. A letter of intent to proceed was delivered in July 1970 to the THTR consortium led by BBK but then Krupp was forced to withdraw from its partnership with Brown Boveri. This did not stop Brown Boveri from pressing on with the pebble-bed and a new contracting company was formed, HRB, with Brown Boveri as majority shareholder and HKG provisionally holding the remainder of the shares. Great efforts had been made by the Government to bring about a merger of all HTR interests in Germany and plans were far advanced when it became appreciated that the proposals of GHH for Geesthacht were unrealistic. This project was abandoned and GHH faded from the nuclear scene. As a result, the action that might have come from a bilateral agreement between the German Government and the Authority, which had envisaged a collaborative programme of HTR development based on the Geesthacht and Oldbury projects, could never materialise, and Germany was left without a direct interest in prismatic HTRs.

On the other hand, in the French national programme, the prismatic HTR had from 1969 onwards, been assuming a growing importance and the CEA was devoting an increasing effort to its development. During 1970, a new industrial group, the GHTR, was created which was to prepare the design of an HTR power station for examination in a few years' time, while the CEA itself was re-orientating part of its existing gas-graphite team and modifying its test facilities for HTR research **(32)**. Work on coated particle fuel manufacture was intensified, contracts were negotiated with Dragon for the irradiation of coated particles in the reactor Osiris, and loops in the reactor Pégase were converted for the long-term testing of fuels in association with Dragon and with KFA. Further HTR work had also been put in hand on physics questions in co-operation with Dragon and the Authority and an agreement signed with KFA for experiments on a pebble-bed configuration. The HTR was no longer just an academic subject in France.

CHAPTER 16

DRAGON'S RELEVANCE IN QUESTION

AT THE Fourth International Conference on the Peaceful Uses of Atomic Energy that took place in Geneva in September 1971 with 4000 participants, the most significant news on the HTR was the revelation by GGA of its "sales" of large HTR power stations to US electricity companies, the most firm being a twin 1160MW(e) station to Philadelphia Electric. Rapid progress was also reported on the construction of Fort St. Vrain. Whatever the hesitations in Europe, it seemed that the USA was convinced that the HTR was now a real competitor to existing thermal reactors. GGA made a deep impression on European atomic energy commissions and electricity companies and was not slow to attack the European market.

Some semblance of cohesion was given to this market by the formation in January 1972 of a new group of electricity companies called Euro-HKG, that crystallised out of a loose association. Its members included VEW, the principal partner in HKG (which had confirmed in the previous October its order for the 300MW(e) pebble-bed THTR power station to be built at Uentrop), a second German electricity company, RWE, and the national electricity undertakings of France (EDF), Italy (ENEL), and England (CEGB). Euro-HKG was set up to provide a forum where experience on HTRs could be exchanged and this is what it remained – a centre of communication and nothing more. Euratom tried to persuade it to take an active role in defining a basic specification for a prismatic HTR power station and the group did have some influence on the standardisation of fuel element design. However, Euro-HKG had no corporate contracting powers and it made little effort to lead the development of the HTR. It was content to follow the sales efforts of GGA in Europe while its members acted independently. No collective attempt was made to exert an influence on Dragon policy, despite the growing influence of the electricity undertakings on national policy decisions.

DRAGON'S FUTURE ROLE

During 1971, Dragon's administration began seriously to feel the effects of inflation and the problems posed by a budget profile that had been drawn up several years earlier with no real provision for indexing and only a small contingency element that had, in any case, been cut to accommodate Norway's reduced contribution. Considerations of the 1972/73 budget indicated that by September 1972, when decisions had formally to be taken on any new extension, the Project would already be in financial difficulties if it maintained its programme. Shepherd was therefore asked by the Board of Management in July 1971 to draw up proposals for a discussion in the Autumn with a view to obtaining a decision by the Signatories in June 1972, so that either an orderly run down could be planned or the programme continued with the certainty that there would be a further injection of funds. Two scenarios were identified: one that assumed that a demonstration HTR would be built somewhere in Europe; the other that no such project could be foreseen.

Shepherd's response was to prepare programmes for three different levels of activity, the most limited being the operation of DRE as an irradiation facility with minimum R & D; the most ambitious, operation of DRE plus a full R & D programme. A third level envisaged a smaller R & D component with no metals research. Should a prototype HTR be built, it was proposed that the programmes should be oriented towards giving maximum support to this project whereas, if no decision was taken on a prototype, Dragon's efforts should be directed towards the longer term development of HTR technology.

Discussions were held first with Marien who, like Shepherd, was in favour of the highest level, and on his advice the intermediate level was dropped from the draft document that was then submitted to the Authority. The Authority's reaction was presented at a meeting held in October 1971 by Moore, who had been appointed Member following the death of Kronberger in the previous year. Level, it was reasoned, should be conditioned by industrial activity and if no prototype prismatic HTR was to be constructed in Europe (as seemed probable at that moment) then DRE should be shut down in 1974 when current sponsored research had been completed. After that, it might be useful to keep a restricted collaboration going between the UK, France and Germany, with each contributing say £¼M.

Euratom and the Authority were evidently not of the same mind and even within the Euratom countries, there were marked hesitations over the level at which R & D should be continued. France for example, let it be known that while it was strongly in favour of DRE continuing, it was critical of Dragon's proposals on the grounds that already too much attention was being given to ancillary activities in comparison with the essential job of running the experiments in DRE and evaluating the exact conditions that obtained in the samples being irradiated. The physics group in particular was the target of this complaint: it was accused of being more interested in hypothetical reactor assessment calculations than in refining the analysis of DRE's own core, a criticism that was voiced subsequently in the PSC when it was supported by both Marien and Thorn. Shepherd defended the group's interest in assessment studies with the argument that staff had to be allowed to do some advanced work, or they would go elsewhere and the Project would be worse off. In fairness to the physics team also, it should be recognised that making such analyses for a core so full of different experiments was much more difficult than for a single purpose core, and it took time to assimilate the effects of changing over from the original designs to block-type elements.

Despite this inauspicious start to discussions on a further extension, when the Board of Management met in November 1971, it seemed that a fair measure of support might be forthcoming for a three-year programme that included a major R & D component, and it was decided that firm proposals should be worked out by a sub-committee that was to meet under the Board's Chairman, Gunnar Holte of Sweden, and comprise one delegate each from Euratom, the Authority, Austria and the Agency, plus the Chief Executive. The sub-committee was also required to bear in mind the overtures that had been made to the Project by GGA and Japan. GGA, it seemed, might be prepared to pay up to $500 000/a for the hire of space in DRE for the irradiation of Fort St. Vrain blocks, and the Board was inclined to encourage such an action. On the other hand, the suggestion that Japan might wish to join the Project was much more coolly received. Marien, in particular, was strongly opposed to European technology being passed to the Japanese on any but the most draconian terms. Only the Authority, which was hoping to open markets for its SGHWR in Japan, seemed at all ready to offer full co-operation.

AGENCY CONCERN OVER FORWARD PROGRAMME

The Agency found the attitudes expressed by the various delegates paradoxical. Williams, since becoming Deputy Director General had been working hard to persuade Japan and the USA to change from being associate members to being full members of the Agency. His reasons were partly financial. Following a revision of budgeting practices in OECD, programmes were funded individually, including overheads, so that both countries had been receiving many of the advantages of the full members without contributing anything to the costs. Williams also regarded the full membership of these countries as highly desirable because it would bring a better balance into the Agency's programmes and make them more authoritative. This was becoming particularly important now that it seemed probable that Britain and some of the Scandinavian countries would be joining the European Communities, following the announcement in May 1971, that agreement between the Communities and the UK had been reached on major issues. Williams saw the Agency's future role as that of grouping the strong nuclear countries while the IAEA formed the global grouping and Euratom a restricted European association.

Already through the Dragon-USAEC collaboration agreement and through the GGA company in Zurich, GGA received a considerable amount of information from Dragon, perhaps more than it gave back. If no prototype HTR with a prismatic core was to be built in Europe, the company would be the chief beneficiary from any advanced R & D programme of Dragon, and if it was also to obtain space in DRE, this was likely to be at very low cost. The Signatories seemed to ignore this, yet were putting up strong resistance to the advances being made by the Japanese, a resistance that was soon to be translated into a demand for an entry fee of £1 million that was evidently out of the question. (Japan did in fact become a full member of the Agency in 1972, at which time the word "European" was dropped from the Agency's title, but never concluded any agreement with Dragon. The USA became a full member of the Agency in 1976 and only in the last days saw any reason to contribute towards Dragon's upkeep when it was already too late.)

The Agency believed that the only justification for a continuation of the Dragon Agreement in the proposed form was the firm commitment in the near future by one of the Signatory countries to build a prismatic HTR power station. If none was to be built, or if commercial exploitation was to go on mainly through GGA, the Project should be replaced by another whose main purpose was to provide fuel test facilities. Space in DRE could be given on a priority basis to Signatory-sponsored organisations and rented to others, the Project providing specialised post-irradiation and instrumentation services. The net cost of such an operation would be markedly less than a self-contained programme with a big research and development content.

PROGRAMME AGREED

Faced with the conflicting view-points of the two major Signatories and the sponsoring agency, Shepherd tried to resolve the divergences for the sub-committee by restating the objectives of the Project in the more general terms of making optimum use of DRE, continuing the accumulation of data to help countries take decisions on HTRs operating on the steam cycle, and working on the longer term applications of the HTR as a high grade source of heat. This avoided the specific distinction the Board had initially made between the two scenarios and so put to one side the fundamental issue of whether or not a prototype prismatic HTR was to be built. It was also not a question that the sub-committee could resolve. Instead, it concentrated its attention on reducing the budget put forward by Shepherd (which proposed an increase of £1.5 million at fixed prices over the previous three years budget of £6.9 million) and cutting down the estimated inflation rate (from 7½% to 5%). A suggestion by Thorn that had previously been put forward by Holte and which would have been welcomed in France, that the general budget be reduced and supplemented by fees for sponsored experiments (as at Halden), was strongly opposed by both Marien and Shepherd, and the subject was not pursued.

Shepherd fought to retain a more realistic figure for inflation but was obliged to pare his estimates down to £7.9M at 1971/72 prices, equivalent to £9.4M after indexation. This was still a sizable increase and he had the satisfaction of seeing his broad programme, including a substantial R & D section, accepted at a Board of Management meeting in March 1972 as the basis for negotiation, receiving strong support from the small non-Euratom countries, which had indicated that any further reduction in the R & D content might deter them from participating. The question of irradiations for GGA did not, at this stage, become an issue as the company did not follow up its original approach.

FINANCING CONFLICTS

So much for the programme, but the problem of finance had yet to be tackled. It now seemed certain that the Communities would be enlarged from the beginning of 1973, and Arnold Allen, who had replaced Peirson when the latter left the Authority in 1970, gave notice that the "UK would, of course, contribute its agreed share as a member state through the community budget. It seems to us that . . . this is the right basis on which any future contribution to the Project would be calculated. But we would accept that there may be some scope for adjustment during the period that is covered by the proposed extension." Norway and Denmark also indicated that their participation was tied to their entry into the Communities. Marien intimated that no increase could be expected from Euratom. Only the delegates from Austria, Sweden and Switzerland were prepared to go to their Governments with a request for funds on former terms.

In Brussels, the proposal to extend the Dragon Agreement for a further three years was successfully piloted through the Commission in the Spring of 1972 and was then passed to the Group for Atomic Questions (see page 285) for its comments. Much play was made of a phrase used by Peirson in an exchange of letters with Jean Rey, President of the Commission, at the time of the signature of the agreement for the 1970-73 extension in November 1969, when he stated that any further prolongation after 1970 would be on similar conditions to those currently applied. Euratom's members chose to interpret this as meaning that the Authority's and Euratom's percentage contributions would remain fixed, irrespective of the size of the Community. It was proposed therefore that Allen's warning be ignored and the Authority be required to continue to contribute 47.24% while Euratom paid 40.4%. Albonetti, speaking for Italy, was not prepared to go this far, protesting that there had been no consultation on the new programme, which was, in any case, too expensive and which should be tied to the decisions still to be made on the Joint Research Centre. He even suggested that Euratom should withdraw, leaving the whole affair to those states with HTR programmes – a proposal that to Marien's chagrin, Caprioglio, recently returned to the Communities as director of the JRC, tended to support. There had been murmurings that Italy might propose that payment should be on the basis of the irradiation space taken in DRE, but this was effectively cut off by Marien who produced an analysis which indicated that 30% of the space was occupied by the Project, 22% by KFA, 16% by the CEA, BelgoNucléaire and AGIP Nucleare, and 32% by the Authority.

By this time, however, the relative interests in the HTR system within the Signatory countries had undergone an appreciable change. In the UK, when the Vinter committee reported in the Spring of 1972, the choice of thermal reactor for the CEGB was left open and its main recommendation was that the industry should once more be reorganised. Although it was not until August that the Government announced its endorsement of such a policy **(33)**, by May it was evident that the Oldbury project had been abandoned. In contrast, France and Germany had come under the spell of GGA which was vaunting the advanced state of Fort St. Vrain and its growing list of orders for power stations in the USA. Licence and co-operation agreements covering fuel and reactor design were being negotiated by the CEA with the American company, the German fuel company Nukem was concluding a licence agreement, and Brown Boveri (Mannheim) was preparing to transfer a major part of its holding in HRB to GGA. German and Swiss electricity companies were alerted to the promise of the GGA designs and it seemed highly probable that orders for HTR power stations based on GGA's designs would be forthcoming.

Receiving unqualified support from its Scientific and Technical Committee, the Commission of the Communities therefore urged its member countries to come to an understanding on a Dragon extension by the end of

June, and proposed as a compromise that the enlarged Community should offer to pay 43% including the UK contribution of nearly 5%. (To maintain the percentage contributions from each country at their existing level, a Community that included the UK, Denmark and Norway would have been required to contribute over 48%). Belgium, France, Germany and Luxembourg were prepared to go forward on this basis and Italy let it be known unofficially it would not in the end oppose, provided there was some advantage to Ispra and to AGIP. It was the Netherlands' turn to create difficulties, insisting that the date for decision was not until the end of September, and intimating unofficially that the longer the decision was left, the greater would be the percentage that the UK could be persuaded to pay. The Netherlands saw no reason why Dragon should be given special treatment when the Council of the Communities had so much to decide – and it was also bent on increasing the income to Petten from external sources. It relented somewhat in July after threatening to veto the Project at a meeting of the Council, and agreed that negotiations could be opened with the other Dragon Signatories. The Commission was finally empowered to proceed at the end of the month.

Clearly, little progress could be made at the June meeting of the Board of Management even though Allen came in a conciliatory mood. Having spelled out the reasons why the UK no longer saw itself as having a leading interest and was therefore on a par with the other Signatories, Allen gave recognition to the special responsibilities of the host country and proposed a combined contribution of the Authority and the national payment through Euratom, of 40%. Marien could still only speak in generalities and the three independent countries, Austria, Sweden and Switzerland, together contributing under 10%, could only stand by and leave it to the big two to try and find a solution.

Even when, in August, Allen raised the UK's total offer to 45%, 7% remained unaccounted for. Marien's limit was 43% and both Denmark and Norway had decided that their contributions would be restricted to their participation through Euratom. For Denmark this meant a reduction from 2.2% to 0.65% and for Norway from 0.73% to 0.44%. Allen confirmed the UK's interest in the Project's continuation as British work on the HTR in the UK had not ceased and the Government was envisaging a possible commercial exploitation with European colleagues. CEGB still thought the HTR had a future and BNDC had gone on with design improvements. At the same time, Allen made no secret of his disappointment over Euratom's stand.

Clearly some move on the part of Euratom was needed and Marien urged the Commission to recommend a contribution of around 43% not counting the UK fraction. The proposal went through innumerable drafts before finally being submitted to the Committee of Permanent Representatives where Italy was the only opponent, suggesting a six-month extension during which the whole Euratom programme could be reviewed. No veto was imposed and the news that the Commission's figures had been accepted was transmitted to Marien while he was attending a special meeting of the Board of Management on 3 October. A condition was that substantial sums were to be set aside for irradiations at Ispra, Saclay, Mol and Petten and that the number of delegates from the two main Signatories on the Board of Management was to be increased from three to five.

A new complication was the vote in the Norwegian referendum at the end of September which definitely excluded the possibility of Norway becoming a member of the Communities. So unexpected was the result, no thought had been given to whether Norway would continue to remain an independent member of Dragon or withdraw altogether.

Marien, presuming that Austria, Sweden and Switzerland would be reverting to their original total percentage contribution of 9.55%, laid before the meeting two offers which he had derived to cover the situation, namely: 47.07% (inclusive of the UK fraction) should Norway withdraw completely and 46.26% should Norway contribute its original percentage. In return, Allen announced that the UK was prepared to come back to its 1970-73 figure with a total payment of 47.24% including its contribution through Euratom of 4.9%. This, however, still left a margin of 0.2% even if Norway should contribute its original 1.65% (which was most unlikely) and there the negotiations halted as both Euratom and the Authority turned the remainder into a question of principle.

Shepherd was by now desperate for a decision as he tried to manipulate his budgets to accommodate both a shut down and a continuation programme. Various devices were suggested for arriving at a conclusion by cutting the programme, but both sides remained adamant – Marien determined that the UK should increase its percentage, Allen with instructions that he should go no further.

No head-way was made at a special meeting of the Board of Management a week later but the break came on 20 October when the UK, unwilling to see the Project founder for such a small amount and conscious of the political repercussions of allowing it to do so, gave in and agreed to pay the outstanding sum, bringing its total contribution to 48.27%. When exact adjustments were made to the UK contribution through Euratom (5.01%) the percentage required from the Communities (less the UK) totalled 42.09% leaving Marien with a comfortable margin inside the maximum figure he had been given. The Council of the Communities ratified the arrangement on 9 November with Italy abstaining.

Drafting the text of the agreement proved to be no easy matter with the change in status of Denmark, the participation of Ireland through Euratom and the omission of Norway, especially in view of the possibility of

Norway's return at a later stage. Adding to the complexity of the problem was the need to use money from the 1973-76 budget to sustain the level of activity up to March 1973 that was contributed by a different set of Signatories. Many drafts were exchanged before a satisfactory wording was arrived at, leaving for further discussion the rights that Norway should have after April 1976.

The 1973-76 extension Agreement was signed at a Board of Management meeting on 8 December, 1972. Its negotiation had occupied just one half of the period covered by the existing Agreement. It was the last such negotiation to be brought to a successful conclusion and the Dragon collaboration terminated on 31 March, 1976.

CHAPTER 17

SUPERVISION BY THE SIGNATORIES

THE AGREEMENT setting up the Dragon Project required the establishment of two international committees to supervise the work of the Chief Executive who carried sole responsibility for executing the programme decided on by the Signatories. Supreme control was vested in a Board of Management which was aided by a General Purposes Committee. In 1969, when a major activity was becoming the irradiation in DRE of test samples produced in the Signatory countries as part of their independent HTR programmes, a Programme Sub-Committee of the GPC was set up with the double objective of determining the experimental charge of DRE and coordinating HTR development in Europe.

Various ad hoc committees were formed from time to time to consider special issues and a Standing Committee on Patents met regularly to advise the GPC on patent applications and the granting of licences. This had not the same status as the PSC and was more an expert group charged with considering which inventions should became project patents, and which information could be communicated under the various association agreements concluded by the Project.

BOARD OF MANAGEMENT

Until 1973 when the representation of the two main Signatories was enlarged, the Board of Management was made up of three delegates from each of the Signatories whose financial contribution exceeded 25% (namely Euratom and the Authority) and two delegates from each of the other Signatories; voting strength on the insistence of both the Agency and Euratom was proportional to the financial contributions. Explicitly, the functions of the Board were defined as being to:

1. Determine each year the joint programme of work and the budget relevant to this programme.
2. Approve the appointments of the Chief Executive and senior staff.

3. Receive progress reports from the Chief Executive and report each year to the Steering Committee of the Agency.
4. Deal with any other matter brought to its notice by the Chief Executive or by the General Purposes Committee.

Formally, decisions of the Board were taken on a two-thirds majority, which meant that neither of the two main partners must oppose a measure for it to be adopted. Only if both abstained did the combined vote of the other countries (whose maximum contribution totalled 13.2%) count for anything. In practice it was in no one's interest to have serious disagreements resolved by a vote and the Board sought always a consensus which everyone could accept. Even though under-privileged in voting power, the independent countries were in no way denied a proper share in the real work of the Project. On the contrary, the proportion of extra-mural research contracts steered in their direction was on the generous side and their share of manufacturing contracts was also high on aggregate. Similarly they had nothing to complain of in the allocation of senior posts or the opportunities for placing other staff. Nevertheless, there was a sense of exclusion resulting from the psychological effect of being relegated to the rank of observer once the problems became serious.

Had the voting been on the basis of one vote per Signatory, as the Authority had suggested, it is possible that the independent countries might have become more deeply involved in policy making. As it was, they tended to leave it to the big two. Either way they would still have been unable to make any impression when it came to a question of relative financial contributions, and once nine of the twelve countries participating were all part of Euratom, no modification to the rules of procedure of the Board of Management could have prevented the essential negotiations from taking place within the machinery of the Communities, remote from the Board's influence.

The Board met three times a year; in February/March when the main item on the agenda was the approval of the coming year's programme and budget (the financial year began on 1 April); in June to approve the Annual Report to be sent to the Steering Committee of the Agency and in the Autumn to approve the accounts of the preceding year. Up to 1966, with the exception of 1961, one meeting (generally the June meeting) was held at Winfrith, one of the other meetings was held in Paris at the Agency, and the third was normally held in the country of the Chairman. After 1966, the Board met only three times in Winfrith, the last occasion being in July 1973; it continued to meet once a year in Paris, and several of the eleven special Board meetings, convened mainly to consider extension agreements (or in the later days, closure) took place there.

Representation throughout the Project's life was, for the most part, at high level considering how small the Project was. Euratom being a composite Signatory was a special case, and the diminution of negotiating power on the part of its delegates in the middle and final years was not due to their level of competence so much as the diminished status of Euratom's programmes and their subjection to political pressures, whereby one national advantage would be traded off against another in a possibly unrelated field. Over the first three years of Dragon's existence, when the structural basis for the whole Project was being laid down, the Commission exercised an almost autonomous role within its budget limitations. Guéron, as General Director for Research and Development and with far more experience in nuclear energy than any of the Commission members, wielded considerable power and, particularly under the umbrella of Hirsch, enjoyed much personal freedom of action.

He was the Chairman of the GPC for its first two years and then, when negotiations on an extension to the Dragon Agreement were begun, he took over on the Board of Management from E. Staderini, Director General for External Affairs. To begin with, everything that went on relating to Dragon passed across Guéron's desk (if at times with some delay. Rennie recalls being ushered into his office, and believed it to be empty until a figure arose from behind the mountain of paper that completely obscured him from view.) With the arrival of Caprioglio in Brussels in July 1959, Guéron was able to delegate more and more responsibility, and Caprioglio became Euratom's project man for Dragon – close to Guéron and on good terms with the Commission – serving on the GPC from July 1961.

When Caprioglio left Brussels in 1967, the first steps had already been taken to curb the authority of Euratom, by cutting its budgets and any long-term planning, a process that was compounded by the merging of the Commissions. Marien, who succeeded Caprioglio on the GPC in 1967 and Guéron on the Board of Management in 1968, reported through a Director General to the Commission – a Commission that had little knowledge of nuclear affairs and was preoccupied with matters of much greater political moment than Dragon. It was also a Commission that had become committed to a heavy administrative procedure for even the smallest of matters. When Marien, who had been Head of the Nuclear Reactor Division left Euratom, his place on the Board of Management (and as Chairman) was taken in November 1974, by Fabrizio Caccia-Dominioni, Director for Research, Development and Nuclear Policy. Caccia was not a technical man but had worked in a legal capacity with CNEN. He also had direct contact with Altiero Spinelli, the Commissioner responsible for Industrial and Technological Affairs who took a personal interest in Dragon. Nevertheless, he had little of the influence that Guéron wielded in the early 60s.

Until mid-1973, the Euratom representative at Board of Management meetings was normally accompanied by two colleagues from France and Germany making up the Euratom delegation of three. These national representatives were for the most part senior members of their countries' scientific administrations but included also from time to time senior technical people from CEA or KFA. The Euratom line to be adopted was worked out beforehand and it was clear that the Euratom staff man was the central delegate. With the modifications to the membership structure resulting from Norway's withdrawal, and the incorporation into Euratom of Denmark, the UK and also Ireland, representation on the Dragon Board of Management increased to five and it was understood that there should be one delegate each from the Commission, France, Germany and Italy plus one from each of the other countries (not including the UK) chosen in rotation.

For most of the Project's life, Euratom was able to present a coherent front at Board Meetings and the Euratom representative was able, without too much difficulty, to take decisions on behalf of his member states within the limits set for funding. Towards the end, there was some spill-over of internal discussions into the Board, as for example, when the Italian delegate would not support the extension programme accepted by the Commission, but by then the essential problem was to arrive at any sort of consensus within the Communities in view of the schizophrenic position of the UK (part of Euratom and, at the same time, principal opponent in the negotiations over percentage contributions). Strictly it was the British Government that participated in the work of the Communities while the Atomic Energy Authority was the Dragon Signatory, but on fundamental issues the Authority was not able to take a line that ran counter to its own Government's policy, even if it had so wished. In consequence, any dialogue on financial issues between Euratom and the Authority within the Dragon Board was, after 1973, still-born.

From the Authority, the Board always benefited from a top level representation – Cockcroft, Penney and Vick, successive Authority Members for Research,* accompanied by Peirson, the Authority's Secretary and *eminence grise,* and Fry, Director of the Winfrith Establishment. Vick was followed in 1967 by Stewart, Member for Reactors, signalling the change from research thinking to power applications within the Authority. After Stewart came Moore, who was also to become a Member, and Thorn,

*The Atomic Energy Authority's terminology is confusing, as strictly the Authority consists of the small group that controls the body that it is generally understood to be. The term 'Member' might be better construed as 'Central Board Member'. Following the reorganisation in 1972, departments of activity for the full time members were no longer designated.

Director of Thermal Reactors in the Authority. When Peirson left the Authority, his place was taken by Arnold Allen, Member and Secretary, and when Fry retired, he was succeeded by the new Director of the Winfrith Establishment, Harry Cartwright, from March 1973. The Authority, it is clear, always paid the Project the compliment of sending top people to Board meetings. These could negotiate in the name of the Signatory as real plenipotentiaries, at least until mid-75 when the Department of Energy took matters into its own hands. Even then it was a Secretary of State who was directly involved.

From the other Signatories, Dragon enjoyed not only the presence of high level scientists and administrators as delegates, who interpreted their role as representing Dragon in their own country as much as representing their country on the Dragon committees, but also a continuity of service that was of great importance for the stability of the Project and of great comfort to the Chief Executives. Pride of place from this point of view must be given to Richard Polaczek of Austria. An original member of the Steering Committee for Nuclear Energy, he represented his country at the first Board of Management meeting and went on with an almost unbroken record of attendance through to mid-1975, twice acting as Chairman. If service with the General Purposes Committee is also taken into account, Gunnar Holte of AB Atomenergi, Sweden, can claim as long a service as any, attending the first GPC meeting in 1959 and the last Board of Management meeting in 1976. The record of Flemming Juul of the Danish Atomic Energy Commission was equally impressive, as even when Denmark ceased to send a separate representative after joining the Communities, Juul represented Euratom or attended the Board as an observer. To go further would require a complete listing of the members of the Dragon committees; it must suffice to reiterate the point that the Signatories did not starve the Project of either high-level or long-service representation.

From the Agency too it was normal for Huet and Weinstein in their time to attend meetings, then Saeland and Williams until 1972, when Saeland delegated full responsibility to Williams. A consistent representative from Paris also was Leslie Boxer from late 1965 until his transfer to the International Energy Agency of OECD in 1973.

For a body that was much concerned with financial and administrative matters, the Dragon Board of Management was unusually well qualified in the technology, which worked both for and against the Project. Early on, the tradition was established of providing a steady flow of well-constructed documents that made the delegates feel that they were adequately informed; the Authority administered the Project in a manner that left no room to doubt either its efficiency or honesty; the more mundane work was taken care of by the GPC, and Rennie adopted the practice, once the Project was running smoothly, of having technical dissertations presented at meetings, which kept the Board familiar with the detail.

As a result the Board had a ready appreciation of the technical problems that were inherent in the HTR and could make a realistic evaluation of the technical success of the developments that were instituted. It was also alive to the interaction between operation of DRE and more general research and development activities and insisted, for example, on the Project doing assessment studies to keep its feet on the ground. It refused to countenance the first efforts to take in gas turbines and turned the Project towards providing an irradiation service when the need arose. It was less well adapted to considering the long term strategy and the more fundamental questions of where the Project was going. All discussions, as one delegate put it, relapsed into technical detail; it was impossible to abstract oneself from the heat exchangers or the fuel or the gas clean-up plant and stand back and think where the Project would be in ten years' time.

Neither Rennie nor Shepherd used the formal sessions of the Board of Management for brain-storming or for long-term policy making. They both considered that it was their duty to initiate action, seeking the Board's approval when the rules demanded. Such a policy suited the delegates also, and excellent relations were established with the Chief Executives who were given wide powers of execution within the broad programme and budget definitions. Both Chief Executives, it must be underlined, treated the overall budget as sacrosanct and although they went to the Board to obtain approval for transfer operations within the budget as little as possible, they made available enough information to convince the Board of the effectiveness of the Project's administration.

At the time of the discussions on a joint HAR project, it had been suggested that the Board should wield greater powers through the medium of an executive committee, chosen from its members, and this committee should have departmental responsibilities within the Project. Then it was the UK that insisted on a single Chief Exectuvie through whom all actions should be channelled; now it would seem that the wisdom of this system is no longer questioned anywhere.

In that preliminary period also, two support committees were proposed according to the CERN system where the Council is advised by a Finance Committee (made up of representatives of member states) and a Scientific Policy Committee (made up of eminent scientists chosen for their individual competence). Such a structure was also recommended by Cockcroft at the high level meeting of experts when the HTR was offered for joint development. Huet and Peirson, however, preferred a single advisory committee as had been adopted at Halden, and with the name 'General Purposes Committee', this became the accepted pattern for Dragon.

GENERAL PURPOSES COMMITTEE

The Dragon Agreement specified that the GPC should be composed of one senior technical specialist per Signatory and ascribed to it two specific tasks, namely to approve contracts over a certain value and the conditions of service of the staff (both of which at CERN are the responsibility of the Finance Committee). It is logical that if there is to be one body only advising a Board that is fulfilling an essentially administrative function in a scientific project, then this body should be composed of technical people. It must also be remembered that Dragon was not a very large enterprise; at the end of the year 1967/68, for example, the total number of staff in post was 90 (comparable figure for CERN was 3200). A heavier superstructure was not, therefore, justified, and although a linear chain of committees risks a certain duplication of interest in practice, the Board of Management and GPC had separate preserves.

The GPC met four times a year typically in February, April, June and November. Until 1967, on average at least three meetings per year were held at Winfrith, whereas from that time onwards, the tendency was to hold them in London and none took place at Winfrith after July 1973 (nor in Paris). With thc completion of the reactor and the fuel fabrication plant and hot cell facilities, there was, of course, less to see at Winfrith than during the opening years, and even if most of the documents (and the staff) were at Winfrith, there was Shepherd's memory to fill the gaps. Also a number of delegates were concerned with extra-mural contracts or involved in other committees so that contact with the Project was a little more solid than might be inferred from the meeting patterns. Nevertheless, the wisdom of dissociating all formal meetings from the Project's centre is open to doubt.

Many of the members of the GPC served at the same time on the Board of Management and in spite of the apparent awkwardness of the different rhythm of meetings and frequent mismatch of date and place, delegates do not seem to have found the arrangement either particularly inconvenient or inefficient. The advantage, in principle, of having the same people attend both meetings was that repetition could be avoided; in practice there was a temptation to go over old arguments and the value of having two committees at all could sometimes be questioned. There was less ambiguity for the Signatories that consciously sent people from different parts of the country's administration to the two committees, so defining more clearly the functions to be expected of each.

The Board of Management was very much occupied with extension negotiations and the GPC was at least able to dissociate itself largely from the political aspects and deal with the day to day affairs of the Project. Being composed of people with active experience in nuclear engineering, it discussed plant aspects and experimental programmes in detail. Other topics making a regular appearance on the agenda were staff, budget and programme matters, contracts, patents and licensing policy, and collaboration agreements with other organisations.

For the most part, the atmosphere in the GPC was constructive and as in the Board of Management, the Project was given considerable liberty to manage its own affairs within the framework of the global budget and the technical programme that had been worked out.

For succeeding extensions, these programmes were generally elaborated in sub-committees composed of Board and GPC members and the technical balance was then examined with great thoroughness. Considering the wide diversity of views expressed at times in the Signatory countries, the effort needed to arrive at a compromise programme was commendably short, which owed much to the knowledge and experience of the participants. It also owed something to the persuasive qualities of the Rennie-Shepherd combination in the early days and the Shepherd-Marien axis in the late sixties and first part of the seventies.

PROGRAMME SUB-COMMITTEE

The Programme Sub-Committee was formed in response to the need to formalise procedures of channelling experiments generated within the Signatories, into the irradiation programme of DRE. By the end of 1968, it was no longer satisfactory to rely on ad hoc arrangements between laboratories and the Heads of the Project's Divisions whose judgements, in any case, could not always be regarded as impartial. Moreover, apart from the number of proposals that could be expected from the Signatories, experiments were becoming bigger and more complex. Decisions had to be taken on their design and, at the same time, on programme schedules many months in advance, to allow core loadings to be defined that were compatible with both the irradiation and safety requirements.

As already noted, a number of the senior Project staff found the idea of an external body determining the programme for DRE unpalatable. They felt more competent than anyone else to judge what was best for the development of the HTR and they saw the PSC as an instrument for turning Dragon into a service facility divesting it of its function as the main centre for HTR development in Europe. Already with the establishment of serious HTR design teams in industry, Dragon's position was being eroded and the formation of the PSC was seen as a further diminution of status.

Within the Signatories which had HTR programmes or fuel manufacturing facilities, DRE was being regarded more and more as a reactor for testing private productions. There was also an element of challenge in the acquisition of space – an exercise in claiming rights. Moreover, the Project's necessary insistence that proposals for experiments had first to be vetted for compatibility and safety by the internal Core Programme Sub-Committee was not always understood, and the PSC, as a result, started life in an atmosphere of competition that was at variance with the coordination functions that had been postulated at the outset. To make matters more difficult, there was a tendency in the beginning for the arguments that had been gone over in the PSC to be re-hashed in the GPC to the point that the question was posed whether the PSC served any useful purpose at all.

It was particularly unfortunate that the PSC should have begun with the Authority, Euratom (and its member countries) and the Project all at loggerheads, because feelings ran high at times and resentments generated then were slow to die and damaged relations over the last period. Meeting for the first three years always at the London office of the Authority, although convenient for travelling, no doubt contributed to the hardening of attitudes as it prevented informal contacts being made with a broader range of staff. The Board of Management also could be criticised for failing to recognise the situation and take appropriate action.

At the working level where the practicalities of the irradiation programme had to be worked out, relations were much more harmonious. Technical aspects had to be examined in detail and there was a greater awareness of the effort needed to mount each experiment. In reality, no waiting list existed and it must be concluded that rigidity of stand-points caused more of the PSC's difficulties than any specific conflict over space.

That it finally came together and performed a useful function owed much to the dedication of its secretary. Alan Saunders, who was core programme co-ordinator throughout the Project. Coming from de Havillands at the end of 1959 he was taken over by the Agency when his company pulled out of the nuclear field. Tall, earnest and vitally concerned with any decisions the PSC might take, in his quiet way he made great efforts to bring the various parties together and smooth the ruffled humours of the contestants.

Neither the Signatories nor the Project tried to use the PSC as a think tank which could help towards the rational development of HTRs in Europe. Its subordinate position to the GPC ensured that it could never become influential in the definition of national or European policies and its work remained largely confined to its function of planning the use of irradiation space in DRE. This in itself was essential, but far removed from a forum where the atomic energy commissions and industry could synthesise a common industrial policy. The desirability of defining, for example, a standard fuel element for steam cycle systems was recognised, but the machinery was not adequate to allow meaningful decisions to be taken. The change-over by the combined UK consortia in 1972/73 to an integral block fuel element, was a unilateral decision taken with a view to establishing a single basic design and only very indirectly owed anything to the PSC.

Nevertheless, useful collaborations were established that brought together experts from different countries working on power reactor studies, to discuss their common problems. The series of Dragon Countries Physics Meetings was one. Started largely on the initiative of Helmut Gutmann, latterly Branch Head of Physics, they provided a centre for the exchange of information on core physics, fuel cycles and cost calculations. Participating in the first meeting held in March 1968 were people from Dragon, Euratom (Brussels and Ispra), HRB and KFA. Saclay was the next to join in and eventually the meetings were attended by representatives of all the research centres, industry, electricity companies, universities and government departments interested in the HTR. Out of this collaboration emerged a co-ordinated method of calculation that replaced methods that were being developed individually or under bilateral arrangements, and the Meetings went on to formulate research programmes on a wide variety of topics including fuel management, environmental problems and in-core instrumentation.

A somewhat similar series of meetings on metals grew out of the PSC. At these, the metals specialists in the HTR field would compare results of their own experiments and draw up plans for submission to the PSC, for research to be included in the Dragon programme.

In addition to these two main collaboration activities, informal working parties were regularly held on fuel to standardise and develop quality control, for example, and fuel specifications; another working party was concerned with the safety of HTRs. All of these contributed to the establishment of broad areas of common understanding of HTR technology in Europe. They were not, however, at a level which could catalyse an effective collaboration in the exploitation of the HTR system. Industrialisation remained a national affair to be defended against outside interference.

FINANCIAL RULES AND REGULATIONS

The Financial Regulations under which the Project operated were finally approved in February 1961 after minor modifications had been made to those that had been in force from the first year. These were drafted very largely by Ralph Reynolds, the Project's first Finance, Accounts and Contracts Officer, under Rennie, working from the principles that the Project's overall budget was predetermined, the life of the Project was to be limited, and for maximum efficiency it was necessary to give the Chief Executive maximum flexibility so as to be able to steer funds into the areas where they were most needed, as indicated by the evolving research and development. During the Agreement negotiations, the Authority had repeated on many occasions the difficulty inherent in costing out a construction and development project that contained so many unknowns, and after taking up his role as Chief Executive, Rennie found that the Authority's latest estimate for the 5-year programme that had been agreed, was nearer £15 million as against the £13.6 million that appeared in the Agreement. With this sort of divergence already in evidence, he would need to be able continuously to manipulate all the funds at his disposal if he were to cover the essential elements of the programme. Even so, he concludes now that it would have been impossible to meet all the conditions had the staff been employed by the Project instead of being seconded to it. The ability to juggle with the staff strength through the secondment mechanism was a key factor in allowing the Chief Executives to meet their commitments.

The first years of the Project were fortunately years of financial stability when inflation was modest and the pound a stable currency. As inflation mounted and exchange rates became subject to rapid and significant fluctuations, new problems arose which created great difficulties for the Chief Executive working within a budget fixed some years before. Moreover, as we have seen, for much of the time he was trying to operate two distinct programmes, one of which provided for a constant level of activity and the other for a run-down to zero by a determined date. This would not have been possible without freedom of action in the internal management of funds and again, an ability to adjust staff numbers without incurring additional costs for compensation.

The original Agreement and each extension determined the global budgets expressed in pounds sterling. Nominally all transactions took place in the same currency but as payments were made to the laboratories of some of the Signatory countries for extra-mural research, and also repayment was made for seconded staff, certain transactions were arranged in the currency of the relevant country to avoid a double exchange loss. Explicit provision for this was made in the regulations. Not having a separate legal personality, all transactions for the Project were carried out through the Authority which kept separate accounts and acted, on the instructions of the Chief Executive, as a service organisation without seeking in any way to impose a policy on the Project. Systems had to be compatible, but both Chief Executives have paid tribute to the impartial and efficient service the Project received from the Authority's administration at Winfrith.

The Chief Executive was required to present an annual budget to the Dragon Board of Management through the GPC, compatible with the total sum provided in the Agreement. Rennie persuaded the Signatories to accept as formal document, a simple breakdown of expenditure under accounting headings, with no reference to function. This, together with a similar breakdown of estimated commitments, was the sole budget document the Board had to approve. Headings appearing regularly included Staff, Construction of the DRE, External Expenditure on Research & Development, Basic Research – almost all relevant to each of the Divisions and Branches through which the Project was administered. Once the budget was approved, the Board was circulated with a document that analysed the budget under technical headings – for information only. Adjustments made during the year would be reported to the Board, and approval from the GPC sought for any modifications to the distribution amongst the various chapter headings. On its side, the GPC maintained a continuous watch on the broad pattern of expenditure but did not waste money by pressing for unnecessary detail and the Board adopted the principle that so long as the boundary conditions were satisfied, the Chief Executive could be trusted to make the best use of his resources.

Monies ear-marked for a specific activity in a given year which were not required in that period, could be transferred to a special fund, and if subsequently shown to be in excess of what was needed, returned to the Project budget, being credited under the appropriate chapter heading to the year in which they lapsed. Consequently, there was no compulsion to go on a spending spree at the end of the year to use up the last penny of the allotted sums, and funds could be nursed along until the most opportune time came to spend them.

Rennie's management of funds was highly personal. Division Heads would present their budget requests for the coming year, on which Rennie based his own budget proposals. When these were agreed, he would allocate sums against given activity headings to be administered by the Division Heads. Close watch on this expenditure was kept by Reynolds and any tendencies to exceed the limits set were notified to Rennie. Apart from the allocated sums, Rennie kept a considerable reserve which he would hand out as the year progressed on presentation of a good enough case. As a result, he could authorise rapid shifts in the direction of a given line of development, or make quick decisions to spend more on the production of a particular piece of equipment by drawing from the reserve that he himself controlled.

Under Shepherd, a more democratic system of management was instituted with the Division Heads participating in the budget allocation decisions to a much greater degree. He gave the same priority as Rennie to restraining expenditure within the boundary conditions, coping with an inflation rate that over the last years was three times the figure calculated when the extension was being negotiated.

CONTRACTS

It was understood that, as far as possible, equipment bought for the Project should have been selected by a process of international tendering amongst the Signatory countries and the Agreement charged the GPC with the task of approving any contracts worth more than £20 000, placed by the Authority on behalf of the joint undertaking. Provision was made for consultation to be in writing provided the value did not exceed £50 000. In the Financial Regulations, the requirement for competitive tendering was made more specific, it being considered normal for all contracts of an estimated value exceeding £200 to have been awarded on this basis. Even so, the Chief Executive was empowered to initiate and to authorise the letting of contracts of value up to £20 000 following competition and up to £10 000 where competition had not been sought. For larger contracts, an obligation was placed on the Project to seek the authorisation of the GPC to go out to tender and to invite the Signatories to propose firms for inclusion in the tender list.

In normal circumstances, when the requirement for an important item of equipment (as defined above) was known, the details were circulated to the Signatories and to the Agency, and submitted to the GPC. A description of the equipment, an indication of the firms being considered, together with an estimate of the value of the contract and the time-table were provided. At this point the GPC would discuss the technical aspects and had the opportunity of commenting on the tender list and proposing additions. Once approval had been given, the Chief Executive had full authority to place the contract, provided that it was with one of the two lowest bidders, reporting afterwards to the GPC, the Signatories and the Agency what had transpired.

Evaluation of the tenders and reduction to a common level often involved a great deal of work, and it was usual for an evaluation team to be assembled. This would consist of the Project's administrative officer and representatives from the designers, inspection branch, and the Authority's contracts branch. During the construction of DRE, additional help was regularly requested from the Agency and the Signatories. Having arrived at an order of preference, the factories of the potential manufacturers would be visited and outstanding aspects of the tender hammered out. If the contract was for a very large sum, or if the firm chosen was not amongst the two offering the cheapest solution, the GPC would again be consulted before the final award. In view of the infrequency of the GPC meetings, the ability to follow a written procedure was of great importance in avoiding bottlenecks.

CHAPTER 18

STAFFING AND STAFF

INTERNALLY, the Project was divided into Divisions whose functions changed to take account of the completion of DRE, the assessment study phase and support for industrial designs, the use of the DRE as an irradiation service to external users and the accent put on research in support of the long-term possibilities of the HTR with greater attention paid to circuit materials.

To begin with under Rennie, three Divisions were formed: Research & Development, Engineering, and Administration. In 1962, when the number of seconded staff reached its maximum, the respective strengths of these Divisions were 102, 103 and 44. The Engineering Division was split into Branches covering different aspects of the construction of DRE, while the work of the Research & Development Division, much of which was in direct support of the reactor construction, was sub-divided into three Branches concerned with Physics and Engineering Research, Chemistry and Chemical Engineering, and Materials, including the preparation of fuels, a task that was to represent a growing fraction of the Project's effort. With the construction work on DRE tapering off and staff numbers beginning to come down, Shepherd was made Deputy Chief Executive in April 1964 as well as Head of a Materials & Chemistry Division; a Physics Division was formed, with Physics and Engineering Research Branches, and the Branches in the Engineering Division were reduced to two, initially labelled simply 1 and 2 rather than with specific titles, then Assessment Studies and Reactor Experiment.

Over the next three years, staff numbers on the Project fell steadily, reaching a low of 89 in post at the end of the 1966/67 year. The divisional structure was simplified correspondingly, to just two technical Divisions called Research & Development, and Engineering, each with just one Branch.

When Shepherd took over from Rennie he retained the two technical Divisions but renamed them Materials Division and Reactor Technology Division. They included a staff of 50 and 38 respectively to which should be added 16 in Administration, including the Chief Executive and Lockett who was made special adviser to Shepherd. A Technical Services Division of just a few people was added the year after as staff complement climbed to 114. Branches dependent upon the Divisions were then: Fuel Element Research & Development, and Fuel Production in the Materials Division; HTR Technology, Physics and Engineering Branches in the Reactor Technology Division, and Reactor Experiment Programme in the Technical Services Division. Later adjustments comprised reorganisation of the fuel element activities into an R & D Branch in the Materials Division and a Fuel Operations Branch in the Technical Services Division. This was the structure that was retained from the beginning of the year 1971/72 until the end of the Project. Staff numbers reached a new maximum at the end of the 1974/75 year, with a staff in post figure of 124.

A few guest scientists made their appearance in the early years and from 1970 onwards great efforts were made to increase the number of people working on the Project whose salaries were not reimbursed. In that year, seven people were attached for one year or more; for the following year, 11 are recorded and the year after, 13. The impact though was shortlived and few remained in the last two years. Over and above the numbers already quoted was the contract labour, typically numbering 60 people, some of whom worked for the Project for several years. They were particularly useful for providing additional secretarial and draughtsman effort. Total turnover of staff can be gauged from the fact that over the 17 years, more than 700 names appeared on the books. To complete the complement of people working at Winfrith on Dragon should be added the Operations Group of about 100 people and process workers provided by the Authority.

REMUNERATION

Grading of the regular staff was in eight levels plus a junior level, no distinction being made between scientific, engineering and administrative functions in contrast to the practice adopted in the Authority. The gradings were, nevertheless designed to be compatible with the Authority's structure. Grade A was reserved for the Chief Executive, B for Division Heads and C for Branch Heads. These corresponded to the senior and 'banded' levels in the later terminology of the Authority, 'banding' being the step into senior management. Section Leaders occupied Grade D and were equivalent to the modern PSO, Principal Professional and Technical Officer, or in administration Principal (formerly Level 1). Lower ranking qualified scientific staff occupied the grades E to G, and junior support staff, such as ordinary draughtsmen, Grade H.

Considerable emphasis was placed by the Authority in the early days on the need to minimise differentials between the overseas and home staff, an objective much easier to achieve in the first half of the Project's life when

UK rates of pay comparable with those operating on the Continent. Comparisons though in this sector are very difficult to make. For example, the starting salaries of the Chief Executive of £3 700 per year and of his Division Leaders of £3 000 per year now seem absurd. Again a Division Head's reimbursement rate of £12 000 per year in 1975 sounds modest in comparison with Continental rates; even more so when compared with the salary scales of other international organisations, but not when compared with the salaries corresponding to the grades in its own staff that the Authority considered to be equivalent. From the staff member's point of view, the cost of living in the Dorset area and such benefits as free medical service have to be taken into account, as well as levels of responsibility for a given grade. Nevertheless it is evident that in European terms, Dragon rates of pay to begin with were quite reasonable but they slipped back progressively, in line with the decreasing salary levels obtaining in Britain relative to the Continent.

On the initiative of Peirson, it had from the earliest days of the negotiations been understood that all staff would be seconded from a parent organisation for a set period. Below the level of the most senior staff for whose engagement the agreement of the Board was necessary, the responsibility for taking on staff lay with the Chief Executive, who would establish the structure appropriate to the programme and recruit people as needed, determining their grade according to their position and functions in the Project. This grade defined the staff member's status and the rate at which his parent organisation would be reimbursed from the Dragon budget. What the staff member was actually paid was a matter between him and the parent organisation, which would frequently find that the reimbursement rate was lower than the person's salary level at home. Not always though, and there were several instances when the reimbursement rate was higher. In these cases it seems to have been the rule that the staff member received the benefit of the difference.

Discrepancies between grade and income inevitably arose and supervisors would have staff members under them who were earning more than they. Particularly was this true in the case of staff on the payroll of Euratom who benefited from a high basic salary and paid no national tax. Yet as far as can be ascertained, very little friction was generated because of this, probably it is felt, because of the complete decoupling of the Project from the administrative departments of the various organisations involved. On the other hand, differences in pay rates inevitably had repercussions on recruitment, as staff were reluctant to move to a post carrying a lower rate than they could expect in their own country even if this did not affect their own pay packet. Status and career prospects were judged by the salary attributed to the grade, particularly in those countries (the majority) which had no national HTR programme and so no clearly defined position to which the person would automatically return once the secondment period was over.

Taxation raised a nice problem as the UK authorities insisted on tax being paid on income 'earned' in the UK yet, of course, taxes were being demanded also in most of the home countries as this was where the salary was paid. To protect the overseas staff from double taxation, the Authority negotiated with the British Inland Revenue an agreement whereby from its own resources, over a period of five years, the Authority paid an annual lump sum rising to over £30 000 per year, in lieu of tax demands being made to the individual staff from overseas. From 1964 onwards, the system was changed, and overseas staff were assessed for British tax on the money that was paid to them in the UK, appropriate measures being taken in the Signatory countries to take account of this in their home payments. The only real losers were the Swiss who had been exonerated from tax in their own country when they came to work for Dragon. For the international staff on the payroll of Euratom or the Agency, tax paid in the UK was reimbursed through their own organisation by the UK, so the new arrangement made no difference to their net income.

OVERSEAS PARTICIPATION

An essential ingredient of a country's participation in Dragon was the attachment of staff to the Project to gain experience from the inside. Documents emanating from the Project would describe the results of the research but in the field of atomic energy it has long been recognised that academic information falls far short of the know-how that accrues from directly working on the problems that emerge, and in having day-to-day contact with the research and development people. In principle, therefore, the Project could have expected the participating countries to exert pressure for their own scientists to be taken on, in order that they could become immersed in HTR technology, and provide a constant source of awareness and experience at home.

So much for the theory. In practice, it was a permanent problem to maintain a high level of representation from the Signatories. As the years went on, the number of senior staff attached to Winfrith from overseas steadily diminished, in spite of the efforts made to reverse the trend.

The problem was less acute during the construction of DRE. By the end of the second year, following an energetic recruiting campaign, almost exactly half the members of the Research & Development and Engineering Divisions (excluding the three lowest ranks) came from overseas, and they included one Division Head and five Branch Heads. The Head of Administration was also from outside the UK. Jumping to the end of the sixth year, with now three Divisions (each with two Branches) plus the

Administrative Branch, two Division Heads were British and one was from Switzerland. Of the seven Branch Heads, three were from Euratom and one from Norway, while of the staff, totalling 115, plus 53 in administration and juniors, 47 were from overseas. From this date on, the problem of senior staff began to be felt and became particularly acute when Euratom was unable to make commitments for 1968 and was unable to replace outgoing personnel. At one point the total number of overseas staff was down to 10, almost all of whom were Austrians. With the regularisation of the situation from July 1968, great efforts were made to increase the participation of the Euratom countries, but even at the end of the twelfth year, with again three Divisions and seven Branch Heads, two of the Division Heads were British, (the other was from Euratom, Brussels) and six of the Branch Heads were British (even if two of them were classified as from the Agency). At the lower levels, the position became much better with 47 overseas staff out of a total of 89 technical staff and this remained true until the Project began to dissolve in 1975.

Had the principle of secondment not been adopted, the Project would certainly have got off to a much slower start. Particularly at a time when the UK held such a dominant position in atomic energy in Europe, there was an undoubted attraction in joining a British inspired project for the experience this would bring, provided the commitment was of limited duration and staff could expect to return with enhanced reputations and the possibility of an important position in the national undertakings, all of which at that time, were short of people with hard practical experience. In its turn, the Project was able to accept people to meet the peak demands during the construction period without being concerned with their long-term career prospects or problems associated with their subsequent redundancy. Moverover, the qualifications of the people engaged could be more closely fitted to the immediate tasks without an obligation to consider their relevance to subsequent operations.

From the Signatories point of view, secondment ensured the return of the staff member once experience had been acquired and, consequently, the feed-back from the Project of this experience. If the Project had been engaging career staff, this feed-back could not have occurred and the country would have been deprived of the services of scarce manpower.

Implicit, however, in this reasoning is the assumption that the Signatory has its own nucleus of interest in the work being done, which can interact with the Project and which can benefit from the feed-back. The national activity in its turn will provide a source of effort which can renew that of the Project as terms of secondment come to an end. In a number of countries this nucleus specialising in HTR technology was never developed and, as a result, there were no specialists who could be detached to occupy senior positions and those returning from Dragon found themselves looking for work on different reactor systems. From the point of view of general training this experience in a different field could be turned to advantage, but in terms of developing a competence in HTR technology, no growth was possible and the individual was made to feel that he had been out of the main stream and wasting time. This was no encouragement to others to follow the same path.

Even in those countries where a national HTR effort was maintained, divergencies in direction meant that a period with Dragon could appear as a step away from the national line of progression. This was evidently the case for staff from Germany where the national system – the pebble bed – was in many senses in competition with the Dragon prismatic system even if so much of the fuel, graphite, and materials technology was similar. In France other forces were also at work, as for much of the life of the Project, relations between France and Britain were not the most cordial and joining a 'British' project was not regarded by many French scientists as an attractive prospect for career advancement.

Again, the uncertainty of Dragon's future for so much of its existence could hardly inspire confidence. In only three of the 11 years from April 1965 until the closure in March 1976, was there an assurance that the Project would live for a further two years. For the more junior staff, often unaware of top-level problems, this was less important and their continuity of employment was assured by their parent organisations. For the senior staff who might have been prepared to gamble on the HTR becoming an adopted system, there was the risk of becoming identified with a project that was running down, unable to complete even its current programme, so leading their careers into a cul-de-sac.

For all the non-British there was, of course, the personal problem of uprooting the family and moving to another country, working in another language, resolving the difficulties of schooling for the children, finding accommodation in a foreign environment, and to many this was a daunting prospect. Mobility has greatly increased in recent years but in the late fifties and even the sixties it was still a major venture to move to another country, and the probable shortness of the stay carried as many drawbacks as it did reassurances.

Secondments of overseas staff ranged from a period of a few months only, to almost the life of the Project; very long periods were the exception and the average was around 2½ years. Authority staff tended to stay much longer and in consequence both the experience and seniority of the British staff rose progressively in relation to those of other nationals. It is interesting to speculate on what would have been the effect of the Authority rotating its staff more rapidly. In the first place, the Project might have seemed a little less British to the outside world which could have provoked a bigger interest in the HTR system elsewhere and modified to some extent the widely held view that it was up to the UK to take the lead in building a prototype. For the staff members concerned (the majority of whom were well content to stay with Dragon at Winfrith for as long as employment existed) the final wrench would have been less traumatic and re-integration into the Authority's ranks less painful.

What is clear is that the progress of the Project would have been very much slower. The amount of work that was got through and the very high standards of both development and production that were maintained were remarkable, when it is remembered that nearly half the qualified personnel were on short term loan and could be effective only after they had become integrated into the existing team and up to the time when they began to plan their return. For this a great deal of credit must reflect on both the top leadership and senior men, notably the Branch Heads and Group Leaders. They were the backbone of the Project and whatever criticisms may be levelled at the proportional representation of the Signatories, the technical success of the programme was guaranteed in large measure by those who remained with the Project for much of its life and assured its continuity.

Without entering into a detailed analysis of the careers of all those who worked in Dragon it is difficult to be dogmatic on how the balance sheet would look for the individuals in comparison with others of similar ability working in other types of organisation. Certainly a number of overseas staff are now in positions of considerable responsibility, and there is little evidence to suggest that any suffered from their period with Dragon. UK staff complain that their promotion was blocked because of the secondment system and there is some justification for this view. Only Lockett, of all the senior staff, received promotion during his period with Dragon. Even Shepherd's appointment as Chief Executive was not acknowledged in any substantive way. For the more junior scientific staff, the freeze was less evident but the engineers suffered because of the Authority's system of treating engineers differently from scientific staff, identifying the grade of a person with his post. Dragon people had a tendency to see themselves also as key people, equating the importance of the Project with that of the Authority as a whole, while inside the Authority, with its staff of over 25 000, Dragon was often seen as just a small operation, no more significant than any other development group of 250 people all told.

THE SPIRIT

Nevertheless, Dragon was not like any other group. From its earliest days, an esprit de corps and a dynamism were developed that were immediately evident to anyone visiting the Project. Dragon people believed themselves an elitist group and elite they were, as their technical record

Figure 4. Compton Rennie (left) and Pietro Caprioglio.

Figure 5. Leslie Shepherd (left) and George Lockett.

Figure 6. Peter Fortescue.

Figure 7. Professor Francis Perrin (left) and Sir John Cockcroft formally start up the Dragon Reactor Experiment at the inauguration on 22 October 1964.

Figure 8. Meeting of the General Purposes Committee with Jules Gueron in the Chair, attended by delegates from the Signatories and members of the senior staff of the Project.

Figure 9. At the dedication ceremony in April 1960, Donald Fry, Director of AEE, Winfrith, presents a silver trowel to Sigvaard Eklund. Compton Rennie is on the right.

Figure 10. DRAGON SENIOR STAFF — January 1962. *Back Row – left/right:* R. A. U. Huddle, Head of Materials Branch; R. E. Reynolds, Head of Finance, Accounts and Contracts Branch; H. W. Müller, Deputy Chief Engineer, Heat Removal and Fuel Element Handling.
Middle Row – left/right: H. de Bruijn, Head of Chemistry & Chemical Engineering Branch; K. O. Hintermann, Head of Physics & Engineering Research Branch; L. H. Prytz, Deputy Chief Engineer, Primary Circuit; S. B. Hosegood, Deputy Chief Engineer, Fission Product Plant; R. Tron, Deputy Chief Engineer, Programming, Test & Inspection.
Front Row – left/right: R. K. Andresen, Head of Administration Division; L. R. Shepherd, Head of Research and Development Division; C. A. Rennie, Chief Executive; G. Franco, Chief Engineer; G. E. Lockett, Chief Designer.

Figure 11. DRAGON SENIOR STAFF — Mid/1973. *Inset.* B. G. Chapman, Head of Dragon Operations Branch.
Standing – left/right. E. Smith, Head of Engineering Branch. L. W. Graham, Head of Fuel Element Research & Development Branch. F. P. O. Ashworth, Head of HTR Technology Branch. H. Gutmann, Head of Physics Branch. R. A. Saunders, Head of Reactor Experiment Porgramme Branch. M. S. T. Price, Head of Fuel Operations Branch. R. A. U. Huddle, Head of Materials Division, S. B. Hosegood, Head of Reactor Technology Division.
Seated – left/right: M. de Bacci, Head of Technical Services Division. G. E. Lockett, Special Advisor to the Chief Executive. L. R. Shepherd, Chief Executive, T. G. House, Head of Administration Branch.

proves. Put on their mettle at the very beginning by such remarks as "What can a few rebels from Harwell and a bunch of foreigners expect to achieve", they set out to prove they had no peers. Staff came and went – some better than others – but whatever the ups and downs of Dragon and the HTR, an astonishing level of dedication was maintained. Perhaps the very uncertainties acted as a spur; undoubtedly the scepticism voiced about the competence of this "heterogeneous mob" acted as a goad. It also made some of the staff (British in particular) very sensitive to criticism; jokes about either Dragon's competence or the HTR were not considered to be funny. Genuine criticisms could be treated as gratuitous hostility and anyone not fully espousing the faith that the prismatic HTR was the best thermal reactor, was branded as an enemy or a fool. Such fervour only went to underline the complete hold Dragon took of its people for most of its life, due in the last analysis to the inspired leadership of Rennie and the unswerving sense of purpose of Shepherd, backed by his omniscient grasp of the technology.

Winfrith and its management and the Dorset people also played their part. It is easy to forget how close in the early sixties were the memories of war in Europe. Especially in Britain, the new fashion of European co-operation was treated with suspicion, and on the personal level there was little experience of working with strangers and their alien techniques. Within the Authority, all had signed the secrecy vows and the whole technology was permeated by notions of confidentiality. Yet from the first there was an integration of peoples and an exchange of friendships that obliterated national differences. Winfrith made no distinction between its denizens – passes were the same for Authority and Dragon staff alike – and within the Project a warm spirit of camaraderie prevailed which still allowed the most rigorous technical debate.

For the staff members, not only was the general ambience agreeable; the detailed working conditions were of a style that encouraged enthusiasm. With a minimum of formalism and red tape, plenty of opportunities to travel and the intellectual stimulus of working with a number of first class people, Dragon was a project of high morale where the job was the thing, and internal politicking at a minimum. Staff responded by transferring their loyalties from their parent organisation to Dragon almost overnight.

During the construction of DRE, such a spirit was to be expected, provided no major catastrophes occurred. It is well known that a project team with a specific task in front of it, perhaps especially when established within a bigger centre with a rather broad programme, develops a close team spirit – the nature of the job requires it and the objectives of all are clear cut. It is altogether different when that task is finished and the explicit objective achieved. Dragon succeeded in carrying on its pioneering crusade after DRE was operational.

Size was also a factor. Dragon never became so big that the management became remote. It has been noted that Rennie gave the impression of being aloof when it came to decision taking; he was nevertheless entirely approachable off-site and the 'Red Lion', a discreetly modernised 15th century public house became the informal lunch club of the Project, where many of the senior staff regularly gathered round the fire, a pint of beer in their hands. Here some of the more nebulous problems were aired and some of the more intransigent clashes of personality resolved. It was a recognised technique in Rennie's day that if you needed to see him, then the best way was to catch him at the Red Lion at lunch time. Shepherd, although more approachable in the office, continued the practice. A canteen was provided on site, of course, and many would eat there with the rest of the Winfrith staff, but it was characteristic of the Project that the senior staff would draw a little away from Winfrith as such, to gather round its own hearth and in an atmosphere of traditional English hospitality, discuss ultra modern technology with a cosmopolitan verve. Ron Smeaton, the landlord of the Red Lion in preserving this atmosphere, made a significant contribution to Dragon staff relations.

THE PEOPLE

Dragon was well sited, well led, of family size, knew where it was going, believed fervently in the technology it was developing, yet all these added together do not quite explain the enormous bounce of the staff or why so many should seem just that little bit larger than life.

One answer is that the original nucleus of people studying the HTR system at Harwell was composed not only of gifted people – Harwell attracting a significant proportion of the country's most able scientists and engineers in the immediate post war years – but of people keen to work on advanced ideas beyond the systems with an immediate prospect of fulfilment. Intellectually they were the frontiersmen and they were joined by people from overseas who brought with them not only their own different training and flair, but that spirit of adventure which had prompted them in the first place to seek experience away from their own homes. A further admixture was the presence of a number of natural non-conformists, especially from the UK, who were out of a mould a little different from their colleagues, and irked by traditional civil service ways of working. The Project was largely successful in distinguishing in these the rough diamonds from the flints. From such diverse elements a close-knit team was synthesised that was endowed with a collective and personal vigour. Fed by the technical success of the co-operation, it was encouraged to develop its full potential in an environment shorn of pernickety rules and regulations.

This administrative structure owed much to one of the diamonds – Ralph Reynolds. Rennie had known him at Harwell, had talked to him about the Authority's administration and heard him express opinions on it that were close to his own. Rennie had also seen him at meetings, putting his points clearly and succinctly and had come to appreciate his gift for setting things out simply and methodically. Hardworking and ambitious, but without a formal university degree, Reynolds had recently been promoted at Harwell (where his talents were recognised) without his new position being determined. He was, in consequence, available and he joined Rennie at once and plunged into the new problems with great energy. By the time that Rear Admiral Andresen from Norway joined the Project as Head of Administration Division, Reynolds already had matters well in hand and he continued to act as the principal administrative executive. Andresen, more familiar with a clear authoritarian structure than this band of individualists, was content to give him full rein and saw no need to extend his own initial engagement of two years.

A contradictory personality, Reynolds could readily be mistaken for an impersonal machine solely interested in applying the formal rules, yet he was adept at bending these rules in the interests of efficiency and it was he who drafted the formal envelope that could foster that simplicity of operation and trust in peoples' competence which Rennie had insisted on. He was in the habit of presenting a cynical attitude towards the outside world, yet no one could have been more dedicated to the Project, and he retains a touching gratitude for the courtesy with which he was treated from the beginning by the members of the GPC (and Guéron in particular) before whom he was at first very nervous. Almost all the early documents came from his pen. Working through Saturday and Sunday he would produce reports at a prodigious rate (he allowed himself one per hour) full of scratchings out and second thoughts but displaying, when typed, a clarity of communication that Rennie regarded as a determining factor in establishing relations between the Signatories and the executive. Professing a strong reluctance to foreign travel which kept him away from his family, Reynolds, nevertheless, continued to guard the Project's interests, accompanying all the visits to foreign firms until he returned to the Authority in 1966. So strongly were people bound to the Project that it was assumed by many that he and Rennie had quarrelled, but this was not the case. He had seen the Project established, the reactor running, and contracts settled, the original aims accomplished, and he saw his own future as advancing only elsewhere.

His replacement as Head of Administration Branch, though not as abrasive as Reynolds, was also not quite the traditional British administrative officer – which was perhaps why he came to Dragon. Exhibiting an independence of mind and temperament, Theo G. House –

referred to often by his first two initials alone – quickly gained the confidence of the delegates and established himself as an effective follow-on to Reynolds, no less assiduous in seeing that contractual arrangements accorded with the Project's rules. More sociable than his predecessor, he was well suited to operating a continuing situation where precedents had been established but where there was still room for innovation. With his heavy glasses, air of permanent wonder, somewhat abrupt manner and sharp sense of humour, he was soon distinguishable as a Dragon man – not a standard functionary. Both he and Reynolds were backed by a succession of highly efficient accountants in the persons of Don Jolly, Eric Weadon and Jim Melville.

Lockett we have already met: universally recognised as an outstanding engineer, universally liked and respected. He was not, however, a Project supervisor bringing together the threads of development, coordinating procurement and installation, fighting the contractors and so on. But this was Franco, the first Chief Engineer (after von Ritter's tragic death). In Rennie's words, Franco was "a tower of strength". When prised away from Ispra early in 1960 he progressively took over the construction of the DRE whilst handing over to his successor in Italy. He found, on studying the general layout of the reactor plant that had already been prepared, work of high standard, forming an excellent basis on which to proceed. Somewhat unhappy though with the design of the fuel transfer mechanism which was to be located inside the pressure vessel and operated automatically, he instituted a new approach with the electro-mechanical parts accessible from the outside, and he still takes pleasure from the fact that his name (together with those of A. N. Kinkead and V. H. Sørensen) appears on the patent covering this device which was required to operate in such novel conditions and which worked with astonishingly little trouble from the outset.

Franco's unique contribution to the Project was, however, his organising ability and his capacity for keeping track of everything that was going on. There was little experience in Europe at that time of building a machine with international tendering and one more success for which the Project and Franco in particular can take credit, is the way in which the various parts came together, more smoothly than in many a national project. Franco, whilst strong, even imperious, was also an engaging personality, radiating an infectious *joie de vivre* and just enough of the classic Latin temperament to enable him to do verbal battle with the best, without losing his ability to switch off the fury and turn on the charm. He got on well with the Authority's Engineering Group – the constructing agent – which was most important and he was a proper match for John Davis who acted as site engineer (under T. G. Williams) and with whom there was many a wordy fight – as became the natural relationship between designer and builder. Davis, a rugged Welshman, first class at his job, would have his viewpoint and the Project would have another, and neither Franco nor he was shy of pressing his arguments. Never in doubt was the motivation of both, namely the good of the Project; both were able, and consequently compromise was always possible with no loss of respect on either side. Handsome, fun-loving, eloquent, Franco's presence left no doubt that Dragon was a European and not just a British activity.

Franco returned to CNEN after four years with the Project to become Director of CNS Casaccia, going back to Dorset at roughly monthly intervals until the DRE went critical. He enjoyed his stay in England and feels that he was able to give of his best, perhaps because of his naval background. Nuclear energy practices of that time owed more to a naval tradition, that was similar in all European countries, than to the universities.

Under Franco, Branch Head responsible for Heat Removal and Fuel Element Handling was Heinz-Wolfgang Müller, a stocky even-tempered German with a natural talent for getting on with people. As well as a good engineer, he too was a good organiser and when he left after four years with the Project it was to take over from Schulten at BBK continuing with HRB when Krupp separated from Brown Boveri. His close knowledge of Dragon and the Dragon staff was an important factor in cementing relations between the Project and the German companies, and it ensured a more open mind towards the prismatic system vis-à-vis the pebble-bed. Like so many others who left the Project to continue work on the HTR system elsewhere, the bond forged with Dragon in the first years had a continuing value so that while there was competition between the pebble-bed and the prismatic systems, at the technical and personal level, co-operation was very real.

Parallel to Müller as Head of the Fission Product Plant Branch was Samuel Hosegood, who remained with the Project from September 1959 through to the end, becoming Head of the Reactor Technology Division in 1968. Assuming the main responsibility for the assessment studies in 1965, he was the originator of many of the innovative designs prepared by Dragon for HTR power stations. As a result, he was in the front line of the efforts made by the Project to promote the HTR as a power reactor system, a crusader with his heart on his sleeve. As the years went on, he developed a strong hostility towards the Risley section of the Authority nourished by such remarks as "when you get into trouble, come to us and we'll teach you engineering". A brilliant and perceptive engineer, whose hands bore witness to his ready contact with things mechanical, he resented strongly the scepticism that met his proposals for downward core cooling, for example, and reacted personally towards any criticisms levelled at the HTR, a system in which he placed complete faith. Towards the end, he took it upon himself to lobby members of Parliament, the press – anyone who might have an influence on the British attitude towards a further extension, seeing the closure as a final betrayal by the Authority, the culmination of Risley's efforts over the years to denigrate the HTR in order to bolster its own systems. His absolute commitment to Dragon made him at times a brittle person to deal with, but he was a thorough organiser and an effective Division Head.

Totally different in character was Tor Midtbo, his fellow Branch Head under Lockett after the Enginering Division was reorganised on Franco's depart. Taking over DRE as it was commissioned, Midtbo concealed his considerable experience gained as chief engineer at the Royal Norwegian Air Force Technical headquarters (following a period working on gas turbines at Power Jets and MIT), under an ebullient, happy-go-lucky exterior. A natural extrovert, he was of the type who injects life into any company and about whom epic stories are woven, as befitted his Viking origins. He did not, however, return to his native Norway, on leaving Dragon in 1966; instead he joined Brown Boveri and made his home in Switzerland.

Epitomising this larger than life characteristic of the Dragon staff, can be singled out amongst the first group of Branch Heads, Karl Hintermann. Prichard recalls his surprise when conducting interviews in Zurich, to be suddenly confronted with a mountain of a man in military uniform. (Hintermann, a fighter pilot in the Swiss air force was at that time doing a stretch of military service). In response to the American Atoms for Peace initiative, he had spent a year at the Argonne National Laboratory in the USA, then returned to Switzerland to work on the national materials testing reactor and to some extent on the pebble-bed ideas emerging from BBC Mannheim. Strictly he was employed by Brown Boveri, Baden, but was on detachment to the national research and development organisation, Reaktor AG. He had been present during Shepherd's talk to the European Atomic Energy Society in Rome on Harwell's work on the HTR, and was so delighted to have an opportunity of questioning Shepherd further, the normal roles of interviewer and interviewee were reversed. He was made Head of the Physics and Engineering Branch and brought to the Project that brand of down-to-earth science that is traditionally Swiss, as well as over 110 kg of exuberant, good humoured intelligence. He also brought a specialist knowledge of gas turbines and with Shepherd and Magnus von Bonsdorff, made the tour of Europe in 1963-64, promoting the direct cycle system. After six years with Dragon, he went to Mannheim, a prismatic disciple in the home of the pebble-bed.

When Hintermann was made Head of the Physics Division in 1964, Erich Schröder moved up into the position of Head of the Physics Branch. Mild, courteous and highly respected, he came to Dragon early in 1960 with a broad background of experience in reactor physics, having performed the safety and shielding calculations on a Magnox reactor that had been considered by the Rheinisch-Westfählisches Elektrizitätzwerk and then on the

Dido and Merlin reactors at Jülich. Without imposing his own personality he had a gift for welding the talents of his Branch into a coherent group that worked together as a unified team. He went back to Germany towards the end of 1965, first to BBC Mannheim, then on to become Technical Director of the nuclear ship development and construction organisation GKSS.

With his departure, physics ceased to be a separate Branch and was included under Marien (see page 115) in the Physics and Engineering Research Branch until Shepherd's reorganisation in 1968 when it became the domain of the Austrians with Gutmann as Branch Head, a post which he held until the end of the Project – the only non-British Branch Head in the team. Of chunky appearance, determined and serious in character, a doughty fighter for his staff, Gutmann was a pillar of the low enriched fuel cycle. Conscious of the political as well as the scientific importance of Dragon, he made great efforts to promote the Project in his own country and to establish a European consensus view on HTR problems.

So far, in dealing with the personalities that made up the Dragon team, no mention has been made of those concerned with that part of Dragon's development that may well prove to be the most far reaching, notably that of coated particle fuels. Its initiator, Roy Huddle, one of Fortescue's original team, joined Dragon at the beginning as Head of the Materials Branch and was made Head of the Materials Division in 1968. Already at Harwell he had made his mark, first, as a gifted metallurgist by inventing the Magnox canning material for the natural uranium reactors, and second as a rebel against authority. In the Dragon atmosphere he was able to give full rein to his virtuosity and soon became known throughout Europe for his mercurial temperament, his changing enthusiasms, his loquacity and in contrast to what could be quite simplistic notions, a quite extraordinarily intuitive understanding of what makes materials what they are.

His approach to materials problems was highly individual, combining a sound academic background with a need to visualise internal mechanisms in a physical way. Since the war, metallurgy has changed from being an art, based on tradition and empirical rules, into a mathematical science whose very complexity tends to obscure the practical implications. Huddle absorbed the science and developed his own models to explain observed behaviour, feeling his way through a succession of simplifications to arrive at a plausible picture which could be quite unsound from a purist's point of view, but which allowed him to seize the essentials. Much of his thinking he did aloud, and a technical discussion with him was more often a monologue, his own conclusions being taken as a consensus opinion. Ideas poured out in a continuous flow and it was vital that behind him there was a body of steady, competent people, able to supply the continuity and attention to detail that transformed ideas into explicit practice. By 1974, Huddle believed that Dragon had gone as far as it could, and that soon fuel and materials development would be much more advanced than more ordinary components of the HTR. He applied for early retirement and left at the end of that year (continuing to work as a consultant to Dragon and KFA). In spite of his being an erratic Division Head, Shepherd recognised his genius and was reluctant to see him go.

To start with, Huddle's attention was concentrated on the reactor core materials, leaving the problems of primary circuit materials largely to John Condé. A great deal of his time was spent in travelling either to the Continent to discuss the tests being made on graphites or fuels in the various Dragon rigs or to Farnborough where much of the graphite impregnation and original graphite coating was carried out. Here Claudio Vivante installed himself almost immediately on his recruitment from Euratom in November 1959, astonishing Huddle by the ease with which he, a foreigner, had become a resident in what was essentially a military establishment. Vivante, quiet, heavy, not at all the thrusting type, was almost the antithesis of Huddle, steady in application, providing the calm to counter the tumult. They became firm and lasting friends and the development of coated particles progressed steadily. Vivante left Winfrith in May 1967 when he considered that the development of coated particles was substantially complete, joining de Bacci in the Commission's services in Brussels.

When the decision was taken to manufacture coated particle fuels within the Project, the task of organising this, elaborating the procedures for producing perfect spheres, coating these with up to four layers of material of carefully controlled thickness and density, was confided to Michael Price. From 1969 onwards he was Branch Head of Fuel Operations, a Branch that started under Huddle but was transferred into the Division of Technical Services under de Bacci who returned to Winfrith from Brussels in December 1969. As Division Head, de Bacci was ready to support Price in his view that the time had come for a systematic approach to be made to drawing up fuel specifications and manufacturing procedures, whereas Huddle's style was to make leaps in the dark that could cut corners in the beginning but made subsequent analysis difficult. Huddle was in any case turning his attention more to the metals research programme and was content to leave fuels to others.

The fuel manufacturing facility of Dragon was something of a phenomenon, a specialised production operation in the middle of a research and development institution. That it could grow to such efficiency was due largely to the systematic, ordered organisation of Price, whose quiet friendly manner and highly developed sense of logical perfectionism was allied to a resolute determination to do things his way. A comment from a reactor engineer required to evaluate the HTR for his commision serves to illustrate the quality of the output from the Dragon 'fuel factory': "There are many brands of good coated particle fuel on the market, but if I had to build a reactor I should see that mine was made by Mike Price''.

When Huddle lost interest in fuels, Leslie Graham took over as Branch Head for Fuel Element Research and Development. Joining Dragon from Hawker Siddeley towards the end of 1961 when that company closed down its nuclear activities, he was initially on secondment from the Agency, and then from the Swiss Office of General Atomic. He had already contributed much of value to the Project particularly in the area of graphite development when with Hawker Siddeley, and the Project was anxious not to lose his services. A brilliant metallurgist (in Huddle's words a ''whiz kid''), dedicated to the HTR and more disciplined than Huddle, he took over where Huddle left off and acted as Head of the Materials Division from Huddle's departure until the closure. The esteem in which Graham was held internationally can be gauged from the fact that since Dragon closed down, he with a small team, has continued research into primary circuit materials in a laboratory at Flight Refuelling Limited, near Winfrith, supported initially by ERDA and then the Department of Energy of the USA, and by KFA of Germany with Austrian and Swiss participation.

Even though reactors are expected to operate safely and without incidents of any significance, it should be noted that DRE was a special case. To begin with, DRE was in itself a novel reactor system whilst latterly the core contained so many different experiments, some of which were included at the last moment, every new start-up constituted a new reactor experiment. Any mistake could not only pose a hazard to personnel but could result in a 'dirty' reactor that was then useless for much of the data taking. The Branch Head responsible for the design of the large variety of fuel elements which made up the reactor core, was Eric Smith an ex-Harwell man with a naval background, always quiet and reliable in spite of the continuous rush implicit in his work. On him and his team rested the responsibility of seeing that the irradiations demanded by the research people could indeed be carried out. Detailed design of fuel rigs does not make the headlines but this was one of those essential, continuing, exacting tasks on which the Dragon programme depended.

Similarly, safety as such is only remarkable when something goes wrong and with Dragon, no incident of significance ever arose. For this a great deal of credit must go to Paul Ashworth, diplomatic, even deferential, easy to get on with, yet clear sighted and consistent. With the able assistance latterly of John de Nordwall, he built up a solid team, thoroughly versed in the complexities of safety requirements and in the behaviour of fission products. Working closely with the Authority operations team under Bernard Chapman (see page 234), equally dedicated to doing things well, the Dragon team earned an enviable reputation for reliability and they left a reactor, that after 10 years' operation, was still amazingly clean.

Naming people in a Project such as Dragon is, however, invidious, as wherever the line is drawn, some will have been omitted whose contributions to its technical or collaborative success were of great significance. In concentrating mainly on the Project Leaders, this does justice only in so far as it would have been they who would have borne the blame should the Project have been a technical failure. Dragon though, was not a group of stars and chorus so much as a company in which all were encouraged to play a determining part, inventing and improvising as they went along. It was no set piece that was being enacted and the result could not have been of the quality it was, if the identification of the staff with the Project's ideals and technical aspirations had not diffused from the Chief Executives throughout the entire personnel, regardless of rank and nationality. Most of those who figured at some time on the Dragon staff list considered it a privilege to work in such a dynamic and innovative environment and responded by making their own contributions to the limit of their capacities.

CHAPTER 19

THE SIGNATORIES

INTERACTION between Dragon and the Signatories was not confined to the secondment of staff and the receipt of information. Even before the Project launched its recruiting campaign, the Agency had initiated a study of the research capacities that were becoming available in Europe and which could be used by the Project for extra-mural research. In several countries, major new facilities were coming into service as the Dragon R & D programme got under way and they provided a vital contribution to the irradiation programme on graphite and fuels, for example. Industrial laboratories also played their part in support of the developments necessary for the construction of DRE. Once DRE was built, extra-mural research continued to carry a major fraction of the R & D load, increasing the scale of the Winfrith effort by 30-50% and giving the Project access to equipment and skills that were not otherwise available. All the laboratories undertaking extra-mural research acted also as communication centres that kept alive local interest in the HTR, even when the system was not included in the national power programme.

In the late fifties, atomic energy programmes were still fluid and research was being carried out on a broad front. Participation in Dragon did not need to be justified by an implicit expectancy of building HTR power stations. Indeed, Dragon was not at first viewed as a central power station experiment at all; it was an exercise in the new technology. At the same time, power stations were still of modest size, and the investment problems associated with a new system seemed tractable, should this particular one prove to have merit.

While DRE was under construction and the HTR was proving to have real potentialities, nuclear power became marginally economic for stations built in large sizes and as part of a series. At the same time the fears of a general energy shortage dimmed with the opening up of more and more sources of oil. Electricity companies were primarily interested in immediate costs and became increasingly reluctant to make risk investments that were not backed by American industry. One by one, the European countries which had nuclear programmes, opted for American light water reactors and outside Britain, the prismatic HTR was relegated to the lowest ranks as a system for central power station use, until GGA in the United States was able to claim that markets were opened.

By this time, the DRE exercise had been completed and Dragon's future depended upon its value as an irradiation and R & D facility to the Signatory countries. Consequently their attitudes towards the collaboration became conditioned more by their power programmes, even though the political will to continue what had been so successfully started, lingered on.

In the following sections, national programmes and policies as they relate to Dragon are summarised **(34)**. Inevitably the picture is drawn with a broad brush whereas view-points in all countries were highly structured and even within an industry or a single Government department, divergences existed which would require much deeper research and understanding to analyse. For Dragon, however, it was the overall situation which counted in the long run and which determined its existence once DRE had successfully operated for a reasonable period.

CHAPTER 20

AUSTRIA AND DRAGON

ALONE of all the continental countries participating in the Dragon Project, more money flowed into Austria from the collaboration than flowed out. Austria's contributions to Dragon budgets totalled £900 000 (1.85%) while contracts placed within Austria were worth over £1M. This no doubt contributed to making Austria the least complicated of all the Signatories. A second important factor was the level and continuity of representation on the Board of Management and GPC.

Reference has already been made to the length of service of Polaczek, Director in the Austrian Chancellery and Secretary for Nuclear Energy Questions. He was a firm believer in the necessity for Austria to gain its experience in nuclear affairs through international collaboration. In line with its policy of neutral internationalism, that gained recognition in the siting of the headquarters of the International Atomic Energy Agency in Vienna, Austria was favourable to the principle of collaborating through OEEC and was, in consequence, an active member of all three of the Agency's joint undertakings. In the case of Dragon, Polaczek had had to contend with the hesitations of the largely nationalised supply industry and the competition for funds from the academic sector, which was pushing the Government to become a member of CERN even though they were ready to believe that a project proposed by the British was soundly based. Industry, although initially needing persuasion, was encouraged by the prospect of manufacturing contracts, and once its opposition was withdrawn, Austria's subsequent participation was never in doubt.

At the scientific level, Austria was represented by firm believers in the Dragon collaboration from the Oesterreichische Studiengesellschaft für Atomenergie (OSGAE), the national atomic energy research organisation which had the task of coordinating Austrian participation in Dragon. Technical Director of OSGAE and Director of its laboratory at Seibersdorf, Michael Higatsberger, with a physicist's appreciation of the merits of the HTR, served on the GPC from its first meeting through to 1971. His place was taken for a short time by Hubert Bildstein, Head of Chemistry at Seibersdorf, followed soon after by Peter Koss, then Head of Metallurgy, who continued the tradition of positive support through to the end.

Austria, with no pressing problems of energy supply, had no intention of striking out on its own in nuclear power developments and did not start construction of any major power station until 1971 (a BWR of 700MW at Zwentendorf that, following a national referendum was, at the end of 1978, blocked from operating). It was, on the other hand, keen to establish an industrial and scientific competence in nuclear energy and also a place in the component market. Strong in quality heavy engineering and in the forefront of materials developments with an accent on such high technology metals as tungsten and molybdenum, Austrian companies were well placed to enter the nuclear market where new exigencies had to be met and quality work was at a premium. In the sixties, the exchange rate of the Austrian Schilling also helped exporters in markets that were extremely competitive. Many contracts for DRE were let at uneconomic prices as companies made their bid to enter the field. Austrian companies were no exception, and the country's share of the DRE construction approached the £½M mark or nearly 10% of the total, excluding conventional plant and site work. Contracts for R & D amounted on aggregate to about £600 000, corresponding to some 3.6% of the overall value of those put out by the Project.

Major items supplied to DRE were primary and secondary heat exchangers and tube bundles, containers for irradiated material, a water cooling system (Waagner-Biro AG); containment vessels and support structures (Simmering-Graz-Pauker); stainless steel supplies (Schoeller-Bleckmann Stahlwerke AG); primary circuit dump tanks (Voest).

When the Project began to look outside for suitable centres within the Signatory countries to which development work could be subcontracted, Metallwerk Plansee was one of the first to be considered because of its reputation in high performance materials. The company became involved thereafter in the development of fission product retaining fuels although there was a certain disparity of temperament that marred the relationship. Huddle's instant improvisation with its rapid changes of direction ran contrary to the company's tradition of meticulous, ordered progression and the two sides were never really able to run in harmony. One of the contracts placed with the company was for the development of an automated coating furnace but this was somewhat premature as fuel specifications had still to be written and manual systems were more appropriate to the development period. As a result, the furnace was not brought into operation; it would have been more useful if the contract, with revised specifications, had been placed at a later stage.

The majority of the R & D work done for Dragon in Austria was performed by OSGAE at the Seibersdorf laboratory, where a speciality was made of detailed post-irradiation examination of fuels and various methods of physical analysis as well as methods of breaking down fuels as an alternative to burning. The reactor at Seibersdorf – a 12MW pool type research reactor – was not particularly suitable for accelerated irradiations, whereas the hot cell facilities and analytical equipment were entirely relevant.

The most striking aspect of Austria's participation was the large number of staff on the payroll of the Project – 28 in all, and the length of time they spent with Dragon – over 4½ years on average. A few were with the Project for a large proportion of its life and the Physics Branch under Gutmann, who was at Winfrith from July 1963 until termination, was, towards the end, nicknamed the Austrian Empire. All but two of the Austrians were seconded from OSGAE (the exceptions coming from Waagner-Biro and Schoeller-Bleckmann) and in consequence, there was a strong bond between the Project and the Seibersdorf Laboratory. Visits between the laboratory and Winfrith took place on a regular basis and when extensions were under discussion the Austrian contingent at Dragon played their part in generating a favourable climate at home. Relations with Vienna were equally warm and Dragon staff were always made to feel welcome whenever they were in the country.

CHAPTER 21

SCANDINAVIA AND DRAGON

EVEN though it is usual to consider the Scandinavian countries under a common heading, particularly in the research field, differences in pace and emphasis in their atomic energy programmes militated against their establishing a common front, despite their mutual interest in boiling heavy water systems. Contact between the countries has been maintained through the Nordic Contact Organisation for Atomic Energy (NKA), a committee which brings together twice a year, representatives of government departments and atomic energy commissions. At such meetings, attitudes to be adopted towards Dragon extensions would be discussed, but joint action was largely confined to expressing support for each other on such questions as contracts, secondments and promotions. Just two combined information meetings on Dragon were held, the first in 1973 and the second in 1975.

In the beginning, it was suggested that there should be joint Nordic delegations to the Dragon committees, but each country preferred to promote its own interests in extra-mural research. A minor effort was made to present a coordinated front on major policy matters and sometimes the Nordic delegates would sit together before meetings of the Board of Management and GPC when they were normally joined by representatives of Austria and Switzerland. Mainly, however, policy was left to the two big partners.

Until the discovery of North Sea oil, there were almost no indigenous supplies of fossil fuel in Scandinavia and the exhaustion of economic hydroelectric reserves in Norway and Sweden was becoming imminent. Denmark had not even this source of energy and had to rely entirely on imports. Nuclear power in the late fifties and the sixties was regarded as a long-term necessity and one which could not be left to the big countries only. At the same time, the effort that could be devoted to research and development was limited, and international collaboration offered a unique opportunity for acquiring the experience needed to evaluate a given system even if not actively to exploit it. Had the HTR become a viable system, it is possible that Norway and Sweden particularly would have found that the scale of the national effort that had been mounted in support of the HTR was below the threshold required to make full use of the participation in the international project. As it was, the question of the national threshold level did not materialise and Dragon is seen now as an important training ground and source of high quality information, whose value lay in the European contacts that were established and in the awareness it created of what had to be understood in whatever nuclear system might be built.

All three countries suffered the experience of seeing the DRE built without a significant plant order being placed with their industries (except for one contract worth about £10 000 placed with Thomas B. Thrige of Denmark for the supply of the fuel disposal carousel). In part this was because Scandinavian companies were not prepared to 'buy their way in' and the impression was created that Scandinavian prices were too high. No great effort was made on either side to establish the contacts that are an essential preliminary to the placing of orders. In compensation, the Project was keen to use the research facilities available and made a point of channelling contracts for extra-mural research into the government laboratories which were, in their turn, prepared to offer attractive terms. For Sweden in particular, the financial benefit of these extra-mural research contracts played a determining role in convincing the authorities to continue their participation. These contracts were important also for another reason. They ensured a more direct association with the technical work of the Project than was possible through the secondment of staff, whose experience was quickly dissipated on their return, through there being no national HTR development wherein it could be exercised. With the research work undertaken for the Project, there were regular visits from senior Dragon research and development people who were able to transmit up-to-date information on the Project and keep alive the interest in HTRs generally.

DENMARK

In the early fifties, before research on the elementary particles had diverged from nuclear developments, Copenhagen was renowned as the centre in Europe for theoretical nuclear physics. Inspired by the presence of Niels Bohr, celebrated as much for his infectious humanism as for his great scientific powers, the Copenhagen school was the Mecca for physicists the world over. Bohr's influence in promoting international collaboration was immense. He was one of the founding fathers of CERN and the theoretical group of this organisation was originally centred in Copenhagen. In consequence, Denmark had an initial spiritual drive to pursue the new sciences and to do so in collaboration with other countries, quite apart from the more practical reasons already mentioned.

The Danish Atomic Energy Commission (DAEC) was set up in December 1955, with Bohr as its Chairman. Its research establishment at Risø, situated on the east coast of Roskilde inlet north of Roskilde, was inaugurated in June 1958. Strong ties had been established with the UK through the personal links between Bohr and Cockcroft, and they led to the purchase of a materials testing reactor (DR 3) with, as part of the deal, a contract to study certain aspects of graphite behaviour under irradiation. Denmark was thereby already in touch with British gas-cooled graphite-moderated reactor work and was keen to join in the further development of the system. Of all the Scandinavian countries, it remained the most steady supporter of Dragon, although some hesitations were expressed about the 1973/76 extension in the light of the decision of the Danish electricity companies to concentrate on light water reactors. However, because of certain safety questions that were being raised in relation to LWRs, and the apparent break-through of HTRs in the USA, it seemed to be an inopportune moment to withdraw. In any case, Denmark was to join Euratom and could continue its membership through that organisation for a contribution of 0.64% instead of the 2.2% that it had been paying up till then.

Denmark avoided the violent changes in direction experienced in so many countries during the sixties, and funding of atomic energy work was not subject to sudden cuts as elsewhere. The economy was reasonably stable and there was little pressure on Risø to diminish its international commitment. Dragon benefited greatly from the continuous support of Hans Koch, Chairman of the Executive Committee of the Danish Atomic Energy Commission and principal delegate to the Dragon Board of Management until 1973. He was an enthusiastic internationalist, had faith in the HTR system and considered the Dragon Project well conceived and run. With his reputation for realism and consistency, he was in a powerful position to protect its interests in Denmark. He was amply assisted by Flemming Juul, Vice-Director of Risø on the scientific side, who served on the GPC until 1973, and on the Board of Management from then on as a Euratom delegate or observer.

Risø was planned as a nuclear research and development centre serving both industry and the natural sciences. It was also to have been the national nucleus for the industrialisation of nuclear power, but in this, it was largely unsuccessful as the electricity companies were suspicious of government involvement and preferred to keep their distance. They were reluctant to commit themselves to the new energy source and Denmark has yet to have a nuclear power station constructed on its soil. No nuclear industry developed, and Risø, without industrial support in the nuclear sector, had latterly to try and anticipate industrial requirements instead of leading a coordinated approach.

After a brief enthusiasm for heavy water organic systems in the early sixties, Risø concluded that a purely heavy water system was the most promising line to pursue in collaboration with other Nordic countries. The electricity companies remained unimpressed. Then in the late sixties, Juul saw in the HTR, the possibility of jumping the first generation of thermal reactors and going directly to a system of high efficiency. Independent assessment studies were initiated and, in parallel with the extra-mural work for Dragon, effort was put into graphite research and coated particle fuel studies. Risø could not, however, take any lead on its own and, disappointed with the slow rate of progress in the bigger countries and the preference of Danish electricity companies for LWRs, allowed its HTR work to run down. Belief in the virtues of the HTR remained, but it seemed clear that the system would not be exploited in Denmark within the foreseeable future. Participation in Dragon continued, as one official expressed it, because it was easier to go on than stop and there was no active opposition within the country.

The extra-mural research provided by Dragon was an encouragement. Denmark's own use of the materials testing reactor facilities of DR 3 were slow to develop and Risø was grateful for the work as much as for the funds it brought in. Prices charged to the Project were reasonable, at one time being one third of those charged by Harwell for similar jobs, and the Dragon experiments received the attention of top quality staff. A wide range of irradiation experiments was conducted on coated particles both as loose particles and as compacts. Post irradiation examination of irradiated specimens in the Risø hot laboratory was developed to a high level and prices for this style of work were kept attractive so that in the later years, specimens irradiated in Sweden were being shipped to Risø for examination. Apart from the irradiation research, Risø was commissioned by Dragon to build a high temperature loop to study the heat transfer characteristics of helium, from which a considerable amount of valuable data was derived. The Dragon work also provided an introduction to HTR research at Harwell and KFA, and Risø supplied irradiation rigs to these organisations and undertook irradiations in DR 3 under contract.

Altogether, research and development contracts worth nearly £350 000 were placed by Dragon with Risø, practically all on the basis of the skills and facilities available without going out to international competition. This represented 2.3% of the total external expenditure in the research and development category.

Twelve people were seconded by the Danish AEC to Dragon for an average of 2¼ years each. The maximum number to be at Winfrith at the same time was five, towards the end of 1961. None had a grading higher than D and only two were placed in this category. To begin with, salaries in Dragon compared favourably with those paid by the DAEC, but with increasing prosperity, the position in Denmark changed and the gap between UK and Danish levels steadily widened. Even though taxes are generally higher in Denmark than in Britain, the few recruited directly into a Dragon position found themselves at a disadvantage financially, and within Risø, there was little enthusiasm to be detached to Winfrith once an established position had been attained. Nevertheless, it was considered to be valuable training to join the Dragon team and several former Dragon staff are now in senior posts which have nothing to do with HTRs.

NORWAY

Norway's per capita consumption of energy has, for many years, been the highest in the world, twice that of the USA, and four times the average of the industrialised countries of Europe. The country had thus a prime interest in new methods of electricity generation as although hydro sources abound, those not exploited are for the most part far from the centres of population; Norway was also attracted by other applications of nuclear energy, in particular for the propulsion of ships. Gunnar Randers, the Managing Director of the national atomic energy institute, IFA, was quick to perceive the potential value of a compact system of high thermal efficiency such as the HTR and was keen to participate in Dragon. However, as the hopes of using nuclear power for merchant ship propulsion receded, Norway's interest in the HTR faded.

In 1964, a major reappraisal was begun of the priorities to be adopted in funding scientific research during which the fraction of the available sums absorbed by IFA's research centre at Kjeller came in for much criticism. An official commission was set up and pending its conclusions, IFA was unable to make any commitments for the 1967-1970 Dragon extension. On the basis of the Commission's enquiry, the Government concluded that Kjeller's funds would not be cut, but would remain at a constant figure with no compensation for inflation. IFA was, as a result, able to confirm its continuation in March 1967 but had to face a major reorientation of its own activities, diversifying in order to attract funds from industry for commissioned research. By this time, it had become clear that for many years to come, the civil application of nuclear energy would be confined to electric power generation (apart from the very limited market for nuclear powered ice-breakers) and with the discovery of oil in the North Sea, Norway's own requirement for nuclear power stations dropped to zero.

Industry's interest in nuclear energy was limited. In an attempt to enter world markets, a centralised marketing agency, Noratom, had been set up to represent a wide selection of companies, but in the case of Dragon at least, it was of no help in bringing custom to Norwegian manufacturers.

Kjeller had no facilities that were of value to the Project and the financial return of IFA from its investment in Dragon was negligible. Yet of all the countries, Norway was relatively the most successful in obtaining Dragon research contracts, the gross value exceeding £450 000 compared with its total contribution to the Project of about £560 000. The contracts did not go to IFA but to another government institute, the Central Institute for Industrial Research (CIIR), and the difference was important. CIIR was set up by the Norwegian Council for Scientific and Industrial Research in 1950 with a charter that made it substantially autonomous, although the Government appoints the management board and provides 18% of the budget, receiving in return a discount on the price charged for the research it sponsors. This constitutes about 50% of the total undertaken by the Institute; the rest is research for industry under contract. The Institute has been suc-efficient business methods. One of its specialities is high temperature corro-efficient business methods. One of its specialities is high temperture corrosion studies and this was particularly relevant to the materials programme of the Project.

First contacts were made by Dragon via IFA in 1960. These led to a contract for research into the outgassing of graphite at high temperatures for which advanced analytical equipment was already available. A second followed on metal corrosion, as the laboratory could claim to be one of the very few able to cope with very large numbers of specimens, and a third on graphite oxidation.

When the time came to consider the 1970-73 extension, IFA was in great difficulties with its own programme, and just could not find the £115 000 necessary. Fortunately, the other Signatories were more settled and an offer of £50 000 was accepted. Shepherd made great efforts to have this figure raised by promising a significant increase in the contracts to CIIR, as the Project was intent on extending its corrosion testing to cover samples under strain. IFA and CIIR, however, worked completely independently and although Dragon benefited from the special government discount, the relations of CIIR with Dragon were purely commercial. Contributions to Dragon had to come from the IFA budget, and for Viking Eriksen, who succeeded Gunnar Randers as Managing Director of IFA, in 1968, the key task was to maintain activity and morale in Kjeller when he was already aware that the worst had yet to come. Efforts made to involve the Norwegian Council for Scientific and Industrial Research in the membership question were also unsuccessful, but a small financial contribution was secured for the metals programme on an unofficial basis, which may well have helped the Board of Management to authorise continuation of the metals contract after Norway had ceased to be a Signatory.

The facilities offered by CIIR and the competence of the people involved, were the reasons behind the big increase in work that Shepherd had indicated. Changing contractors once a long term research and development project has got under way is time consuming and expensive and although some of the corrosion work was hived off to Fiat in the later years, a new contract for over £100 000 was placed, even after Norway had withdrawn from the Project.

Norway's complete withdrawal was not intended. During negotiations on the last extension, IFA knew it was to suffer further cuts in 1973 which would involve a 15% reduction of staff, even without any contribution to Dragon. It was with great relief that the Institute viewed the coming Norwegian link-up with the European Communities, which meant that participation in Dragon could be taken care of through its membership of Euratom and funded by other sources (and for less money). The result of the referendum, firmly excluding the country from the Communities, came as a complete surprise; no allocation had been made for Dragon in IFA's budget and the Government refused to provide other funds. CIIR made some representations through the Department of Industry and attempted to influence the Chairman of the Industrial Committee of Parliament. Income and expenditure were, however, completely divorced, and the Government was not prepared to increase the subvention to IFA, just to help bring business to CIIR.

Norway's waning interest in Dragon is reflected in the number of people seconded to the Project. For a short period towards the end of 1961, eight Norwegian staff were resident at Winfrith out of a total of thirteen whose length of stay averaged 3¼ years. They included not only technical staff, but the first Head of Administration, Rear-Admiral R. K. Andresen and the first Technical Secretary, the late Bjarne Aabakken (tragically killed at Tehran Airport after he had left the Project, when returning from a mission in aid of the developing countries). The last remaining secondment from IFA left Winfrith in June 1969. The longest serving member was Tor Midtbø, who was with the Project for seven years, latterly as Head of the Reactor Experiment Branch. He took over from Lars Prytz, also a grade C, and left to join Brown Boveri rather than return to IFA. Three others attained grade D, which indicates that the general level of representation was usually high.

North Sea oil transformed not only the energy supply situation in Norway, but also the investment demand. From the middle sixties onwards, nuclear energy development was of low priority. Only through international participation has the Halden project been able to keep going and its continued existence is entirely due to the flexible way in which IFA has managed the Project, adapting the programme continuously to meet the requirements of the users. Halden remains a resource; Dragon became a drain.

SWEDEN

Enthusiasm for the potentialities of the HTR played little part in deciding Sweden to join the Dragon Project and, as time went on, even less in its continuing as a Signatory. Initially it was support for OEEC and the style of collaboration it represented – without political involvement – that was the determining factor. Later, it was the value of the research contracts that Dragon brought to the country, measured partly in financial terms and partly in terms of the high level technical activity that resulted.

In the early fifties and sixties, Sweden was one of the most vigorous nations in the pursuit of an individual competence in nuclear energy and only in the seventies, under the impact of the persistent anti-nuclear campaigns that had been waged over the previous half decade, did resolution waver. Twenty years ago, however, ecological problems were not to the fore and the country, under the stimulus of the national research organisation AB Atomenergi, focused the majority of its attention on the development of a natural uranium fuelled, heavy water moderated reactor system. This led to the commissioning in 1963 of a combined district heating and electricity generating stations at Ågesta, and in the same year, Government approval for a 200MW(e) power station based on a boiling heavy water reactor to be built at Marviken on the network of the State Power Board. In the longer term it was believed that the future lay with fast reactors.

At the 1955 Geneva Conference, Sweden's scientific leaders were prominent in the discussions and Eklund was invited to chair the OEEC Working Group on Experimental Reactors. It is significant that after Shepherd had given his paper to the Rome meeting of the EAES late in 1957, he gave the same paper again in Stockholm before the meeting of top level experts in Paris, at which Cockcroft offered the Harwell HTR as a joint undertaking. Carrying his country with him, Eklund played a major role in the negotiations that led to the Dragon Agreement and was unanimously elected first chairman of the Board of Management. He continued in this position in the following year and was then appointed Director-General of the IAEA.

It was of quite secondary importance at that stage that Dragon was an HTR. Sweden would have been very happy with – and indeed for some time expressed a preference for – the homogeneous aqueous reactor. Participation was seen as a political gesture with the additional merit of giving access to another technology should it be required. No effort was made to establish a national group corresponding to the international Project; the Project was expected to lead a separate life and to provide all the information necessary while the country got on with the development of a boiling heavy water system.

Sweden's power industry has a polyvalent structure involving industry, local authorities and the Government. The State Power Board has no monopoly and generates under half of the country's electricity needs. The remainder is divided roughly 5:2 between the private companies and the municipal undertakings. Even the manufacturing industry (particularly today) involves partnerships between Government and private industry.

With this tradition of mixed public and private enterprise, it might be reasoned that Sweden was in a better position than most countries to resolve the problems posed by the new energy source which required such large public funding before a viable commercial structure could become established. In practice, however, the creation of a stable partnership was achieved only after a lengthy period of trial.

ASEA, the leaders of the manufacturing industry, saw the future as lying with light water reactors and, whilst acting as main contractor for the Marviken station, went ahead on its own with the development of light water technology. The first commercial order for a nuclear power station was placed with ASEA in 1965 by a group of private and municipal electricity undertakings, and the system chosen was a boiling light water reactor. AB Atomenergi, while giving technical support to this project, including some redesign of Marviken to cater for test loops for light water fuel, continued with its heavy water system. An internal competition was thereby generated which could not go on indefinitely. The first extension of the Dragon agreement passed without incident, but when it came to the second, to continue the collaboration beyond 1967, no one in Sweden, was giving thought to the HTR and the Government felt that its resources were already being stretched enough.

Rather than abandon the Project, it proposed that its percentage contribution be reduced, pointing out that Sweden's share of Dragon work was small in comparison with both Austria and Switzerland amongst the independent countries and its technical interest was no bigger than that of the Netherlands, whose contribution was much less because of its participation through Euratom. Only by stressing the importance of the irradiation contracts placed by Dragon with AB Atomenergi and the promise of substantial contracts to come, was the Government persuaded to modify its attitude. Although Dragon was funded directly by the Government under a budget heading separate from that covering AB Atomenergi, the two were still related, and the Government found the argument meaningful; for the latter, this meant work for its reactor R2, an income for its research centre and no direct charge on its own budget. Sweden agreed to continue as before.

On the Coast of Studsvik near Nyköping, AB Atomenergi had built a research establishment with, as central feature, a 30MW light water MTR (R2) that became fully operational in November 1961. R2 was one of the most advanced installations in Europe and was ideally suited to the fuel research programme that the Project was planning. It was also supported by extensive ancillary facilities and a highly competent staff and there was an evident match between the Project's needs and the centre's capabilities.

As the sixties drew to their close, the balance of industrial opinion swung heavily in favour of light water systems and led to the constitution on 1 January, 1969, of a joint company of ASEA and the Government, ASEA-

Atom, to design and build light water reactors and manufacture fuel. The following year safety studies showed that a substantial redesigning of Marviken would be necessary if it were ever to become operational, and it was decided then to convert it to an experimental reactor for LWR safety investigations. From then on AB Atomenergi concentrated on fuel cycle studies and research and development in support of the LWRs. Research into coated particle fuels for possible application in a gas-cooled fast reactor, that had gone on in parallel with Dragon's programme, lost much of its drive at the same time.

More than ever, the HTR was then regarded primarily as a source of customers for irradiation services. Studsvik had made a speciality of measuring fission product emission on line during irradiation and this capability was of direct relevance also to KFA's development of pebble-bed fuels. Indeed, as a result of its Dragon experience, Studsvik found in KFA a more important customer than Dragon itself. The Government for its part saw this business as adequate return on its investment and, in spite of making cuts elsewhere in research and development spending, supported the subsequent extensions to the Dragon Agreement.

Altogether, contracts to the value of a little under £850 000 were placed with Studsvik (of which only those to a value of £22 000 were the subject of international competition) and although they were used as an argument to keep Sweden in the Project, the Project maintained that they were justified by the high quality service given and the value for money. Sweden's share of research and development contracts was 5.5% and, of the total contracts placed, 3.4%. These figures may be compared with a contribution as Signatory of 4.4 – 4.46%, that totalled a little over £2M over the life of Dragon. During the second half of the Project's life, Sweden was receiving back in contracts about half the payments it was making.

The HTR as an electricity source did have one staunch supporter in the country, J. E. Ryman, Managing Director of the Stockholms Elverk, but he was practically alone and there was little encouragement from any of the other companies when AB Atomenergi organised an occasional information meeting on Dragon. Ryman had negotiated in the early sixties an exchange agreement with GA and was hopeful that advantage could be taken of the HTR's safety characteristics to build a station near Stockholm. Also some exitement was generated in the early seventies by the apparent breakthrough of the HTR and the second largest utility in the country, Sydkraft AB, announced in the Spring of 1970 that it was considering plans for an industrial chemical complex centred on an HTR, inspired no doubt by a study their technical director had just completed in San Diego. Nothing came of these ideas.

Effectively, the HTR was never in the running for power production and latterly, any remaining interest was in its very long term possibilities as a heat source for chemical conversion. The direct cycle was followed from afar with only minor support. In consequence, Sweden made little attempt to steer Dragon policy but would favour the broad research and development programme rather than any concentration on the steam cycle.

Only with difficulty could staff be persuaded to join Dragon, as only at the senior technician level could it be represented as a futherance of career prospects, although AB Atomenergi did try to present a secondment as a reward, not as a rejection. Salary scales in the UK were significantly lower than in Sweden so that salaries had to be made up, and this did not help. Twelve people altogether were seconded to the Project for an average period of a little over three years. None attained a grade above D. The longest serving member was Jan Blomstrand who was first with the Project from 1962-1966 and then returned in 1968 with the object of writing a book on HTR physics. AB Atomenergi was sponsor also to one Finn, Magnus von Bonsdorff, who was involved in assessment studies at Winfrith, particularly concerning the direct cycle system. Following his return to Sweden he became managing director of a utility that has BWRs on its network.

In view of the minimal effort Sweden expended in support of Dragon, its return could be considered as satisfactory, at least from a financial point of view. In particular, the Dragon-inspired use of R2 for testing HTR fuel was of importance for Studsvik and continues to be even now. It is recognised also that considerable benefit was derived from the contacts that membership stimulated, and from the opportunity given for a complete and open exchange of views on fuels, safety, and techniques of varying sorts, not necessarily directly connected with the HTR: in short, from the intellectual stimulus that Dragon provided.

Only at the very end did AB Atomenergi give serious consideration to the HTR being a system that might be exploited in Sweden in the light of its inherent safety characteristics. When Dragon closed, a new evaluation was being made to see whether it might provide the answer to the objections of the anti-nuclear factions to the existing power programme. As a result, Sweden's support for a further extension would probably have been more enthusiastic than in previous years.

CHAPTER 22

SWITZERLAND AND DRAGON

WHEN considering the interaction between Switzerland and the Dragon Project, account must be taken of the country's constitutional structure of a confederation of States and the strong attachment in the sectors concerned with industrial development to the principle of private enterprise. Over a thousand companies are engaged in the supply of electricity and, whilst the great majority operate very small systems, the number able at some time to contemplate the installation of a nuclear power station is still large for a country with a population of six million people. Moreover, although there is a great deal of interconnection between the networks and extensive provision for exchanging power with neighbouring countries, the spirit of competition is very live, and in the late fifties and early sixties it could be expected that the electricity companies would opt for different nuclear power systems.

Partly as a consequence of the segmentary nature of the network, and partly because of its need to export in order to survive, the manufacturing part of the power industry has had to put itself into a position where it could supply systems or components for whatever markets emerged. Toughened by world competition, manufacturers had learned to stand on their own feet and to rely mainly on auto-financing for their research and development, trusting their own judgment and resources to keep them in the van of new technologies. It is estimated that even today, 75% of the research and development conducted in the country is funded by private industry. Faced with the variety of systems under development across the world in the early sixties, the big companies, such as Brown Boveri, Escher-Wyss and Sulzer sought to identify themselves with specialised plant components and conventional systems rather than with complete nuclear systems. The project for the 30MW(th) gas-cooled heavy water reactor at Lucens, nationally inspired and 50% financed by the Government, can be seen as just a temporary departure from this industrial policy.

Of particular relevance to Dragon was the strength of the industry in the fields of heat transfer by gases and high quality rotating machinery, including gas turbines. The HTR, as presented in 1958 with its possibilities for direct cycle conversion was thus instantly of interest to the manufacturing industry as many of the out-of-core components corresponded closely to its own specialisation.

Government in Switzerland saw in Dragon an occasion for inter-government co-operation of a style that conformed well with the national policy of neutrality and the desire to remain independent of bloc action. Dragon was a means of supporting the OEEC, widening contacts with other European countries, without involvement in any politically motivated union. Given the support from industry, the Government had little hesitation in deciding to join. Part of the funding was also to be provided from industrial sources, although in practice most of the finance came from the Confederation at the start, and all of it for the subsequent extensions.

Participation was seen in the beginning as essentially a Government-industry partnership and this was reflected in the representation on the Board of Management that until 1969, consisted of K. Niehus of Brown Boveri and Urs Hochstrasser, Delegate of the Federal Council for Atomic Energy Matters. Natural gravitational centre for the collaboration was the Federal Institute for Reactor Research (EIR) with its research establishment at Würenlingen. Founded in 1955 as a joint venture of industry and Government, it was restructured in 1960 and has since been operated as an association of Government with the Federal Institutes of Technology of Zürich and Lausanne. EIR was represented on the GPC from the beginning by its Director, Andreas Fritzsche, who served continuously until 1971 and then from 1973 by Peter Tempus, Deputy Director of EIR.

The departure of Niehus signalled the end of the most active phase of industrial attention to the HTR and representation on the Board of Management was from then on limited to a delegate from the Federal Office for Science and Research. From 1969-1973 this was Hans Enzmann (a member of the Dragon staff from November 1959 to October 1965 and also GPC delegate following the retirement of Fritzsche) and from 1973, Jean-Michel Pictet. Pictet was Board of Management Chairman for the last year of the Project, during which time he made every effort to gain agreement on a further extension.

Support for Dragon in Switzerland was continuous and consistent but this in no way implied a choice of the HTR as a national system or the rejection of other systems. Industry needed to establish a broad base of knowledge and experience in order to be able to offer equipment for whatever systems the world chose to adopt. As soon as the dominant position of light water technology became apparent, industry adapted its efforts to fall in line with the evolving world markets. At the same time, the Swiss electricity companies, for the most part, also concluded that their safest future lay in the light water systems that were so heavily underwritten by US technology and which were being adopted by the majority of European countries also. The success of the Swiss industrial policy can be gauged as much from the proportion of world orders obtained in the sixties as from the consistent performance of the LWR stations at Beznau and Mühleberg which have been in operation since 1969 and 1972 respectively.

Early on, there was a certain optimism that the direct cycle HTR system would become a commercial success in the medium term but it became evident in the middle sixties that generation via gas turbines associated with large reactors was for the long term only, and interest concentrated therefore on the possibilities of the steam cycle and a fuel cycle based on low enriched uranium. As such, the change in emphasis in the Dragon programme was supported with enthusiasm by Switzerland, but this did not mean that the prismatic HTR had been adopted, and the efforts of Brown Boveri, Baden in this area were widely misinterpreted, even in the Dragon Project itself.

The Brown Boveri company in Mannheim had, through its commitment to the AVR project and then the THTR Association, devoted the bulk of its effort to the pebble-bed system with the highly enriched uranium-thorium cycle. As the Dragon Project matured, and the General Atomic prismatic reactor at Peach Bottom began to show such promise, Brown Boveri, Baden felt it necessary to initiate an independent evaluation of the two systems, and in the Summer of 1966, established a design study department to work on prismatic HTRs. In due course, a team of about 50 people was built up, supported by other departments within the company, so that although the number engaged on the study was still far short of an effective design and development organisation, it was still of significant strength when related, for example, to the Dragon Project itself.

As Switzerland was a signatory to the Dragon Agreement, Brown Boveri had access to all the information distributed by Dragon and it had complete freedom to propose staff members or to attach guest scientists. As a result, its demand in 1966 for a special consultancy agreement with the Project, apart from worrying Euratom so much, was interpreted as a commercial statement that the company had decided to develop its own prismatic HTR. This was not the case. Brown Boveri was used to conducting its studies privately and in its own way. It wished to pose its own questions in private and have its answers given in private. At that time, it would not have considered working through its Government and throughout the Project's life communication with the relevant Government departments was minimal. The agreement with Dragon was concluded finally in June 1967.

By 1970, the HTR group at Brown Boveri had completed its study and had concluded that under the prevailing conditions, the HTR steam cycle system could not be commercialised without enormous investment. Attention was then turned to the direct cycle system, but in 1972, the company took a new look at its nuclear activities and decided to disband the HTR team in Baden and to leave to Brown Boveri, Mannheim any further exploitation of the system.

By this time also, Mannheim had enlarged its interests and was negotiating with GGA for a partnership in HRB, in order to be in a position to offer GGA's prismatic HTRs on the European market as well as continue with the development and construction of the THTR pebble-bed, seen then as a longer term proposition. Moreover, in France, the Brown Boveri group CEM, was involved in the consortium of GHTR which also was negotiating an agreement with GGA for rights to the company's design of a very large prismatic HTR. There was no need for any separate Baden effort.

GGA's activities in the early seventies had a profound influence on the policies of many organisations in Europe. In Switzerland, however, the GA impact dated back as far as 1960 when it established a company in Zürich. This was essentially a research outstation, remote from San Diego, with Peter Fortescue as its Technical Director. Designed to tap European scientific talent, its attention was focussed on various aspects of gas-cooled reactor technology and the company was given considerable freedom by San Diego to develop its own ideas. It was also an important base from which contact with European nuclear bodies could be made, generating an early enthusiasm for HTRs in Switzerland and acting as a continuous source of publicity for GA's competence in the field.

Being formally registered in Switzerland, GA Zürich had access to the information that was available to the Signatories on Dragon's work, on the same footing as any other Swiss company. Patent rights were not transferable, but information could be transmitted across the Atlantic without restriction, which meant that the various limitations on exchanges with the USA imposed by Dragon could readily be circumvented by the American company. Illustrating the level of GGA's current awareness of Dragon's activities, a senior staff member of Dragon ruefully recalls being questioned in San Diego on the meaning of a particular paragraph in the confidential Monthly Progress Statement to the Dragon Board of Management, that had been sent out only a few days earlier to the members of the Board of Management and the General Purposes Committee.

Regardless of the lead role that Brown Boveri had played in HTR work in Switzerland and the presence of GA in Zürich, the preparation of the first actual offer for a Swiss HTR power station was begun by a French consortium. This was in response to a call for tenders issued by the Lausanne centred company, l'Energie de l'Ouest-Suisse, which announced in September 1974 its interest in an HTR station for a site at Verbois outside Geneva. Troubles with GA's designs and Fort St. Vrain intervened and, as orders for HTRs in the USA were cancelled, the French consortium felt obliged to limit its offer to an outline design only. As a result, the only HTR power project that came on the books in Switzerland died in the study stage.

Switzerland was, in the main, well content with its participation in Dragon and the evolution of the Project's policies. It approved the accent put in the late sixties on the low enriched cycle and supported the later trend towards the long-term development of the direct cycle and the possibilities of using the HTR for process heat. It recognised that as a small country obliged to spread its interests, it was necessary to collaborate with other countries and was disappointed that conflicting national objectives caused the larger countries to limit their contributions to Dragon, to the point that in the later research programme, the total effort expended upon advanced concepts was below the level which would be effective. Realisation of this was a major reason for the country's agreeing, in 1973, to participate in the German HHT project for the development of a direct cycle HTR to which both the Federal Institute at Würenlingen and Swiss industry are contributing.

Support for the Project was not influenced by the possibility of doing extra-mural research. At the beginning, some irradiations were conducted in the EIR materials testing reactor (Diorit) at Würenlingen, but this was a 20MW natural uranium fuelled heavy water moderated reactor, best suited to large volume, medium flux irradiations and Dragon was mainly looking for high flux irradiations of small samples. Switzerland recognised that the Project was obliged to spread its outside work as evenly as possible and the country had done well enough in the supply of components. It could not expect as well a significant fraction of the R & D work. In any case, EIR became fully engaged in doing hot cell work for the THTR project and was not short of funds to maintain a healthy level of activity.

Industry's share of the contracts for equipment for DRE (up to March 1965) was valued at over 15% of those placed outside the UK and included notably the charge/discharge machine (SECA), the variable frequency generating sets and the primary gas circulators (Brown Boveri), the company contributing a great deal of its own money to the development of gas bearings. The equivalent figure for contracts for R & D (other than EIR) was 3½% placed mainly with Battelle in Geneva. Altogether, including the work done at EIR, the total value of orders placed in Switzerland was £735 500, equivalent to 9% of the value of contracts placed outside the UK. These figures may be compared with the country's total contribution to the Project of £1.6 million, representing a maximum of 3.33% of the budget, or 6% of the contributions coming from outside the UK. Moreover, as participation in Dragon provided some of the necessary background to doing design work and making components for the German and American HTR projects, there was an immediate return on investment through the traditional channel of industrial exports.

While Switzerland remained a keen supporter of Dragon to the end, its direct interest in the HTR system should not be exaggerated. It can be noted that while 10 people were detached to the Project in its early days, the total number amounted to 15 only, out of an aggregate overseas staff over the Project's life of 234. In part this reflected the policy of Brown Boveri to hire staff once they had gained experience rather than detach existing staff to learn the work, but it was also a commentary on the time-scale in which the HTR was viewed.

If there had been a possibility of continuing Dragon, Switzerland would have been among the first to give its support. Real regret was felt on the termination of the technical programme.

CHAPTER 23

THE ORIGINAL SIX EURATOM COUNTRIES AND DRAGON

FOR EURATOM, the OEEC initiatives to establish European joint undertakings appeared as a challenge and the offer of the British to internationalise their HTR work, coming less than three months after Euratom came into existence, was a major test of the Community's will to coordinate activities in the civil nuclear field. Had the member countries elected to act independently, the authority of Euratom, already dimmed in comparison with the hopes that were expressed 18 months before, would have been seriously compromised. At the same time, members of the Commission were reluctant to support a project on foreign soil before any comparable project could be launched within its territories. With uncertain leadership, the Commission was swayed by the opposing pulls of wishing to represent its members and be regarded as the proper channel through which their participation should be directed, and trying to limit support for the Project to a level that was considered inadequate by the other partners.

For political reasons the member countries of Euratom were ready – in some cases insistent – for Euratom to act on their behalf and there was equally the financial argument that having agreed to a substantial budget for Euratom, the countries were reluctant to find additional funds for co-operative ventures in nuclear energy. Hesitations on the part of the Commission nearly resulted in the UK deciding to go ahead with the Project alone and it was paradoxical that while Euratom was so determined that it should act for the Six, it was the offer of independent contributions by France, Germany and Belgium that finally allowed agreement to be reached on the financial provisions.

With the appointment of Hirsch as President of the Commission, relations inside Euratom became more stable, whereas soon the tensions between the Commission and the Member States began to make themselves felt. The replacement of Hirsch in 1962 by Pierre Chatenet marked the end of Euratom's hope of coordinating in any real way, the nuclear programmes of its member countries, and by 1966, the Community was leading a hand to mouth existence, operating on yearly budgets with no background plan, whilst promoting a reactor system (Orgel) that had everywhere else been rejected. Participation in the German THTR was stopped in 1967 so that from then on, Dragon was effectively the only reactor project of promise in which the Community was deeply engaged. Even when the member countries were no longer enthusiastic, Euratom staff still hoped that Dragon could be kept in existence as one example of a successful European co-operation.

In spite of big differences of opinion between the Community members, for example on the accent to be placed on R & D vis-à-vis operation of DRE, Euratom was reasonably successful at blending a common policy and presenting within Dragon a coherent front, at least until 1973 when the arguments within the enlarged community spilt over into the Dragon meetings. Guéron, Caprioglio and then Marien all developed a strong personal commitment to the Dragon collaboration and were able to pilot the Project through its crises in the Community. These were most serious when discussions on an extension to the Dragon Agreement coincided with disputes over Euratom's own forward programme. Then Dragon became embroiled in the wranglings over general Community policy which resulted in 1967, in a complete break-down of the decision-taking process. Between the periods of negotiation on extensions Euratom delegates were able to speak for the Six with a single voice. After 1973, the situation was much more confused and the presence of national representatives in Board of Management meetings dispelled the notion of a corporate stand.

The Project took care to respect the role of Euratom as the Signatory and when dealing with laboratories in the countries of the Community, either in preparation for an extension or in regard to research contracts, would invite a member of the Commission's staff to be present. Even when the interaction both technically and financially was directly between laboratory and the Project, Euratom was not by-passed. Such a policy paid dividends in the easy relations that existed even in the early years when Euratom was especially sensitive about its position as representative of the Six.

Out of the Six countries, four had no national HTR programmes – Belgium, Italy, Luxembourg and the Netherlands. They regarded Dragon as a long-term research and development activity which, in the years to come, might have commercial implications, and in the immediate, could provide interesting research tasks for the national laboratories and development and manufacturing work for industry. Industry would thereby be able to acquire a background experience should a market for components open up. Contributions were not onerous, the overall cost amounting (in round figures) to £1.9M for Belgium, £1.4M for the Netherlands and £4.6M for Italy. With the exception of Belgium, the Euratom countries paid roughly two thirds of the contributions they would have paid if assessed on their national revenues. Belgium, because of its high contribution to the research budgets of the Six until 1973 (9.9%) paid, on aggregate, about the same.

Participation as measured by the number of secondments would suggest a reasonably active level, but in practice, many of the staff who worked at Dragon returned as Euratom staff members and ceased to make a direct contribution to their countries' nuclear efforts. From the total of 234 secondments from outside the UK, 13 came from Belgium, 2 from Luxembourg (which had virtually no nuclear programme), 8 from the Netherlands and 39 from Italy. A number of the Belgians were on secondment from CEN, but otherwise the great majority from the four countries were recruited through Euratom and remained as Euratom employees. Only the few detached from organisations doing work directly for Dragon such as BelgoNucléaire, AGIP Nucleare and Comprimo had an established position to which they could return.

BELGIUM

One of the most consistent supporters of Dragon was Belgium which, as host to the Commission, but conscious of a wider Europe, saw in Dragon an expression of its traditional internationalism. The presence of Marien was no doubt a catalyst also.

From the early 1950s onwards, the country maintained a vigorous nuclear programme, an energetic industry being supported by the national laboratory whose name was changed in 1957 to the Centre d'Etude de l'Energic Nucléaire (CEN). At the main centre at Mol, an early gas graphite research reactor BR-1 was followed by a 50MW(th) light water MTR, BR-2, that went critical in 1961, and a 10.5MW(e) PWR supplied by Westinghouse, that began operating the following year. This set the country on a PWR course from which it has never diverged. By 1976, Belgium had an installed nuclear capacity of 2000MW(e) which provided in the following two years, some 25% of the total electricity generated in the country. Industry has become self-sufficient in the technology and has been active in seeking markets for components and fuels outside its own territory.

Nuclear research was given a strong impetus by the country's exceptional role as a uranium producer and also by the special arrangements that were made after the war between the Government and Union Minière (the company exploiting the Congo uranium reserves), for the development of the Mol centre, which incidentally was chosen for the site of the joint reprocessing plant of the Eurochemic company, set up by the Agency under a Convention signed in December 1957. BR-2, which is a particularly powerful materials testing reactor of unusual form, has been run as a CEN-Euratom association within the context of the Joint Research Centre. The core of the reactor is in the shape of a diabolo and it was designed to give the highest neutron flux of any facility in Europe. Access to the central irradiation positions is rather restricted but loops can be installed and Dragon found the reactor to be an excellent tool for making accelerated tests on small specimens. Emphasis was placed on the study of graphite corrosion and the transport of carbon that could take place from the core of an HTR to other parts of the primary circuit. Altogether, contracts worth about £165 000 were placed with CEN, including some out of pile work, which included the request to adapt, for Dragon fuel, a dry process the Centre had already developed for making spherical kernels – a job that was done very quickly and regarded as quite minor, but which nevertheless formed the basis for Dragon's coated particle fuel production from then on.

Other fuel work was undertaken by Dragon in collaboration with BelgoNucléaire, a company that made a speciality of plutonium handling. It prepared for Dragon plutonium bearing fuels in connection with the low enriched uranium fuel cycle research, and also supplied enriched uranium driver fuel for DRE – the first time it was acquired from a source other than Dragon's own fuel laboratory. In the Inter-Nuclear enterprise, BelgoNucléaire was in charge of fuel design and development, collaborating respectively with TNPG and SNAM Progetti. The company equipped itself for the fabrication of full-sized HTR fuel elements on a pilot scale (10t/a) that could handle both highly enriched uranium and plutonium.

Overall, Belgium's share in Dragon contracts was relatively small. No plant orders for DRE were placed in the country and the only other significant contract not so far mentioned was with SERAI and worth £32 000 for the study, out of pile, of the growth of carbon deposits on metals such as could occur in heat exchangers. Dissatisfaction with this situation was expressed on more than one occasion but it was never made a determining issue in establishing the country's broad policy.

THE NETHERLANDS

The Netherlands saw in Dragon a political instrument for involving the UK in continental Europe and over most of the Project's life, was very positive towards its continuation, although no more willing than the others to dissociate Dragon from Euratom's overall programme in the late sixties. Its most negative period was before the 1973-76 extension when the UK was virtually certain to join the Communities. Then it felt no need to hurry into any decisions on Dragon, particularly when government expenditure on scientific activities was under pressure at home and the funding of Petten was uncertain.

The Netherlands' power reactor programme has remained amongst the most modest in Europe, just two reactor plants operating in 1976, a BWR of 50MW(e) output and a PWR of 450MW(e). Nevertheless the country was not slow to start research, initially in collaboration with Norway. A national commission was set up, the Reactor Centrum Nederland (now renamed the Netherlands' Energy Research Foundation) and a research centre established at Petten which in 1963 was partly taken over by Euratom as a unit of its Joint Research Centre. Principal research tool there is a materials testing reactor, the HFR, which began operation a few months after BR-2. HFR is a 20MW(th) light water reactor substantially identical to R2 at Studsvik. It was a natural candidate for rig experiments of various sorts, and initially the irradiations planned for it were concerned with the accelerated testing of fuels. Later, Petten specialised in the measurement of dimensional changes in graphite over a wide range of irradiation conditions. Altogether contracts worth nearly £0.5 million were placed by Dragon with the laboratory which also carried out a significant amount of work for KFA and did joint work for Euratom, KFA and Dragon together.

Contracts were also placed with the Foundation for Fundamental Research on Matter at Utrecht (FOM) for studies on the application of thermal diffusion to the concentration of impurities in helium to facilitate their measurement, and with KEMA for the development of the Sol-Gel process for making fuel kernels with a thorium and uranium loading. This independent research organisation of the electricity companies had continued, in association with Euratom, research into the homogeneous aqueous reactor (regardless of its rejection by the OEEC countries as a joint undertaking) and had developed the Sol-Gel process of Oak Ridge, for the manufacture of very fine spherical particles to be used as fuel for its own reactor Suspop. Although the work for Dragon was technically successful, it was not used, because when the liquid route to kernel production was eventually adopted, it was on the basis of SNAM Projetti's process which was more advanced.

The Netherlands' share of DRE construction contracts amounted to 13% of the Euratom total as compared with a 6.9% participation in the relevant Euratom budget. Major items included support structures for the pressure vessel (NV Neratoom), fission product delay beds and iron shielding (Wilton Fijenoord) and the pilot plant built to develop the helium purification system (Comprimo NV), perhaps more properly regarded as an R & D contract.

ITALY

The first nuclear power station to be ordered in Italy, in mid-1958, was a 200MW(e) Magnox plant that was built at Latina by the Nuclear Power Plant Company of the UK, in collaboration with AGIP Nucleare. It was followed soon after by projects for a BWR and a PWR. Growth in electricity consumption in the country has been consistently amongst the highest in Europe, hydro resources have also been completely exploited and in the late fifties, Italy was considered to be one of the most important markets for nuclear power in Europe. Internal strife, however, which surrounded the eventual nationalisation of the electricity supply industry in 1963, occurring over a period when oil was plentiful, brought nuclear construction to a halt. After the first three stations, only one major power reactor (a BWR) was started during the life of Dragon. The national electricity authority (ENEL) has shown strong leanings towards heavy water systems but in the long run, US industrial pressure has won out and Italy's nuclear power programme has become centred on light water reactors.

The national nuclear research body, the Comitato Nazionale per l'Energia Nucleare (CNEN) was also at the centre of a great deal of political controversy, finally coming into existence in 1960 after years of argument over its statutes. On its formation, the research centre at Ispra was handed over to Euratom as the main laboratory in the Joint Research Centre. CNEN's own laboratories are at Casaccia and Frascati near Rome, the former being devoted largely to technical developments and the latter to fundamental research. In the nuclear power sector, CNEN has become committed to a hybrid heavy and light water system as thermal reactor and in 1967 embarked on a project to build a 35MW prototype experiment (also at Latina) named Cirene, that is expected to start up in the early 1980s. CNEN's attentions therefore, have been directed towards this system and its fast reactor programme that includes a third share in the French FBR project at Creys Malville.

In all this there has been little room for any initiative in the HTR field. A keen promoter of joint action in nuclear energy to begin with, Italy's position has become more and more conditioned by its sense of injustice over the return from its international investments and by its dissatisfaction with the support given to Ispra which has had a most unsettled existence. Home of Euratom's Orgel project whose main reactor experiment ESSOR began operation in 1967, Ispra was always in the front line when Euratom's programmes were under fire. This did not encourage the Italian Government to continue its support for Dragon. Italy saw no point in Dragon going on once the host country lost interest, and was strongly opposed to any long term R & D programme that was unrelated to specific industrial needs. Unfortunately in the last years when Albonetti, Director of International Affairs and Economic Studies at CNEN (and a regular spokesman in the Community's Group for Atomic Questions) was representing Italian interests in the Board of Management, it was widely assumed in the Board and in the Project that his criticisms of the Project's programme stemmed from personal antipathy and did not represent the views of his Government. Outspoken to the point of brusqueness, his style was not one to engage sympathetic understanding. He made no secret of his impatience with the irrationality he saw in Europe's nuclear development generally, and although always articulate in his pleas for a *juste retour* for Italy, he castigated other European countries for their insularity and their failure to come together in nuclear energy at the level where it would count. Alone of the delegates he had called for industrial participation in the Dragon Project, but this was taken for rhetoric and it was put forward in the crisis days of the late sixties when there was no time to institute such a fundamental change. He was not alone in believing that once Britain was no longer ready to take the lead in HTR developments, there was no real sense in continuing a collaboration based on a British project, but he was alone in being prepared to grasp the nettle and say so bluntly.

Politically, Italy was reluctant to see a collaborative programme disappear and at the special meeting of the Board of Management called in July 1975 to deal with the hiatus that had developed in the extension negotiations, it was Albonetti who, after running counter to the other delegates on almost all subjects and expressing himself astonished that the Project could be facing a financial crisis eight months before the formal closure date, proposed establishing a survival budget to carry the Project through to December 1976, and so give enough time for negotiations. Ironically this was seen at the time as an unnecessarily desperate measure and when it became the final survival ploy, too many conditions were attached for it to receive support. Despite disagreeing with the programme, the budgets and decommissioning charges, it is more than probable that had the other members of Euratom been willing to continue Dragon on similar terms to those already in force, Italy would not have imposed a veto. Instead it would have taken the same stand as in 1972 when, having exercised pressure to ensure a maximum *retour* to the country, Italy abstained in the crucial vote of the Council of Ministers.

Not until 1973 was the Project able to commission irradiation work in Italy, due to the absence of spare reactor capacity there. Then, to honour promises made during the extension negotiations, Dragon capsules were installed in ESSOR which had been converted to undertake general work. The contract was worth about £100 000 up to the end of March 1975.

Italy had not been ignored in other areas: R & D contracts in the second category, which excluded irradiation and other major facilities, were worth in total over £⅓million. In fact, Italy's share of this group of contracts placed in the Euratom countries was over one third, which can be compared to its contribution to the Euratom budget (until 1973) of 23%. Of particular importance was a contract for almost £180 000 for experimental and theoretical studies of fission product migration in fuels, placed with SORIN – the industrial research centre originally set up by Fiat and Montecatini, which later became a wholly owned subsidiary of Fiat. Over the last few years, Fiat became increasingly engaged on metals research for the Project, and one of the inducements for obtaining agreement to an extension beyond 1976, was the promise of a substantial continuation programme in this area.

It should be noted that the industrial structure in Italy has gone through a number of permutations as new groups have been formed and new companies have come into nuclear energy while old ones pulled out. To begin with, the main gas cooling interest resided in AGIP Nucleare, the subsidiary of ENI (the Ente Nazionale Idrocarburi) that was co-operating in the construction of the Latina power station. After this was finished, AGIP was dissolved and the nuclear engineering activities of ENI were grouped under SNAM Progetti. Later a new AGIP Nucleare was constituted as a nuclear fuel company. SORIN's nuclear research was taken over by Fiat and extended to cover the study of the failure of materials under mechanical stress.

The old AGIP collaborated with the French group Indatom in the first assessment study contracted out by the Project, and then a number of experiments were made for Dragon by SNAM, notably on coolant properties and reactions with graphite. SNAM was very active in the fuel field having acquired rights to the original patent on the Sol-Gel process and then gone ahead with its own development of the liquid process. In its partnership with BelgoNucléaire, SNAM made the kernels and these were coated by the Belgian company. Essentially the initiative for SNAM's fuel work came from the industry rather than the Project, which had some hesitation about supporting too many fuel fabrication facilities.

CNEN became involved in thorium cycle fuel reprocessing studies, partly as a result of a collaboration with Dragon and the Authority to evaluate systems and costs, and partly because of its research and pilot plant PCUT, built at Rontondella near Taranto to establish recycling techniques. Under an agreement with the USAEC, the plant which was completed in 1967 was to process thorium bearing fuels used in the 12.5MW(e) BWR of Allis Chalmers at Elk River. (This reactor began operation early in 1964, and would seem to be the only light water reactor plant that has incorporated thorium in its fuel.) Because of this experience, it looked at one time near the end of the Dragon collaboration as if CNEN would be taking over irradiated fuel from DRE for processing. The offer was, however, made conditional upon the successful outcome of tests that were being conducted at Saclay and, as there was no further extension, was never followed through.

Fiat in particular would have been well content to see the Dragon collaboration prolonged, if only to continue the metals programme, but interest on the whole was muted. CNEN itself seconded no staff to the Project – leaving aside its subsequent Director General, Franco – and once the GGA-inspired HTR projects were cancelled, industry recognised that markets for HTR components would be non-existent. Only mild regrets were therefore uttered when the Dragon programme ended.

FRANCE

Almost certainly the first person outside the UK to know that Cockcroft was proposing to internationalise the Authority's research on high temperature gas-cooled reactors was Francis Perrin, High Commissioner of the French Commissariat à l'Energie Atomique and Chairman of the OEEC's top level committee of experts that recommended the Project as suitable for study as a joint undertaking. Despite the absence of any real co-operation between Britain and France, there were close personal links between Harwell and the CEA, cemented by the European Atomic Energy Society. There was also a considerable measure of mutual respect.

Even if interesting from a scientific point of view, there were doubts in France at this time, as over the next decade, that the Project could lead to an economic power reactor. Strong reservations were felt about the uranium-thorium cycle and the need for highly enriched fuel. The need to close the fuel cycle pointed strongly to a concentration on uranium-plutonium in preparation for the day when plutonium would become the fuel for a fast breeder programme. Natural uranium had to form the basis for the first thermal reactors and very quickly, as in Britain, both civil and military arguments led to the adoption of a gas-cooled graphite-moderated reactor as the source of power and plutonium. Second in the line of thermal reactors was the heavy water system that had both a technical and national appeal in view of the early pioneering work of French scientists. But whilst it was acknowledged that the uranium-thorium cycle enlarged the fuel supply market, its adoption would mean the establishment of a completely separate fuel processing technology that was considered to be inherently difficult technically, and would impose a severe additional burden on the nation. This opposition to a second fuel cycle has remained, being only temporarily submerged during the period of co-operation with GGA in the years 1972-75.

Equally objectionable in the early HTR design concepts was the need for a fuel that was highly enriched. During the formative years of both Euratom and the Agency, considerable efforts had been made to create a European enrichment organisation, but the UK was unable to co-operate because of its restrictive agreements with the USA, and the resolution of other possible partners crumbled in the face of the American offer to open its stock-pile to European markets. France finally in 1957 decided to build its own diffusion plant at Pierrelatte. This could not be in operation for many years and even when military requirements for highly enriched material had been satisfied, could not produce such material for civil purposes at a cost comparable with that obtainable from the much larger American plants. France was not prepared to embark on a civil power programme that was dependent upon outside suppliers for its fuel, while a power programme based on nationally produced highly enriched uranium was unlikely to be economically attractive.

Technically there was not much fear about the feasibility of building an HTR experiment. An active primary circuit was not seen to present insuperable engineering difficulties and the basic technologies of graphite moderation and gas cooling were already understood. Even helium as a coolant gas did not cause worries as France had access to supplies outside the USA. The basic problems were the fuel feed and the fuel cycle that did not fall within the established French nuclear strategies and so raised objections of principle. France, it might be said in international affairs is prone to examine situations in terms of fundamentals and to adopt policies and attitudes that are consistent with these basic considerations. The HTR was not a valid contender in a French nuclear power programme. Dragon qualified for initial support because it was a technically interesting exercise that would allow co-operation with the most advanced nuclear country in Europe.

Politically the French attitude towards international co-operation was ambivalent. Prime mover in the negotiations that led to the Messina agreement and the setting up of the Common Market, there were strong reservations in regard to the role of Euratom even before de Gaulle came to power in May 1958. Within the CEA, which had been given an unusual degree of autonomy that it was not inclined to see eroded, support for a strong centralised European organisation was less than whole-hearted. Attempts to include military activities in the Euratom Treaty had been successfully thwarted and there was strong resistance to any encroachment upon what the CEA regarded as its industrial preserves. Under de Gaulle, nuclear self-sufficiency became national policy, even though it was still necessary to support some programmes of Euratom. These were least controversial when they concerned research projects which had no equivalent in France and which had no immediate commercial application. Opposition to the element of supranationalism in Euratom's functions created a predisposition towards OEEC projects. Personalities also played a role. Dragon benefited on all counts. Britain's HTR was a sensible project for Euratom to be involved in; it was sponsored by the nascent Agency and it cost the CEA very little as Euratom's budgets were paid out of a separate ministerial fund. Perrin was ready to back the Project and it was his offer of a supplementary credit should this prove necessary, that finally unblocked the negotiations over the initial percentage contributions and allowed Euratom to match the Authority's payment towards the first £10M.

Over the next decade, the CEA's attitude towards the Project was characterised by its readiness to go along with an international venture once begun, and by the low priority it accorded the HTR in comparison with the FBR in its programme for advanced reactors. Its initial enthusiasm for working with the Authority waned as Britain's dominance diminished, and was tempered by a suspicion that the Authority was manoeuvring to draw the maximum advantage from the Project while giving away as little as possible. The situation was not helped by the marked deterioration in the political relations between France and Britain, highlighted by the vetos imposed by de Gaulle in 1963 and 1967 on Britain's joining the Communities with, as corollary, an active policy of preferential collaboration with Germany.

The CEA had no formal HTR section – its own Réacteur à Haute Température was never more than a paper study even if a great deal of research and development was devoted to materials, notably beryllium and its compounds. As a result there was no established nucleus from which staff could be seconded to the Project. Those who did join in the early period did so on their own initiative. They were in a way the square pegs, but were nonetheless, for the most part, good quality people and their contribution was important.

Negotiations over the first Dragon extension took place against a background of rising disaffection with Community policies. Money was not so much a problem as basic disagreements over the principle of supporting American projects in Europe and the Commission's plans for fast reactor development. Neither of these touched Dragon directly and France was prepared to follow the Commission line on the Project. For the second extension from 1967-1970, Dragon did not escape so easily. With nuclear power programmes everywhere having suffered from the highly competitive position of conventional power, France, in the middle sixties, was engaged on a reappraisal of its external spending and was intent on trimming this to a lower figure. While Euratom was trying to get agreement on its third 5-year plan to start in 1968, France had become so disenchanted with the Communities, in particular the Common Market, it was refusing to participate in the meetings of the Council. Moreover, by insisting that Dragon could only be dealt with in the context of the complete nuclear programme of the Communities, a moratorium was placed on further negotiations.

The first move away from this rigid line allowed the Community to fund the Dragon programme up to the end of the second five-year plan and then following initiatives of the Agency and the OECD Steering Committee for Nuclear Energy, France offered provisionally to fund the Project on a national basis – a proposal that was opposed by the other Euratom members as they were not willing to see Dragon taken out of the Communities' programme and given special treatment. It will be recalled that the deadlock on funding the Project for 1968 was broken by the Authority assuming temporary responsibility for Euratom's share. The full extension up to March 1970 was just agreed by the deadline of 31 July, when the objections of several countries (including France) to giving separate consideration to Dragon were withdrawn. It was France, in the search for lower expenditure abroad that insisted on a reduction in Euratom's 46% share, agreeing finally to a 40% contribution instead of the 30% it had first proposed. The Authority took over the difference.

It would seem that at no time did France have any intention of forcing Euratom to withdraw permanently from Dragon, to be replaced by individual State membership as with Halden in 1963, even though the technical side would have been in favour of such a move as it would have made for a more direct link with the Project and eliminated the need to pass through Brussels. Not that the delays in receiving reports were particularly significant, but there was a psychological reaction against the formal position. Caprioglio was very conscious of this, which was why he made regular visits to Saclay to lecture on progress at Dragon. During his time at Euratom, however, there was scarcely anyone who wanted to listen.

Only in 1973 did the CEA deem it worthwhile negotiating a special exchange agreement with the Project, which would have been quite easy to arrange in view of the KFA precedent if communication problems had been anything more than an irritant. Then it was for administrative convenience in view of the size of the joint irradiation programme. At the working level, it would seem that contacts between Winfrith and Saclay as well as the industrial contractors (even l'Air Liquide) were cordial and easy. In effect, whatever the emotional reaction, the interposing presence of Euratom did not have any material effect.

NUCLEAR PROGRAMME CHANGES COURSE

The years 1968 and 1969 saw great changes in the French nuclear programme. Until that time, policy was determined by the CEA and followed by Electricité de France (EDF) and industry. This was fixed on a line of natural uranium gas-graphite reactors which would in due course be complemented by plutonium fuelled fast reactors. Jules Horowitz, Director of the Direction des Piles Atomiques at the CEA had firmly resisted the pressures towards an enriched fuel that could give a system of lower capital cost at the expense of higher operation costs as in the AGRs. EDF was not, however, willing to play a passive role indefinitely and had been gaining experience of water systems through a collaboration with the Belgians at the

260MW(e) Chooz station, started in 1961 under the US-Euratom agreement. In the Spring of 1968 an agreement was concluded for a 50-50 participation in the Belgian 880MW(e) PWR station at Tihange and discussions were also taking place on a joint LWR project with the Swiss at Kaiseraugst. Impressed by the success of the US light water programme, EDF, supported by industry, pressed for France to swing into line behind the majority.

National pride was saved by the fact that in its military programme, France had been developing a light water reactor for submarine propulsion and following the successful operation of the land-based prototype PAT in 1964 at Cadarache, had embarked on a programme of construction of nuclear submarines powered by PWRs. The first, the *Redoutable*, was launched in 1967. Pierrelatte, the national enrichment plant came into full operation in 1966 and from that date, France was no longer dependent upon outside suppliers for enriched uranium.

The clash of policies between EDF and the CEA over the civil programme came to a head during the debate on the sixth national plan that was to run from 1971 to 1975. The outcome was that in 1969 when the advisory commission for the production of electricity from nuclear energy announced its conclusions, the switch to light water reactors was firmly recommended.

Meanwhile Dragon, having initially been conceived entirely as a highly enriched uranium-thorium system, had from 1966 onwards been concentrating more and more on a low enriched uranium-plutonium cycle that Rennie had been determined could compete with the British AGR in terms of fuel. France's fundamental objections to high enrichment and a second fuel cycle were being removed and there was a growing appreciation of the qualities of coated particles. In the CEA's Annual Report for 1968, the HTR was for the first time raised to the rank of *filière*, and in the Advisory Commission's report the system was explicitly recommended for intensified study in collaboration with industry. From this point on, the HTR figured in the formal nuclear development programme of France and the basis for its development was the Dragon low enriched uranium system.

An additional incentive to the CEA to take the HTR seriously was the growing availability of qualified staff released from the natural uranium gas-graphite system who had relevant experience and who were being made redundant by the turn-over to light water reactors. Official secondments to Dragon ensued and France became for a few years a strong supporter of the Project. Extension of the Agreement from April 1970 to March 1973 was actively approved and the CEA placed a special emphasis on collaborative experiments with the Project in parallel with its own irradiation experiments. Contracts from Dragon were vigorously sought and resulted in the largest ever placed by the Project for irradiation work – £200 000 for irradiation experiments in the Colibri rig in Osiris and £400 000 for irradiation experiments in the Pégase reactor. In addition, using fuel provided by Dragon, a zero energy reactor Marius was rebuilt as an HTR physics facility and a joint programme of physics was worked out in collaboration with Dragon, EDF and the Authority.

Publication of the CEA's expenditure on specific systems began only with the 1972 Annual Report following reorganisation of the Commissariat the previous year. By then the annual sum allotted to HTR work was approximately 88M FF – over three times the budget of Dragon, and this was for R & D work alone. In the reorganisation, Perrin retired as High Commissioner of the CEA (to be succeeded by Jacques Yvon) and a new Division of Reactor Research and Development was created under Georges Vendryes, former Deputy to Horowitz. Vendryes was prepared to take a personal interest in Dragon and whereas previously the French member of the Euratom delegation had normally been André Messiah, Deputy to the Head of the Reactor Design Department, from June 1972 until November 1973 Vendryes came himself.

INDUSTRIAL EXPERIENCE

Even if during the construction of DRE, the CEA had adopted the role of distant observer, it had still made sure that French industry played a significant part. It was pressure from Guéron, however, that resulted in the contract for the helium purification plant being awarded to l'Air Liquide with the Nuclear Chemical Plant of the UK as a sub-contractor, although the latter company had put in a marginally lower price. A special meeting of the GPC had to be called before this contract was placed – against the Project's recommendations – and it proved to be the one contract that went seriously awry financially. Placed at a figure of about £240 000, to which was added further work amounting to £100 000, the final settlement made at the end of 1968 after years of negotiation was for a little over £500 000. The company contended that its cost topped the million pound mark and for a long time it looked as if a settlement out of court would not be possible. In the negotiations, the Project, supported by the Board of Management (and Guéron) adopted a firm line, which French delegates did not oppose. Original criticisms voiced by the Project against l'Air Liquide centred on the belief that the company did not have the necessary experience in the fabrication of plant that could not be repaired once in operation. In the event, the company put in so much effort that the standard of design as well as fabrication surpassed the Project's specification, and it was particularly ironic, therefore, that even before the plant went into operation, coated particles should have diminished certain of the requirements by over three orders of magnitude.

Additional major components of DRE provided by French companies were the pressure vessels for absorber rod drives, SFCM; main valves and drives, Ets. Neyrpic; main compressor units, Ets. Corblin; the overhead crane in the containment building, Coupe-Hugo–Soretex-Levage; and circulators for subsidiary systems, Société Rateau. This last company was also commissioned to develop pad-type gas bearings and manufacture a set of main coolant blowers held in reserve against troubles developing with the Brown Boveri design. The development was successful but their use was unnecessary, and they were subsequently sold off.

Péchiney participated in the fabrication of the first core, undertaking the graphitisation and purification of partially processed material. Throughout the life of the Project the company maintained a close collaboration in the development of graphites suitable for high temperature cores, bringing to bear its experience in the production of graphites for the French gas-graphite programme that was backed by a sound scientific appreciation of the problems to be solved. A contract placed with the company for the development of high density graphite valued at £170 000 was amongst the largest placed by the Project for R & D work that did not involve an irradiation facility. Towards the end of the Project's life, Péchiney was in a position to offer material that did not depend upon the single US supply that was the basis of UK manufacture, a fact that could have had great importance for HTRs in view of the temporary cessation of production by the US company following a fire.

In collaboration with AGIP-Nucleare of Italy, Indatom, a consortium of French industrial concerns, performed in 1963 the first assessment studies in Europe, outside the UK, on the design of an HTR power station with prismatic fuel elements, introducing at the same time the concept of reinforced concrete pressure vessels for HTR applications. The French reactor industry was thus well aware of the problems and potentialities of the HTR system at an early stage, but with no follow-up being contemplated in France, little further work was done until near the end of the decade. Then when the sixth 5-year plan was announced, a new industrial group was formed to be a design organisation that could in a few years' time be in a position to present tenders for HTR power stations. This was the Groupement Industriel Français pour les Réacteurs à Haute Température (GHTR) comprising the Compagnie Electro-Mécanique (CEM), that is 41% owned by Swiss Brown Boveri, the Compagnie des Ateliers et Forges de la Loire (CAFL), the Société des Forges et Ateliers du Creusot (SFAC), Péchiney, and the Compagnie pour l'Etude et la Réalisation de Combustibles Atomiques (CERCA).

CERCA had begun development work on particle coating in 1967 and detached a staff member to Dragon the following year. The company showed exceptional initiative in entering the coated particle field, not waiting for

CEA contracts and organising directly with Huddle the first irradiation of compacts that it had produced. CERCA made the first fuel pins of the tubular interacting form which were loaded into DRE in 1971, and the teledial fuel which was taken to the highest burn-up of any in the reactor. After Dragon closed, this was subjected to a very thorough examination by the CEA and shown to be still in excellent condition. Other companies involved in fuel fabrication were the Société de Fabrication d'Eléments Catalytiques (SFEC) which was developing two kernel preparation processes for the CEA – one based on the Dragon formula and the other on its own techniques – and Le Carbone Lorraine which was offering graphites that were suitable for fuel compaction. The fuel industry was, in consequence, well set to launch itself into coated particle production.

FRANCE TURNS TO GULF GENERAL ATOMIC

In the early seventies, the HTR had taken a firm hold within the CEA which wanted to sell its experience and become involved in a real HTR power station project, provided this could be arranged under someone else's sponsorship. Despite its own efforts and that of industry in gas cooling, concrete pressure vessels, graphite behaviour, fuel handling and so on, the 1969 reappraisal was too recent for new initiatives to be taken outside the light water–fast breeder programme. The whole of French resolve – that of EDF and the CEA alike – was tied to the success of these two systems, and industry was certainly in no position to strike out on its own. Industry had never been encouraged to take the lead in a new system or even become strong enough to take the responsibility for complete stations. EDF contracts were for parts of a station and EDF acted as its own architect-engineer.

Dragon had nothing to offer in the industrial area; the engineering studies conducted by the Project were regarded as trivial and in any case, the Project was being turned towards longer term activities and direct cycle applications. For the translation of research experience into concrete project terms, GGA appeared to have all the answers: Fort St. Vrain nearly complete, a growing list of orders for very large stations; the fearful jump from laboratory to commercial plant had been made. Uncharacteristically, the CEA abandoned its principles concerning the fuel cycle, allowed itself to be seduced by the designs of the 1160MW model without going into them in real depth and committed itself to the GGA line.

A series of contracts was signed in the Autumn of 1972 between the CEA and GHTR on the one side and Gulf Energy and Environmental Systems International on the other, whereby in exchange for information on GGA's know-how, the CEA would undertake a programme of research and development in support of the design. Concurrently, GHTR, supported by the company Technicatome, which was created in June 1972 out of the former reactor construction department of the CEA (with a 10% participation of EDF) began the preparation of a commercial offer. These contracts were subsequently modified, first in 1973 to take account of Shell joining Gulf (q.v.) and again in 1974, when the GHTR was converted into the Société pour les Réacteurs Nucléaires HTR (SHTR) that was owned 40% by Creusot Loire, and 20% each by CEM, CERCA and Péchiney-Ugine-Kuhlmann.

From mid-1972 the whole of the French HTR effort was mobilised round the GGA agreement. The first integral block fuel element of the CEA began irradiation tests in DRE in June 1973 and that same month a special fuel company was formed to manufacture American style elements – the Société pour les Combustibles de Réacteurs à Haute Température (CORHAT) with a 32.5% holding by GGA and a 67.5% by CICAF, a company controlled by the CEA with also industrial interests including CERCA and SFEC.

The 1973-76 extension of the Dragon Agreement was supported for the use that could be madc of DRE in the fuel development programme relevant to the construction of HTR steam cycle power stations. This was now the sole technical justification for keeping the Project running. Even on the irradiation programme, France had reservations as there was dissatisfaction with the accuracy of the data covering experimental conditions in DRE and there was competition from its own research reactors. Moreover, coated particles had been taken to the point where the fuel was by far the most proven aspect of an HTR design; the manufacture of fuel giving an entirely satisfactory performance was no longer a problem.

The opportunity to exploit the HTR was presented in 1974 when l'Energie de l'Ouest Suisse (EOS), after extensive preparatory discussions, and encouraged by SHTR's outline designs that were given a wide dissemination earlier in the year, put out a call for tenders for the station at Verbois, just inside the Swiss-French frontier near Geneva. EDF would be able to participate as a junior partner on foreign territory. Designs prepared by SHTR and Technicatome would have CEA backing for the core, while for the rest of the plant they would have the security of several American stations pioneering the way. French industry would thus have much of the design responsibility off-loaded on to other shoulders and stood to gain a commanding experience in the manufacture of the new system with minimum risk.

The Verbois project did not materialise. Continuous delays at Fort St. Vrain followed by the cancellation of orders for the large stations destroyed both the bolster to morale and the technical guarantee. SHTR, having found that the American design did not form a satisfactory basis for tendering, felt unable to present more than a technical proposal to EOS in the

Spring of 1975. With this, France withdrew to the HTR sidelines, retaining its 250-strong team of experts on gas-cooling at Saclay for the time being, but under pressure to cut costs. The science was available; it was the industrialisation structure that had collapsed. Dragon could do nothing to rebuild it.

In the field of advanced HTRs, the CEA had undertaken some studies on the direct cycle (in which EDF was not at all interested) and with Gaz de France on hydrogen production and then coal gasification. These never got beyond the stage of paper studies and in no way could be regarded as a basis for the HTR programme in France. They could, however, form the basis for a collaboration with the German programme which was again being encouraged politically and it was expected that a series of agreements would be concluded with Germany that were complementary to the American agreements. In these too, Dragon had no place.

The only motive as far as France was concerned in keeping the Dragon collaboration going, once the industrial power programme dissolved, was to avoid breaking up an international project, a view-point consistently presented by French delegates to Dragon and to the Communities during the discussions on an extension beyond 1976. If the other partners had been enthusiastic for a continuation along existing lines there is little doubt that France would have gone along with the majority, as a gesture of international goodwill. On the other hand, France was not prepared to increase its contribution at all, nor prepared to make any diplomatic effort to prolong the life of the Project in the face of opposition. Dragon's closure was regarded as logical and preferable.

STAFF

Considering the brevity of the French interest in Dragon and the traditional reluctance of French people to work anywhere other than in France (or even Paris, if that is the region of origin), the participation of staff in the Project was surprisingly high and closely comparable with the German. Altogether, 42 people joined the Project, 20 from the CEA, 11 as Euratom staff and 11 from six industrial concerns. Their average duration of stay was 2½ years. They came for the most part in two waves – the individualists in the first few months of the Project and then the more formally seconded staff in the period 1968-1971. None came in either 1966 or 1967 and in the latter part of this period there was no one from France at Winfrith. Just one made Grade C and this was Robert Tron, who was Head of the Programming, Test and Inspection Branch during the construction of DRE. He left a year before DRE went into operation.

GERMANY

Traditionally, technical developments in Germany are led by industry, and nuclear energy, at least in the early years, was not considered to require any fundamental change in the established practice. Development is orientated towards an essentially engineering approach and new technologies are viewed in terms of the realisation of complete plants or customer-directed products. The various power companies, even those largely financed by local authorities, foster a spirit of independent enterprise and they expect to contribute to the introduction of new techniques by intensive interaction with the manufacturing industry.

This independence and the federal structure of the administration, militated against the establishment of a single national policy in nuclear energy, and development proceeded as a series of parallel activities. In more recent times, aid from public funds has come to be recognised as a necessary ingredient of innovation to the point that most of the extra cost of introducing the new technology now falls on the Federal Government. Government does not, however, seek to impose a specific technical policy on industry but rather to support and coordinate its efforts. Industry is considered to have the best machinery for determining the lines of development that will prove to be the most sound commercially. Long-term research is funded both from federal and state sources and is carried out mostly in the Universities, a broad range of institutes and, in the nuclear sector, particularly at the two centres of KFA Jülich and Karlsruhe. Jülich, while being the site of the AVR plant and the national centre for HTR research, comprises also several institutes with a strong academic bent, where the accent is on interesting research more than industrial realisation.

Lines of communication are imprecise and Government, its institutes and industry find difficulty in keeping closely in touch in rapidly changing situations. For the Project this was to present problems, as its contacts in these various areas were motivated by different interests and did not act in a concerted way. In the final months of Dragon especially, it was not easy to obtain a full appreciation of the situation and because of the tendency to listen mainly to supporters, the depth of official feeling towards keeping the Project in being was exaggerated.

While the majority of nuclear power stations ordered in Germany have been LWR's, with 6800MW(e) in operation and 11 000MW(e) under construction at the end of 1976, by the same date, Germany had invested between 1500 and 2000 million DM in the HTR. These figures can be compared to the country's total contribution to Dragon of around 60 million DM. The cost of Dragon has thus represented but a small fraction of the total expenditure on HTR work, the majority of which has gone into the pebble-bed system. It seems apposite, therefore, to review the evolution of the two main pebble-bed reactor projects – the AVR and the THTR – before examining further the country's participation in the Dragon collaboration.

AVR AND THTR

As described earlier (page 59), a group of electricity companies led by the Stadtwerke Düsseldorf, in April 1957 commissioned the associated companies of Brown Boveri & Co., Mannheim, and Fried Krupp, Essen, to prepare the design study of a 15MW(e) experimental power reactor based on the pebble-bed principle that Rudolf Schulten had elaborated. The reactor was to be gas-cooled and fuelled by a mixture of highly enriched uranium carbide and thorium oxide contained in balls (pebbles) of graphite that would serve as moderator. Fission products escaping from the fuel would be extracted in an appropriate clean up system. In the course of the following year, the study was presented and on 3 February 1959, three weeks before the Dragon Agreement was signed, the Arbeitsgemeinschaft Versuchstreaktor AVR GmbH was formally constituted with a view to ordering the proposed plant. The contract for the construction of the plant, which was to be sited close to the new nuclear research establishment of the Land Nordrhein-Westfalen at Jülich, was signed on 13 August, 1959. Total cost was estimated at 40 million DM of which AVR would contribute one half, the remainder being provided by Government sources. Additional costs (which came to a further 50 million DM eventually) were to be contributed 80% by the Government and 20% by the main suppliers. Actual construction was started in August 1961 by the newly formed company Brown Boveri/Krupp Reaktorbau (BBK); criticality was achieved in August 1966, full power in February 1968, and the plant has handed over to AVR in May 1969.

In the meantime, the Jülich establishment had become a largely federal funded organisation and been renamed the Kernforschungsanlage (KFA) Jülich. During the first half of the 1960s, the primary aim of KFA was to bring into full operation its research reactor (Merlin) and materials testing reactor (DIDO). Various reactor systems were being studied and there was a natural interest in the construction going up next door. It was, however, Schulten's appointment as Director of the Institute for Reactor Development (and at the same time Professor for Reactor Engineering at the Technical University of Aachen) in 1966 that finally polarised KFA into adopting the pebble-bed as a main research theme.

Schulten was in favour of other countries participating in the development of the pebble-bed, notably France, and although he had earlier declined an invitation from Guéron to join Euratom, the two men kept in close touch. Guéron, despite his reservations over the fuel design until the advent of coated particles, saw it to be Euratom's duty to be involved and offered to pay for the first fuel charge of the AVR reactor. In return, Euratom would require full information on core performance for transmission to its members, a condition that AVR could not accept. For AVR, the problem of the fuel concerned not so much the finance as finding a supplier. The facilities of the German company Nukem could not be ready in time; Dragon's were fully occupied with the Project's own requirements. AVR was forced to go to Union Carbide in the USA and accept, against the strong opposition of Guéron, a clause in the contract that excluded from the core any fuel from other suppliers; otherwise Union Carbide's guarantees would be void. Nukem was thus dependent on DRE for testing its first production fuel balls and in the Autumn of 1966, four elements in the core of DRE were given over to this work.

Once construction of the AVR reactor was well on its way, BBK began to look to the future and in 1963, launched the Thorium Hochtemperaturreaktor (THTR) project which was to be a 300MW(e) pebble-bed demonstration plant. Appreciating by then the costs involved in such a development, and encouraged by the Government which was keen to see some return to Germany from the contribution the country was making to the common Euratom budgets, BBK agreed to negotiate on Euratom's participation. Patent problems, which had presented insuperable barriers in the past, were simply resolved by making declarations that if difficulties arose, discussions would be held. This opened the way to the formation at the beginning of 1964 of the THTR Association comprising Euratom, BBK and KFA as contract partners, with Nukem as sub-contractor. The agreement covering the Association was, in the first instance, to cover a period up to December 1967 corresponding to the end of Euratom's second 5-year plan. Renewal was expected after that date but could not be agreed owing to the failure within the Community to agree a programme for the following five years. Many of the Euratom staff who had been working for the Association joined BBK or KFA which continued the project independently.

Throughout the study, BBK kept in close touch with the electricity companies and, convinced by the good record of the AVR reactor in its run up to power, a new utility group was formed early in 1968 called the Hochtemperatur-Kernkraftwerk GmbH (HKG), with the specific objective of commissioning the construction of a prototype pebble-bed power plant. The exact composition of the group changed with the years, gaining in particular the participation of the Vereinigte Elektrizitätswerke Westfalen (VEW), but the name remained. HKG received the preliminary study from BBK in the middle of 1968 and following the choice of a site offered by VEW at Uentrop, near Hamm, was presented with a firm offer at the end of 1969. For the construction, a THTR consortium was formed of BBC, BBK and Nukem, to which a letter of intent was sent in July 1970.

All was not well, however, with BBK. Krupp's commitments to the THTR were in the nature of risk investments that on re-examination appeared likely to incur substantial losses. These the company was no longer willing to accept, in view of its current financial difficulties and in June 1971 it withdrew from BBK. BBC stood firm and began to look for a partner willing to take about 40% of the share capital, intending this time to keep a 60% holding instead of its former 50%. In order to maintain the work rhythm of BBK, HKG agreed to participate to the tune of 10% and in October 1971 BBK was transformed into the Hochtemperatur-Reaktorbau GmbH (HRB). All parties considered this to be a temporary measure only but sufficiently solid for a firm contract to be placed that same month by HKG with a consortium comprised then of BBC/HRB/Nukem. Main financial assistance came from the Federal Government and from the Land Nordrhein-Westfalen, and the project was recognised by the Communities as a joint enterprise from which it would derive tax concessions in return for making information available to the Euratom countries. KFA was no longer directly involved in the contract but later returned to the scene to take responsibility for the fuel cycle as well as supporting research and development.

Even before the break-up of BBK, exploratory talks had been opened with GGA on a merger of HTR interests and these were renewed with more purpose at the September 1971 Geneva Conference. With GGA's prismatic system gathering momentum in the USA, Brown Boveri deemed it vital to gain access to GGA's experience and agreed to transfer to the American company 45% of the shares in HRB, HKG withdrawing as had always been understood. Since January 1973 when the formal transfer took place, until the time of writing, no further changes had taken place in the THTR project structure, and the plant was expected to start major commissioning tests in the early 1980s. Even ten years for construction after seven years of study makes a long time scale. THTR was, however, like Fort St. Vrain subjected to a rapidly evolving pattern of new safety regulations that have required continuous 'back-fitting', and so an unending series of changes in the plant design. Also, soon after the start of construction, the pebble-bed was to be relegated to second place and only in the second half of the seventies come back as the main HTR system in Germany.

COLLABORATION AGREEMENTS WITH DRAGON

When the AVR project was launched, it was quite independent of any other HTR work going on elsewhere in the world. Conversely, Government support for Euratom's participation in the Dragon Project was neither motivated by the consideration that it would provide relevant information for the pebble-bed nor moderated by the thought that it would be competitive. With two exceptions, German staff who joined the Project in the first four years were recruited individually through Euratom and were not seconded by parent organisations interested in the HTR. Nevertheless it was soon evident that there was a considerable overlap of interest between the two projects and following personal contacts between Dragon and BBK staff, negotiations were begun in the Spring of 1961 for a continuous exchange of information. In principle, BBK through Euratom had access to all Dragon's published information but the route was long, and could be unacceptably cumbersome. Rennie was not opposed to such an exchange, but the Board of Management hung back and delayed so long, that no direct agreement was ever concluded, the need being superseded by the agreement with the THTR Association and KFA.

Board objections centred on the fear that a single company would have better and swifter access to Dragon information than the Signatories, even though it was acknowledged that terms similar to those used in the agreement with the USAEC could be applied. Another difficulty was that BBK did not have sole control over the information pertaining to the plant items sub-contracted to other companies – notably Nukem for the fuel, and Linde for the fission product extraction plant. More important though was Euratom's anxiety not to be by-passed. Close personal links were nevertheless established between BBK and Dragon that were simplified by Dragon's contract with BBK for the building and operation of certain experimental rigs and by the fact that both organisations were to effect similar irradiations in BR-2 at Mol and HFR at Petten. In due course also, BBK was to hire from Dragon, certain facilities installed at Risø and Studsvik. Whatever the objections put forward by the legal representatives on both sides concerning rights to patents and subsequent exploitation, scientific information flowed reasonably smoothly between the two organisations.

With the creation of the THTR Association, Euratom, as a member, had no longer any reservations about an exchange agreement being concluded with Dragon and proposed that an initial request for irradiation facilities that had been put to the Project in the Autumn of 1964 should be widened. The Board of Management agreed and negotiations began in earnest to resolve the problems posed by patents arising from work undertaken in collaboration. After innumerable meetings and the preparation by the Agency of a large number of drafts, an agreement was finally reached in July 1965. Active collaboration had begun earlier. Several guest scientists from BBK and AVR, had spent some months at Winfrith during the commissioning of DRE, the Board agreed in March 1965 to provide irradiation space in DRE in so far as this could be made available, and the first formal information meeting was held at Jülich in May.

According to the agreement, all Dragon information that the Project had a right to disclose was to be made available to the Association and in return, the Association was to make available to the Project similar information regarding its own research. Detailed manufacturing techniques were excluded as well as specific information on component design and economic studies. Through the agreement, relations between Dragon and KFA became very close and the flow of information was even freer than the exact terms might imply. When the THTR Association dissolved, the collaboration was continued with KFA.

With Dragon's development of coated particles, Nukem, too, had felt the need for a closer contact with the Project and early in 1963 had asked through Euratom for information on coated particle production in return for subsequent information on the performance of Nukem manufactured material. Nukem had grown out of Degussa's nuclear division, becoming independent in 1960 with, as shareholders, apart from Degussa, Rio Tinto of the UK and Mallinckrodt Chemical of the USA. (Later it was restructured and is now owned by RWE 45%, Degussa 35%, Metallgesellschaft 10% and Rio Tinto 10%). Nukem had an understanding with BBK so was not bothered by competition and was wise enough to understand at an early stage that academic knowledge was only a small part of the total competence necessary to make and deliver highly enriched nuclear fuel. As a consequence it was not inhibited by notions of commercial secrecy and welcomed Euratom help (including staff) to build up the know-how necessary for the manufacture of the new fuel.

Nukem understood the importance of coated particles for the pebble-bed, having realised that the original concept of a fuel pellet resting in a cavity in the middle of a graphite ball, could not be satisfactory as it would lead to a primary circuit activity that was unacceptable. BBK, although initially sceptical about the reliability that could be achieved with coated particles, also saw the simplification that could arise from having a fuel that intrinsically retained fission products and changed its approach accordingly. Discussions between Nukem and Dragon, smoothed by the setting up of the Standing Committee on Patents and Information, finally ended with the signing of the agreement on 19 March, 1964. Through most of 1963, Nukem had been following Dragon progress closely and equipment designed for installation at Winfrith was, in a number of cases, duplicated for erection at the company's works at Hanau. In addition, a member of the Nukem staff, Dieter Kern, was seconded to the Project in March 1963 (unhappily never to go back to work in Germany as he died in a road accident in May 1965 while still at Winfrith).

As already noted, Nukem found the time needed to develop and prove its production processes to be outside the timescale for the start-up of AVR, but from 1967 onwards, became the main supplier of fuel for the reactor.

During the first years, the flow of information was very largely from Dragon to Nukem. Then paths diverged and Nukem began to make its own innovations, developing production processes that ran counter to Huddle's ideas. Nukem, it should be noted, was able to benefit not just from Dragon experience but also from the KFA and THTR irradiation work in Studsvik and Petten, and from the operation of AVR. As a result, the company was able to establish a competence in fuel technology independent of Dragon. One of its important developments was a moulding technique for making fuel balls, a process that Huddle was very reluctant to believe in. Nevertheless, this did not deter the company from sending its inventor, Milan Hrovat, to Winfrith in 1969 and 1970 to gain experience on fuel for prismatic systems. Working in the fuel production group under Price, with whom he had a close rapport, he formulated his ideas on a moulded block for producing prismatic elements with minimum graphite wastage.

HTR FUEL POLICY

Concerned only with the use of highly enriched uranium in HTRs – mainly fuel balls for AVR and THTR but also latterly, driver fuel for DRE and developments for a GGA type reactor, Nukem was spared the excursions into the various types of fuel that Dragon studied for the low enriched system, and which involved years of work. In Germany, Dragon's preoccupation with low enriched uranium from 1966-1972 was regarded as an aberration whereas Rennie saw the low enriched route as the only way to get the system launched. For all sectors in Germany, the HTR was associated with the highly enriched uranium/thorium cycle and the shift towards a system that resembled an AGR carried no support. Industry, having concluded that natural uranium reactors could not be economic, did not consider a low enriched system to be any less dependent on outside suppliers than a highly enriched system and, up to 1979, the commitment to the highly enriched uranium/thorium cycle as the reference system for HTRs was maintained.

Yet at no level was any serious effort made to modify the direction of Dragon's fuel development. The Project was regarded as an important source of quality information and of people. DRE was a valuable irradiation reactor – complementing the statistical results from the irradiations in AVR with detailed information on particle behaviour. So long as this continued there was no reason to interfere. The Authority which ran the Project (in German eyes) was determined upon the low enriched development and seemed ready to exploit it; Euratom and especially Marien was not only supporting the idea but through Inter-Nuclear, trying to involve other countries; perhaps something might come of it and Dragon was allowed to go on as a parallel insurance.

FIRST PRISMATIC HTR ACTIVITIES IN GERMANY

The AVR and THTR projects were both concerned with the development of a pebble-bed system coupled to a steam cycle. They were not the only HTR projects launched in Germany. All the big groups were looking for openings in the late 50s and early 60s, and Gutehoffnungshütte Sterkrade (GHH) was no exception. The GHH nuclear department was created in 1962 with Egar Böhm at its head, a former Deutsche Babcock and Wilcox man, whose early work had brought him into contact with gas-cooled graphite-moderated reactors. Considering the USA to be the main nuclear country, and having no opportunity to enter the light water and fast reactor fields, he turned to General Atomic. The Peach Bottom project was proceeding well and GHH negotiated a licence for this reactor design which was offered to the Nordwestdeutsche Kraftwerk (NWK) for a station at Wiesmoor. By early 1964, however, it had become evident that Peach Bottom was already out-moded and NWK decided not to go ahead.

GHH had in the meantime, partly as a result of the joint studies it had been making with GKSS, the ship research organisation, become interested in a direct cycle system and found in Dragon a corresponding enthusiasm. This was the period when the Project was promoting a policy of R & D into gas turbine applications culminating in the Paris Colloquium in May 1965. Even though, as we have seen, the Signatories elected not to follow this course as a main theme, the Project's effort was not damped out completely and GHH was encouraged to continue by both Dragon and GA.

Through its connection with both organisations, GHH eventually opted for a prismatic core. The company had considered the pebble-bed in passing, but its main attention was really devoted to the plant design for which there were many different ideas (all incidentally different from GA's) involving different detailed engineering. Only a small effort was available for core evaluation and it was assumed that the core could be defined by the scientists working with Dragon data when the need arose. The company grossly under-estimated the extent to which core designs – still in a state of flux – and plant engineering interacted once it came to producing manufacturing specifications.

GHH's study was focussed on the development of very large gas turbines suitable for full scale power stations and the design of a 25MW(e) direct cycle plant as a system demonstration. A potential customer existed in the form of the Land Schleswig-Holstein, and in the Spring of 1966 an offer was made in competition with groups proposing other systems, of a direct cycle HTR at a unit price comparable to a large light water reactor. This offer was accepted and led to the formation of the Kernenergiegesellschaft Schleswig-Holstein (KSH) to own and operate the station that was to be built at Geesthacht. Towards the end of the same year, GHH as representing

KSH made an approach to Dragon requesting a collaboration agreement including consultancy. This was received favourably by the Board of Management provided that assurances could be given that information did not thence pass to GA through the medium of the GHH–GA exchange (associated with the licence agreement between the two parties) and that a clear indication was given of the consultancy services that were required.

In spite of the precedents set by the similar agreement concluded between Dragon and Brown Boveri (Baden), discussions with GHH proceeded slowly. The agreement was finally signed in April 1968 but it was not until the end of that year that Dragon received a formal request for advice. Before that, Lockett informally had been helping in the design of the reactor and regular visits between the GHH centre and Dragon had taken place; the first formal meeting, however, did not occur until February 1969. An explanation may lie in the company's involvement in the negotiations on setting up Inter-Nuclear leading to the view that connections with TNPG and the other companies taking part would be more important. GHH was also temporarily distracted from its gas turbine activities and participated with TNPG in the designs of a 600MW(e) steam cycle station that was eventually dropped.

RESTRUCTURING OF THE INDUSTRY

Over the latter part of the sixties, the Federal Government, finding itself assuming an increasing proportion of the cost of nuclear development, had been seeking ways of consolidating the national HTR effort, and under its persuasion, negotiations were opened between BBC, BBK and GHH with a view to setting up an equally owned joint company that would take over all the HTR Projects, and which could also enter into co-operation agreements internationally, notably with GGA and the British consortia. Discussions proceeded as far as choosing the headquarters (at Cologne) and even establishing an organigramme. It then became clear that the new company would be assuming financial responsibility for the Geesthacht project that had been offered at a fixed price, a price that clearly could not be met. Because of this, merger negotiations were discontinued and, for the same reason, late in 1971, by mutual agreement between the contracting parties, the Geesthacht project was abandoned. GHH then closed its nuclear department.

Despite the fact that the formal agreement with Dragon had terminated in March 1970, contact was maintained, as under another development programme, GHH was designing a 50MW closed cycle helium, oil-fired generating station that has since been built at Oberhausen. In the development of helium technology, the specification of materials and circuit techniques, the Dragon contribution to the company was therefore significant, even if in the end, the nuclear side collapsed.

The year 1971 was doubly difficult for Germany's HTR industry with the withdrawal of Krupp and the abandoning of Geesthacht. For KFA too it meant a change as, with the signing between HKG and the THTR consortium of the commercial contract for the construction of the THTR power station, KFA ceased to play a leading role. The contract marked effectively the end of KFA's preoccupation with the HTR as applied to the steam cycle and from that time on, KFA concentrated on advanced concepts. Certain responsibilities for THTR remained, notably fuel and graphite testing plus other research and development but KFA's main HTR line became the direct cycle and process heat linked to the pebble-bed.

In 1971 the underlying clash of HTR ideologies also came to a head, stimulated by Philadelphia Electric's selection of GGA's design for a twin 1160MW HTR station. National pride was tied up in the pebble-bed which had its passionate disciples, for whom no other system was comparable. Nevertheless, Brown Boveri and HRB, the Director of which, Wolf Müller, as already noted was an ex-Dragon man, had kept in close touch with prismatic developments. The companies had been impressed by the performance of DRE and Peach Bottom and were well aware of the market importance of Fort St. Vrain and the new orders to GGA. The prismatic system appeared to be ready for commercial exploitation in large sizes whereas there was still much to do on the pebble-bed.

Within KFA and the Government also there were those who felt that the pebble-bed should be regarded as a long-term proposition and that collaboration on the prismatic system for the shorter term should be encouraged. Having catalysed the consolidation of the light water industry in the Kraftwerk Union (KWU) in 1969, the Government still hoped, in spite of the failure to bring about the BBC/BBK/GHH merger, to arrive at a strong HTR combine as a second national group, while closer ties with the USA were desired from a political point of view. The HTR was seen as the most promising instrument for bringing this about. There was thus both a political and commercial inducement for HRB to link up with GGA and the acquisition by the American company of 45% of the shares of HRB following the negotiations that took place during 1972, was seen as a major commercial breakthrough – indeed something of a coup for Brown Boveri, giving the company access to the GGA 1 160MW design in exchange for rights on its experience in the pebble-bed development.

Brown Boveri investigated the market for a large prismatic plant in Germany and succeeded in winning the interest of VEW, by then the principal partner in HKG, and leader of Euro-HKG, the international group of electricity companies formed in January 1972 to pool experience in HTR operation. At the instigation of VEW, HRB was commissioned by Euro-

HKG (without financial commitment) to prepare detailed plans for a station based on GGA's designs to be sited at Schmehausen near the THTR plant. In consequence HRB became firmly committed to a prismatic development, relegating the pebble-bed to a secondary position. Nukem, in its turn, had seen the change in market potential and had felt compelled also to look to the USA for support, concluding through its wholly owned subsidiary HOBEG specialising in HTR fuels, a licence agreement with GGA in July 1972 – six months even before GGA formally became a partner in HRB.

Dragon, having expended its main effort over the previous years on the low enriched cycle, was supplanted as the main source of experience in prismatic core development, the UK was no longer giving a lead in prototype construction and the German effort became wholly oriented towards US technology. Dragon could still be useful to the German project as DRE was the only reactor in which representative fuel block sections could be irradiated and its contribution in the materials field was also recognised. Otherwise, any further impact it might have on the steam cycle HTR was regarded as being of small consequence.

ADVANCED APPLICATIONS

So far, in considering the German HTR scene, attention has been focussed exclusively on electric power generation, and the direct cycle application has been treated only in the context of GHH. From the beginning, Schulten had promoted the HTR both for direct cycle operation and also for process heat, for which there exists a special requirement in Germany. Quite apart from having a strong modern chemical and steel making industry, Germany possesses large deposits of hard coal which are expensive to mine. Politically it is necessary to keep the coal industry in business and from a resource point of view desirable to exploit indigenous reserves. Theoretically, they may be used economically by converting the coal to a gaseous fuel, and the HTR is widely considered to be the most suitable generator of the heat needed in the processing operation.

When the Government began in 1969 to try and bring together all the HTR activities, it had in mind not simply co-operation on the construction of steam cycle stations but a coordinated development of the whole HTR field including the direct cycle approach and process heat. Its first positive result came with the setting up in 1972 of the Hochtemperaturreaktor mit Helium Turbine Grosser Leistung (HHT) project to undertake the development and design of a large direct cycle system including the construction by GHH of the closed cycle helium station fired by fossil fuel, already mentioned. This brought together BBC, HRB, KFA and Nukem with whom were associated the Oberhausen electricity company EVO and the Technical

Universities of Aachen and Hannover; other participants were expected to come in later. KFA provided the focal point and Hermann Krämer was nominated Project Leader. Contact with Switzerland was renewed and in 1973 Switzerland agreed to take a 10% share, bringing in EIR, Brown Boveri (Baden), Sulzer and Alusuisse. The year after, the HHT project entered into an exchange agreement with GGA, further locking the German and USA programmes together.

Parallel with HHT was a process heat project involving KFA and coal interests that attracted the attention of KWU (the light water reactor company of Siemens and AEG) as a long-term development exercise that was not competitive with its existing activities. In the Spring of 1973, with the encouragement of the Government and KFA, KWU decided to establish the Gesellschaft für Hochtemperaturreaktor Technik (GHT) with the object of concentrating on pebble-bed reactors for process heat applications. Engineering services were to be provided by the KWU subsidiary, Interatom (engaged on fast reactor development), and to bring itself up-to-date on pebble-bed technology, GHT entered into a licence agreement with HRB. Although the HHT and process heat projects remained separate for a long time, KFA provided a common element and gave a semblance of cohesion to the whole advanced programme.

INDUSTRY'S CONTRIBUTION TO DRAGON

The most important contribution to Dragon's development programme from German industry was made by Sigri Elektrographit in the preparation and supply of fuel element bodies, and other graphite components. In the manufacture of the first charge for DRE, Sigri participated in that thoroughly European circuit in which Morgans (UK) supplied partially processed material that Pechiney (France) graphitised and purified. The material was then returned to the UK where it was machined by the Authority or Graviner, before being transported to Sigri for impregnation and medium temperature treatment, finally completing its journey at Winfrith where outgassing and high temperature treatment was applied. It should perhaps be emphasised that the companies for each phase were chosen on technical grounds and not political. The Project had no wish to complicate the fabrication procedures, nor put at risk valuable material by unnecessary transport.

Sigri's share of the work increased and it supplied many of the graphite bodies for later fuel elements, for a time being the only company in Europe able to machine Dragon's complex block elements to the dimensional accuracy demanded.

Leaving aside fuel supplied by Nukem (worth over £100 000 in 1974-75) the total value of contracts placed in Germany for services and research and development was fairly low in comparison with the other Euratom countries, amounting to only about 16% of the total for the Six. For the construction of DRE the fraction was higher, amounting to about ¼ of the Euratom share. Major items of plant supplied were the reactor pressure vessel, Mannesmann-Export (a contract which cost the company dear but elicited no claim for compensation); valves for the cooling circuit, FX Stöhr; fuel element assembly plant, the upper load facility, spent fuel element transit flask and a production sintering furnace for fuel element assembly, WC Heraeus. The value of contracts placed with the last company amounted to about £250 000 in all.

Germany had no spare capacity in its materials testing reactors to offer, and even for its own HTR programmes made parallel and joint use with Dragon of reactors in other countries.

Of the total of 46 staff seconded to the Project, nine were from BBK, two from AVR, six from GHH, three from Nukem, seven from KFA, and the rest from Euratom. Not all returned to work on the HTR in an industrial capacity; a number of physicists returned to research establishments or into the academic world.

EVOLVING ATTITUDES

Germany's initial participation in Dragon came through its support for Euratom's general programme. Once the Project got under way, it was seen to be complementary to the German HTR development, a useful adjunct that gave access to British experience in gas cooling as well as specific R & D information that was relevant to the pebble-bed.

The three reactors of AVR, DRE and Peach Bottom were regarded as each having separate and distinct roles **(35)**. AVR was to demonstrate the feasibility of the pebble-bed core concept, the satisfactory irradiation behaviour of the fuel in a statistically significant way as well as the ability to operate an HTR safely and reliably; DRE provided a test bed for fuels under operational HTR conditions; Peach Bottom was there to generate electricity at a steady rate.

The design and manufacturing industry, as represented by BBC, BBK, HRB, GHH and Nukem, respected the talents to be found in Dragon and recognised the wide knowledge in the Project of core performance, fuel manufacture and materials. For a long time, the expertise assembled in Dragon was significantly greater than in any of the German project groups and there was a greater awareness in Dragon of the problems to be resolved. The industry nevertheless had little faith in the assessment studies made by the Project (and which were so strongly supported by Euratom) believing that these had to be carried out in an industrial environment to be of practical value. Moreover, pebble-bed engineering was a discipline distinct from the prismatic. When the main attention turned to the prismatic system, this was under the tutelage of GGA, and although DRE remained the only reactor in which representative fuel block sections could be irradiated, this was not regarded as being of great significance compared with the wealth of experience obtainable from the Americans.

At the working level in KFA, Dragon's skills coupled to its drive and imagination were much appreciated. The Dragon team provided a continuous intellectual stimulus as well as a constant flow of information. The special exchange agreement with the Project that was taken over from the THTR Association cemented relations between the two organisations and KFA became the natural gateway through which Dragon information was disseminated into Germany.

Bearing in mind that with the exception of the UK, Germany stood to gain most from the international effort, yet was paying only about 14% of the Project's costs, it might have been expected that this would have been reflected in the position adopted by the country in regard to extensions to the Dragon Agreements. However, even in 1965, before construction of the AVR reactor was finished, Germany was suggesting within Euratom that operation of DRE should be stopped in 1969 and R & D work run down a year earlier. As negotiations on the 1967-70 extension progressed, Germany made no attempt to unblock the jam caused by Euratom's inability to take decisions; support for the THTR Association was considered more important.

Failure to obtain this support no doubt contributed to the hardening of attitudes towards international projects discernible in the last years of the 1960s. Although liberated from external controls in the middle fifties and enjoying a dynamic economic recovery, Germany had still to throw off its burden of guilt left over from the war. With the election of K.H.F. (Willy) Brandt as Chancellor in 1969, the remaining inhibitions fell away and Germany began to assert its authority, not least in scientific and technological circles. International projects continued to receive support, but it became established policy to seek to centre new initiatives on home territory and attract the participation of other countries into German-run ventures.

Dragon's effort to widen the Project's operations were considered to be an encroachment on German preserves. Despite the dependence of KSH on Dragon for the Geesthacht plant, the German member of the Euratom delegation resisted the inclusion of gas turbine studies in the Dragon programme for the 1970-1973 extension "because of the substantial programme in Germany". KFA also wished formally to exclude direct cycle work from its collaboration agreement with the Project.

Shepherd's latter-day efforts to coordinate European work on advanced systems including the construction of a Very High Temperature Reactor gained no support from Germany's representatives. They expressed the view that any contribution from Dragon would be of marginal value in comparison with the extensive German programmes and would represent just a diversion of funds. Germany, although perfectly prepared to welcome external participation (Switzerland and possibly France), was not willing to see control pass into other hands. Dragon was under the general direction of an international Board of Management and had the reputation of being very independent in carrying out its R & D work. HHT and process heat were German programmes upon which large funds were being expended and would remain German.

Similarly Shepherd's proposal that the Project begin to divert some effort to studying, as a serious research activity, the small scale reprocessing it was obliged to do in order to reduce the volume of its irradiated fuel to manageable proportions, elicited a negative response. Closing the fuel cycle, while admitted to be of the greatest importance in Germany, was considered to require a large financial investment that would be better made in the appropriate industries. Dragon had no really relevant facilities and building these up would be again just a diversion of funds that could be more profitably used at home. Co-operation with other countries could be organised as necessary outside the Dragon Project. (It can be noted that the Agency joint undertaking for fuel reprocessing of Eurochemic was being closed down for lack of support.)

In effect, although at the working level there was a desire that Dragon should continue, at the policy making levels from about 1972 onwards, Dragon was not seen to be essential. Steam cycles were being looked after by GGA, advanced systems by German inspired programmes. It was assumed that Dragon would, of course, go on after March 1976 and rigs were prepared for installation in DRE. Krämer certainly viewed the impending closure as a disaster but KFA as a body, with its many institutes and different study programmes, was very slow to move. Only when any hope of an extension had died was there a reaction and a major effort was deployed in the last months to fund an information retrieval operation, thereby running the Project down a little more smoothly than would otherwise have been the case. The rapidity with which funds were made available for this purpose shows clearly that it was not shortage of money that closed Dragon, it was shortage of will.

The reactor industry was even less mindful of Dragon's predicament as negotiations on an extension seemed less and less likely to produce a solution; it had other more pressing problems.

As HRB worked through its design for the Schmehausen plant (the HTR 1160) it became aware of deficiencies in some of the major plant components as conceived by GGA and found it necessary to do much more work than merely adapt the American system to German conditions. Moreover, tenders submitted by prospective sub-contractors were running higher than had been expected and the overall cost began to look unfavourable in comparison with competitive systems. Even allowing for the fact that this was a prototype, it became more and more difficult to produce an attractive offer. Troubles at Fort St. Vrain confirmed doubts about certain of the original design concepts and with orders for HTRs in the USA lapsing, HRB restricted itself to sending to Euro-HKG a detailed technical description of the plant instead of a firm bid.

VEW for its part was up against a deadline. In order to stimulate activity in the electric power field, the Government had offered major tax concessions to companies ordering plant before the end of 1975. Neither VEW nor HRB was prepared to be responsible for the first off in a new series, and VEW felt obliged to renounce its plans for an HTR at Schmehausen and order a light water reactor instead. (Ironically, five years later the licence to build this reactor had still not been granted.) Nukem abandoned its development of moulded blocks, and as the HTR programme in the USA collapsed so too did the German HTR programme on prismatic steam cycle systems.

More than that, Germany's whole HTR programme was propped up by agreements with (by then) GAC, not just the HTR 1160 plant. When the prop broke, the whole programme fell into disarray. On top of this, the Government was engaged in changing its system of funding, so that contributions to the Communities would not be treated separately but would come under explicit project headings. Any separate contribution to Dragon would have run into departmental difficulties and might well have involved cuts in (say) KFA's share. On a wider front, Germany was leading the fight in the Communities against rising government expenditure and the high rate of inflation, so that from the point of view of fiscal policy, an increased share of Dragon costs was not to be entertained.

Above all the Government needed time to consider the situation created by the GAC debacle. If the next Dragon extension had been a simple case of continuing at the same rate there is little doubt it would have been approved by Germany without difficulty. On the other hand, if the various HTR programmes had been pushing forward steadily, it is perfectly possible that Germany would have considered taking a larger share in order to keep DRE and the metals research running. For a brief moment indeed, it seemed as if Germany would make a special diplomatic effort, but other programmes of the Community were under discussion and Germany was not prepared to compromise its bargaining position on these, to bolster a Project that had seemingly run its time, and had such little support elsewhere.

CHAPTER 24

BRITAIN AND DRAGON

CONCEIVED in Britain, born in Britain and nurtured in Britain, Dragon was for all the Signatories with one exception a British Project. The exception was the United Kingdom Atomic Energy Authority which fostered it, cared for it, encouraged it to stand on its own feet and acted as its legal guardian, but never regarded it as a member of its own family. Yet throughout Dragon's life, the fortunes of the Project were dominated by UK policies on the HTR. These have already been considered in some detail and what follows is in the nature of a résumé and a supplementary note in which other aspects of the relationship will be explored.

In broad terms, within the Authority, basic research is carried out by Harwell, while engineering studies and the design and construction of experimental reactors is the responsibility of Risley. When a system is ripe for commercialisation, industry takes over the design and construction of the plant while Risley manages the back-up research and development and through its fuel organisation, the definition of the fuel cycle.

National policy, while requiring Government authorisation was, until the middle sixties, largely determined by the Authority with the compliance of the (nationalised) generating boards and industry. Since then it has become the subject of wider consultation in various ad hoc and permanent committees in which a large number of interested bodies participate. This, it might be noted, has not led to a greater consistency nor speed in taking decisions.

It is evident that when Cockcroft first proposed that the HTR should be developed as an OEEC joint undertaking, he was expecting that Risley would have a much greater role in the design of DRE than materialised in practice. He foresaw the Authority continuing to be responsible for the detailed design under contract to the international project, and even as late as the beginning of 1959, he spoke of the Risley design office under Jack Tatlock, moving down more or less *en bloc* to Winfrith. It is unlikely that this would have been acceptable to the other Signatories and in any case, Tatlock would not have been able to make the move. After von Ritter's death, the question was again raised, but Thorn, who was Tatlock's Deputy, was not willing to make the transfer, particularly if he was to take responsibility under an unknown and possibly unseasoned figure-head. The appointment of Franco who was both gifted and experienced, resolved the dilemma allaying at the same time the Authority's fears that the Project would suffer if Dragon did not have a reputable reactor builder as senior engineer.

Risley continued with the design of DRE until the international team could take over. The Engineering Group of the Authority then acted as the construction agent, in which capacity its experience was of great value, and was recognised as such. DRE was nevertheless Dragon's own design and the Authority was careful not to impose its own ideas. For the Authority, Dragon was a research experiment on a system which might or might not prove to be interesting and which might or might not find a place in the country's programme. There was no advance commitment to commercialise the system whatever its virtues; the HTR would have to prove its potential if it was to move beyond the experimental phase. Risley's task, apart from its contract work, was to evaluate that potential in the light of its own and Dragon's results and the changing nuclear scene.

DRAGON AND WINFRITH

Within Winfrith, Dragon held a privileged position. Fry, aided by Cyril Hart-Jones, the Deputy Administrative Secretary of the site, was intent on seeing the Project a success, and they enjoyed the continuous support of Peirson until he left the Authority in 1971. Relations with Rennie and Shepherd were cordial and positive, neither side being weak or aggressive when it came to fixing the rent of buildings for example – perhaps the liveliest issue between them. Accommodation before the Dragon offices were built (later known as A.10) was not easy and it says much for both sides that there was little friction. Excellent work was done by the Winfrith/Dragon Liaison Committee, the formal instrument for resolving problems affecting the two parties, while outside the committee, much was achieved on a personal basis. Winfrith administrators were not hide-bound and it seems that the challenge posed by the novel relationship was met with a great deal of good will and initiative.

At the working level also, Dragon people were welcomed with friendliness, and the overseas staff were assimilated rapidly into the community. Conflicting priorities created the minimum of problems, certainly no more serious than are to be expected within any site. Here again the absence of bureaucracy was a contributing factor. If someone from Dragon urgently needed a job done in the workshop that had not been scheduled, he would find his own way of pushing it through and the cost attributions would be sorted out later. Throughout most of its construction period, Dragon had the advantage of being the most important project on the site, stepping down to second place only when work was started on the SGHWR in May 1963.

The pattern of cordial relations established by Fry was continued by Cartwright when he took over as Director in 1973. Naturally courteous and reasonable, he continued the tradition and could contribute his own broad experience in the reactor development field, having been in charge of experimental reactors over Tatlock in 1958/59 and so in contract with the Risley DRE work. Afterwards, he became Director of Industrial Power in the Development and Engineering Group and then the Reactor Group when it was formed, and subsequently Director in charge of water reactors and then fast reactors. Similarly Allen, when he took over from Peirson, brought to the Project a sensitive and sympathetic understanding. He was well aware of the inner workings of the Authority, having been Plowden's private secretary in 1956/57, and after a short period in the Treasury, Director of Personnel and Administration of the Engineering Group and then of the Reactor Group at Risley. He left the Authority again in 1963, but returned to the London Office in 1968, following which he was appointed Secretary on Peirson's departure, and later Member.

The success of the integration into the Winfrith scene is epitomised by the smoothness with which the Authority assumed responsibility for the operation of DRE. Elementary considerations of liability showed that the Authority had to take the ultimate responsibility for the reactor and indeed that safety on site could only be in the direct charge of the Director of Winfrith. It was then a short step to concluding that the correct way of delegating such responsibility was to have the DRE operated by Winfrith staff on behalf of the Dragon Project. The danger was that Winfrith might seek to exert undue influence on the operating programme or would make less effort to ensure a high work load than would Dragon people. Neither aspect in practice ever gave rise to the smallest qualms.

The Project retained the right to second people into the Operations Group, the size of the Group had to be agreed by Dragon, and individual appointments were also subject to Dragon's approval. Staff costs appeared in Dragon's overall salary bill and other operational expenditures were broken down as if they were incurred directly by the Project. To outsiders, the members of the Group were indistinguishable from Dragon staff and devotion to the HTR system became as solid as in Dragon itself.

Estimating that two years' experience of the plant would be needed before operation began, assembly of the team began in 1961/62. In terms of executive responsibility, the Operations Manager came under the Winfrith engineer in charge of reactor operations, who in turn, was responsible to the Director through the Chief Engineer. In parallel with this route of delegated authority, the Operations Manager was required to report to the Dragon Safety Working Party which examined in detail the safety aspects of the programme and reported on them to the Site Safety Committee that through its Chairman, Ray Cox, advised the Director on the conditions under which operation should be authorised.

Plant, as it was completed by the Construction Group, was handed over, not to the Dragon management, but to the Operations Group, and it was the Operations Group that went through the process of commissioning, learning the details of the plant in the process and helping to correct the faults. Dragon people were also closely involved and solutions to difficulties were worked out together. Franco set up a routine for the recording of modifications that was adhered to throughout the Project's life so that unusually, in an experiment continually under development, the plant drawings were largely up-to-date and accurate. This was of considerable importance in view of the big turnover of staff. Drawings are, however, only a part of the picture and as the Project progressed, the detailed knowledge of the reactor's components came to reside mainly in the Operations Group where continuity was preserved.

During commissioning, the Operations Manager was Frederick Barclay, energetic, meticulous, but perhaps a little brittle for the special relationship that was needed. From 1965 onwards, his place was taken by Bernard Chapman, appointed to the team originally as technologist. In this position he had come to know the plant well and, temperamentally, he was ideally suited to the job. An inveterate pipe smoker, he gave the impression of having stepped out of a publicity film for a special leaf, blended to bring out that balanced, good humoured, unflappable judgment that manufacturers of pipe tobacco love to project. A good technologist – a physicist by training and one of the early Harwell pioneers – his technical competence was valued. He saw no ambiguity in a position where, being primarily responsible to the Director of Winfrith, his task was to provide Dragon with the best service that he could, and as a result his motives were never in question. When he opposed a given innovation he was listened to with respect. Flexible in his approach to people and to problems he still kept his independence and could be firm when necessary. Particularly important, he liked the DRE and he liked the Dragon people. Complete trust reigned between him and the Project and as such he enjoyed the confidence of everyone. He was invited to all planning meetings of Dragon, including the PSC, and he had access in both Dragon and the Authority to all information on HTRs that could be relevant. When Dragon came to an end, his knowledge of prismatic HTRs must have been amongst the most extensive in the world.

Standards adopted for DRE were higher than those that pure safety considerations would have dictated. Contamination of the primary circuit at perfectly safe levels could spoil the experimental integrity, confounding measurements such as those on fission product migration. DRE stayed clean because coated particles were so good. It also stayed clean because of the mutual comprehension between the Project, the Operations and Safety staff, who were all motivated by the same desire to do things well.

It should also be said that they had a good reactor to operate – well designed, manufactured to high quality and put together to rigorous standards, for which the Dragon staff, industry and the Risley construction team can all take credit. The one 'error' that stands out, concerns the means used to control acidity in the water in the intermediate cooling circuit. It had been decided to dose the water additionally with lithium to make sure that the reactor could not get out of control should water pour in. Unfortunately, the combination of chemicals was not compatible with the steel used for the heat exchangers and corrosion occurred. A fairly mundane affair in a system with so many novelties.

AUTHORITY'S HTR PROGRAMME

Apart from the operators and the safety experts, who were to all intents and purposes Dragon people, the only staff working on the Authority's own HTR programme at Winfrith were the physicists and engineers concerned with the experimental programme of the high temperature physics reactor Zenith. This reactor, which was initiated by Shepherd at the end of 1956, was one of the first to be built at Winfrith when the site was opened. For two years after the reactor became operational, Zenith was rented to the Project by the Authority but even when it was returned, Dragon could still attach people to the research team working on it, and in the first phase all results were openly available. The reactor structure was rebuilt in 1970-71 to take full sized fuel elements for physics tests on the HTR prototypes that were being designed for Oldbury, and from then on Dragon's presence was not encouraged.

At Harwell, work that had been started, for example, on the chemistry of the system, graphite behaviour and fuel, continued throughout the life of the Project, but for some years was largely confined to supporting Dragon. When the first extension was negotiated, the Authority agreed that its own programme should be integrated into that of Dragon and it was understood that most of the work that had been going on would be paid for by the Project. While the Authority's annual report for 1960-61 stated that "the major part of the Authority effort in support of the HTGC concept has been directed towards the 20MW DRE", the report for 1962-63 uses the expression "all the Authority effort" was in support of Dragon.

In the majority of its extra-mural research contracts, Dragon received much more than a strict cost accounting would have indicated as reasonable. Authority contracts were no exception. As with Zenith, a complex and expensive loop in one of the Harwell materials testing reactors was hired to the

Project for a time and then when it became too expensive to be continued, was run by the Authority and Dragon was given full access to the results. Over and above the contract work, Harwell was engaged on work designated as fundamental and had relevance to HTR development and this too was made freely accessible to the Project.

The total value of contracts for research and development placed within the UK came to roughly £9½M of which £7½M was paid to the Authority. Major items in this sum were £400 000 for Zenith, £500 000 for irradiation work, £700 000 for post-irradiation examination and a similar sum for basic research at Harwell. Some £400 000 was spent on graphite and its machining and £1½M on uranium and thorium – substantial sums in Project terms but relatively insignificant in comparison with the Authority's annual budget that in 1968, for example, was of the order of £70M.

A break-down of the Authority's annual accounts shows no independent expenditure on gas-cooled reactors beyond the AGR until 1967-68 when a figure of £0.7M is quoted against the heading of the Mark III reactor – the designation that was used when there was competition between the Dragon system and an AGR with silicon carbide canned fuel **(36)**. From that year onwards, the figure climbed rapidly to a peak of £5.2M in 1970-71. Most of the 1967-68 expenditure would be on the silicon AGR but after this was abandoned and the HTR was chosen for the Oldbury plant in 1968-69, all the Authority's efforts in this domain were directed towards the HTR. Over the two years, 1971-72 and 1972-73, expenditure decreased to £5.0M, then to £4.0M and by 1974-75 was no longer being recorded separately. In the last two years of Dragon's existence it can be imagined that the independent HTR work of the Authority was running at about £2M per annum. (The figures quoted above do not include the Authority's contributions to Dragon which are shown separately in the accounts as grants to international organisations.)

With the adoption of the Dragon HTR system by the UK, plant design became the responsibility of industry and development was directed by Thorn at Risley. Fuel manufacturing facilities were built at Springfields (the Authority's and then British Nuclear Fuel's plant for Magnox and AGR fuels near Blackpool), test irradiations were carried out in Dragon, Peach Bottom and the UK materials testing reactors. It might have been thought that the Authority would also have set up a programme office at Winfrith, particularly as the establishment had been transferred to the Reactor Group in 1960 and so was answerable to Risley in place of Harwell. This was not the way the Authority worked however; engineering development had always been centred on Risley and moreover, the Authority would have regarded it as improper to appear to lean so closely on an international undertaking.

PERSONAL INTERACTIONS

At the senior level, there were fewer staff 'loans' from the Authority to the Project than the numbers suggest because there was little recycling. Of the total staff of 720, 472 were from the UK of whom 313 were from the Authority*, a few from the reactor consortia and there were 122 engaged as contract labour. In terms of man-years on the Project, the UK share was significantly higher than the bare figures indicate, and the Authority's fraction higher still. Once attached to Dragon, Authority people tended to stay and few of the senior technical staff returned to the Authority, particularly to work outside Winfirth. It was the long secondments of Authority staff that gave the Project its stability.

British staff employed by Dragon were available, of course, for consultations on the UK programme, and both Rennie and Shepherd took an active part in the technical comparisons of the Mark III proposals; the basic asessment report on the HTR was drawn up at Dragon. What was missing between Dragon and the British engineering teams was a direct interaction which might have created a greater identification of purpose and avoided the confrontations that characterised exchanges in the crucial years at the end of the sixties. A larger turnover of senior staff also might have led to a better understanding and to a more united stand in the terminal years.

Personal relations outside Winfrith were not always of the best. Huddle and latterly Hosegood, for example, were at loggerheads with the Authority, feeling the reverse of loyalty towards their actual employer. The antagonism between Rennie and Kronberger went to the limit of what could be expected from the promoters of two competitive systems. The reasons lay in a conflict of view-points. It is not uncommon for there to be a lack of communication between the R & D side and the engineering side, with the research people accusing the engineers of taking decisions by rote and the engineers accusing the R & D people of promoting ideas without regard to the practical implications. This had become exaggerated in the Authority in the days of Cockcroft and Hinton and the old Harwell staff had a tendency to preserve the traditional distrust of Risley methods. Overseas staff were astonished at the malignity (to quote one senior member from the Continent) reserved by some of their British colleagues in Dragon for Risley when they themselves met nothing but co-operation. In their turn, Risley people could give the impression of being scornful of these southern theoreticians inviting the conclusion that they suffered from the 'Not Invented Here' syndrome.

---

*Operations staff are not included in these figures.

This may well have conditioned certain attitudes but the reaction was probably more fundamental. Dragon staff were so convinced of the superiority of the HTR over the AGR, they could not accept the slowness with which the system was first adopted, and they regarded the later rejection as mischievous and stupid. Being British they felt personally engaged in the debate. Other countries could take their decisions in their own way without the British members of the Project feeling slighted; Britain could not.

The Authority could also be accused of being maladroit in its relations with the Project and with the other Signatories. A strong impression was left by the negotiations over the first extension that Britain was wishing to terminate the collaboration. Few could believe (even if they had been told) that the Authority had concluded, as a result of Thorn's assessment, that the fuel that Dragon seemed intent on using in DRE could not lead to a competitive system, and so the Authority was wishing to try another tack. Had they done so, it would only have served to reinforce the Authority's reputation for secrecy in any matter which might have industrial significance. On a number of occasions the Project was warned not to stray too far into commercial application but to stick to R & D (which it should be noted was what the Project wanted to do). In Britain, nuclear information is treated as a national asset to be commercially exploited. Companies are given access under licence agreements and it was at one time seriously suggested that information on Dragon in the UK should be limited to those companies with access agreements on gas-cooled technology. The practice was not adopted but added to the picture of the Authority as a self-seeking and arrogant body.

Again the Authority was seen to be insular in the collision in 1969 with the Project and other Signatories regarding the allocation of irradiation facilities in DRE. Thorn, confronted in the PSC with opposition from the Project and the Signatories, quickly earned a reputation as an aggressive critic of the international programme, while the Project was labelled in the UK as unco-operative and some of the Signatory countries as egotistic. In the last year also, the UK seemed unable to convince its partners of its real views regarding further work on the HTR nor that any further support for Dragon would be in the nature of a gesture of goodwill only.

As the only big Signatory representing a single country, the Authority was permanently in the position of appearing either to be asserting an undue influence or running away from its responsibilities. Euratom could always retreat behind its complex machinery when it came to negotiations on extensions for example, whereas the Authority was totally exposed, having until the last year the power to take rapid decisions unrelated to some fixed 'plan'. If circumstances had been reversed in 1968, Euratom could not have stepped in and assumed the load of the UK; the Dragon collaboration would have ended then. Nor can one see another Signatory in 1972, when

Oldbury had been abandoned, agreeing yet again to raise its percentage contribution (to compensate for Norway's withdrawal and Denmark's diminished payment). When the UK did finally take a firm stand on the size of its contribution and insisted on a reduction, the Project closed.

UK VIEW OF THE HTR

Even if on the surface the HTR was the logical extrapolation of the gas-graphite system, as it was first presented, it introduced two fundamental departures from current technology: the fuel cycle and helium cooling. For a country with reasonably assured sources of uranium supply, only in the very long term was the thorium cycle seen as being necessary and even then only in the event of the plutonium fuelled fast reactors running into problems. Even if, from a physicist's point of view, the thorium cycle was to be preferred for thermal systems, it still meant the setting up of a completely new processing cycle and Risley was far more conscious of the effort required than was either Harwell or Dragon. Whatever the fuel, the development of a special process for breaking down the fuel element would be necessary and that alone would require a new approach. Thorium involved a new procedure throughout – not necessarily more expensive than uranium/plutonium, as the costing prepared for the Project by the Authority's Engineering and Production Group and by CNEN were to show, but a major effort in money and man-power would be needed and this was only justifiable if an extended HTR power programme was to be adopted.

The problem with helium was its limited availability. Only in the USA were significant quantities extracted (as a by-product of natural gas production), and supplies to overseas users were far from assured, as helium was classified as a strategic material. Moreover, the main markets that would determine the price lay outside the nuclear energy field. For nearly a decade helium supplies were a source of acrimony between the Risley gas-graphite group and Rennie, Risley exaggerating the problem as helium represented a threat to the AGR system with its carbon dioxide cooling and Rennie refusing to admit the realities of the situation. In the Harwell days, helium had been seen as the coolant for the experiment without further commitment, and in the original report to OEEC it was acknowledged that the problem of its availability "throw(s) doubts on the feasibility of employing it in an extensive power programme". The Authority's concern was thus not of recent invention. Before embarking on a power programme it was clearly essential to be sure that a key material would not become unobtainable or available only at an unacceptable price. Carbon dioxide for Magnox and the AGR was available in unlimited quantities at low cost while helium was a rare and expensive element. In addition, Risley was sceptical of the Project's estimates of leakage rates, and hence make-up demands for a power reactor, whereas the Project interpreted the doubts as another artificial defence of the AGR. Both sides tended to take extreme positions but Rennie now admits that it was only the discovery of helium in North Sea oil that convinced him that the HTR really could form the basis for a national power programme.

In practice, the concentration and quality of the helium from that source has proved to be so low, that extraction would not be worthwhile. A more telling argument was that even a large HTR programme would have little impact on the world's demands for helium. Furthermore, there are, in fact, many sources that could be exploited and in the last resort, a helium/neon mixture could be extracted from the atmosphere or obtained as a by-product from liquid gas production. In the early years this was not so clear and as late as May 1966, a Dragon paper presented to the British Nuclear Energy Society's Symposium on 'High Temperature Reactors and the Dragon Project' spoke of the cost of extraction from the atmosphere as "prohibitive" **(26)**. Soon after, the conclusion was shown to be misleading as the additional unit power cost from a programme of HTRs designed for a noble gas mixture extracted from the air should be no more than a few per cent. Substituting a mixture in a reactor optimised for helium would increase costs substantially, but it should be possible to guard against the necessity for such an action.

When the CEGB came to examine the whole question, the conclusion was drawn that helium availability was not a serious problem and from about 1967 it was no longer an issue. Nevertheless, the stringent requirements for leak-tightness that the cost of helium imposes on a reactor system, was a constraint that the Authority would have preferred to dispense with. Moreover, with its own unhappy experience of corrosion by carbon dioxide, the Authority had still to be convinced that contaminated helium in a reactor is really as inert as its proponents make out, despite the experience in DRE.

To start with, the HTR was just an experiment, not to be considered on the same time-scale as the AGR or even necessarily on any time-scale at all. With the emergence of coated particles, that evaluation began to change and the Authority was quick to see the importance of the development for fuel technology in general. Further research on coated particles became largely a triangular collaboration between Harwell, Dragon and RAE Farnborough and it was the Harwell group which patented the RAE's coating process for silicon carbide that provided the essential complement to the pyrolytic carbon coating. In fact, the Harwell group's action caused much resentment in the Project and some embarrassment in the Authority which gave assurances that the Signatories would enjoy the same rights as if the patent had been their own.

The real turning point in the Authority's attitude came with the demonstration that an HTR could be built with a fuel enriched to 5% only, operating on a uranium/plutonium cycle, i.e. when the HTR became a direct challenge to other thermal reactor systems. The subsequent adoption of the low enriched HTR as Britain's Mark III thermal reactor was a unanimous choice of the three partners in nuclear energy, the Authority, industry and the CEGB, wherein the CEGB's insistence on a fully replaceable core was a major element in the decision.

The design consortia's participation in HTR development was vital but ambiguous, as was their influence on Dragon, partly because the Authority was the Signatory and partly because of the structure of nuclear energy in Britain. In the immediate post-war years, industry's role was solely that of a component supplier to the various branches of the mushrooming Authority. However, as the prospects for economic nuclear power grew brighter in the early fifties industry pressed the Government to hand over responsibility for the design and construction of nuclear power stations. First four and then five competitive consortia were formed to prepare individually their own designs for a Magnox reactor. Each had its merits and so the UK embarked on a nuclear programme that consisted entirely of prototypes, first with the Magnox system, making the same mistake again when the AGR programme was instituted.

The main customer was the Central Electricity Generating Board (created out of the Central Electricity Authority) with a duty to provide a reliable supply of electricity at minimum cost from an investment programme largely controlled by the Government of the day. It had no mandate to make experiments or to take risks and insufficient authority to decide its own nuclear system or force rationalisation on an industry, too numerous for the market, and too thinly spread to cope with a new technology, a rapid advance in unit size and a moving price target to compete with conventional fuels.

An irregular ordering sequence compounded by managerial and technical inadequacies brought the nuclear industry into a crisis at regular intervals and each time some reorganisation took place. It was not, however, until near the end of Dragon's days that the nuclear design and construction industry was reduced to one organisation, and the principle of competition that had never been allowed to develop its own rationale was formally abandoned.

When the HTR first came to be considered for central power station generation, three consortia remained **(37)**: The Nuclear Power Group (TNPG) (formed in 1960 out of the Nuclear Power Plant Company and AEI – John Thompson); the English Electric – Babcock & Wilcox – Taylor Woodrow Atomic Power Group; and Atomic Power Constructions (APC). Eighteen months of discussion on the future structure of the industry ended in July 1968 with the announcement of the Government's plans for regrouping. The outcome was that the English Electric consortium joined with the Authority (acting for the anticipated fuel company) to create a new organisation that the following year became known as British Nuclear Design & Construction Limited (BNDC). APC was to join up with TNPG. By then, however, APC was running into grievous trouble with the first AGR station (ordered in 1965 for Dungeness B) and was no longer empowered to act as a reactor company able to consider new business. Its merger with TNPG was abandoned and TNPG continued with a 20% holding by the Authority. The next upset came soon after with the divergences of view on the extent and size of the forward nuclear programme. The Vinter committee was unable to resolve the problem and recommended in 1972 a further contraction in the number of design consortia. As a result, in the following year BNDC and TNPG were shuffled into a single complex called the Nuclear Power Company, a subsidiary of the National Nuclear Corporation that was created as a holding company with GEC owning 50% of the shares, the Authority 15% and British Nuclear Associates the rest. This last company represented the interests of all industrial companies other than GEC. This was the position in 1976.

So much reorganisation with the inevitable internal manoeuvring for positions of responsibility in the latest set-up, created distractions that militated against a harmonious evolution of even basic activities; they were not designed to create a climate in which bold initiatives could prosper. Britain has not found it easy to establish a smooth industrialisation procedure. The contrast between the swift and successful programme for the construction and operation of the Windscale AGR, and the long drawn out and costly process of bringing on stream the commercial AGRs brought home the enormous gap that exists between the demonstration of an experiment like DRE developing 20MW heat, completely independent of any external constraints, and the construction of a power plant developing one hundred times the power and integrated into an electricity supply system. The parallel between the AGR and the HTR is not necessarily exact, as the basic design of the commercial designs of the AGR were so very different from that of the Windscale AGR experiment, whereas the core performance of DRE should be superior to any demanded in a commercial version. Nevertheless, in the UK, it was difficult to avoid seeing the two systems in the same light.

Although the industrial consortia embraced the HTR, they remained essentially the followers of fashion, not the makers. Competition, however artificial, obliged them to try and anticipate future policy before the opposition, whilst avoiding commitment that was not underwritten in advance. TNPG, the first company to take up the HTR, was seeking to establish a territorial gain when it put in its bid for an HTR with the tenders sent to the

CEGB in 1967, for the station at Hartlepool. This marked the start of the industrial study of the HTR in a dedicated way. By the late sixties the two remaining groups, TNPG and BNDC were each engaged on the designs of an HTR for Oldbury and they had every reason to believe that one would be chosen. Their position was nonetheless equivocal and lacked coherence. At the Nuclex exhibition in Basle in October 1969, the British presentation could be summarised as: we are building AGRs for the CEGB, we are trying to sell SGHWRs outside the UK; the system we believe to be the most economic is the HTR.

Quite apart from the disarray over the choice between the indigenous thermal reactor systems, sections of industry were promoting the abandonment of them all in favour of light water systems so that Britain could align itself with the majority of the rest of the world and participate in world-wide markets. Only by the narrowest of accounting margins, based on questionable premises, had the AGR been chosen in 1965 in preference to LWRs for the next round of nuclear power stations, and when the AGR construction programme ran into difficulties, pressure to go over to LWRs, was renewed. Leading the LWR lobby was Sir Arnold Weinstock, Chairman of GEC and under his powerful influence, the CEGB was converted, presenting to the Government as its counter to the 1973 oil crisis, a plan for a massive programme of construction of LWRs. In contrast, the South of Scotland Electricity Board, supported by the Authority, urged adoption of the SGHWR. Once again, the Government was faced with conflicting counsels and the HTR was relegated to the ranks of a non-combatant. Within the CEGB there were still senior people who believed that a place should be found for it but like other electricity organisations caught out in the past by the difficulties of coal production, the deceleration in the demand for electricity caused by economic recession, and the changing price of oil, the CEGB had insufficient reserves of credibility to sponsor whole heartedly another untried system. In its eyes, the HTR was still far from proved either technically or economically, and it would have required a great deal of courage as well as unqualified support for it to launch out into a new system, whatever the apparent initial success of the American HTR programme.

The subsequent delays and troubles at Fort St. Vrain had less effect in Britain than in Germany and France as by then, irrevocable steps had already been taken. They were a source of ammunition for the opponents of the HTR, but in effect the UK abandoned the system when it opted for the SGHWR in 1974. Once that decision was taken, it was argued that it was logical for the Authority to stop work on a competitive thermal reactor. Risley was adamant that it could not do both systems and it was necessary to deploy all the effort available on to the SGHWR. This became the keystone of Authority policy that the Chairman, Sir John Hill presented to the Government from the beginning of 1975 onwards.

It was unfortunate that Anthony Wedgwood-Benn should take over the Department of Energy from Eric Varley in June 1975, not because he instituted a fundamental change in policy but because he allowed several months to go by before defining that policy clearly to the other Signatories. With his known opposition to Britain's membership of the Communities and his doubts on the necessity for a nuclear power programme, it was natural that Benn should later be accused of sabotaging the one nuclear energy project in which Britain was associated with other members of the Communities. Effectively though, the Authority had taken its stand well before then – a stand which gained the official support of industry and the generting boards. The Government of the time was primarily concerned with establishing some sort of nuclear policy centred on a British system. Its attitude towards Dragon was determined by this, moderated only by its unwillingness to be the instrument through which the collaboration was terminated.

CONTRIBUTIONS AND CONTRACTS

In terms of finance the closure of Dragon meant only a marginal saving to the Authority. Its contribution to the Dragon budget in 1975-76 was £1¼M while revenue from the Project in terms of rent and overheads alone amounted to £330 000. This can be compared with an expenditure of £6.5M on the AGR and £49.1M on the fast reactor. Dragon's budgets over the years totalled £47.3M of which less than 18% was spent overseas in contracts for research and development and the purchase of equipment. Of the remainder, about 22% was paid out in salaries, of which about 60% went to Authority secondments and a further 20% to other UK staff. Contracts to British firms, excluding the Authority, totalled nearly £8M, say 17% of the gross.

Significant sums were paid to Winfrith for consumed items such as electricity and water (£1.14M) and it can be argued that charges such as £2.5M for plant maintenance and £650 000 for accounting services were at cost and so contained no element of profit. Rent and overheads (£3.6M) are not so lightly disposed of. Real costs and contributions are in practice very difficult to analyse but the bare fact remains that compared with a total contribution of £20.3M paid to Dragon by the Authority (plus about £470 000 paid from Government funds through Euratom) almost double that sum was spent in the country, out of which over £20M was paid to Winfrith for services and staff. It is nevertheless, worth reiterating that the Authority's contribution was not limited to its bare subscription and the Project benefited from a great deal of parallel experience, the original background effort in know-how and equipment and generous terms for such items as the hire of hot cells. Dragon also benefited significantly from the fact that it had no capital investment to make, nor on-going charges to pay for such facilities as workshops. When the Project stopped, the other Signatories just walked out.

Outside the Authority, the contribution of RAE, Farnborough was both technically vital and financially advantageous. The sum of £53 671 paid to the establishment for the development of fission product retaining fuels and graphites was a nominal price to pay for the experience put at the disposal of the Project, the personal contributions of William Watt and Robert Bickerdike and their collaborators and the open house policy adopted towards the Project.

In the supply of equipment, British companies had a marked advantage, partly for geographical reasons and partly because of their previous experience in the gas-graphite system. Considerable sums were spent in the UK, for example, on the development and purchase of graphite and its machining. In the beginning, Morgan Crucible was the only source of suitable material and Graviner (Gosport) was one of the few companies with machining facilities already available. Anglo Great Lakes also came into the picture but as the Project went on, competition from the Continent increased and the initial near-monopoly was ended.

One of the largest single experiments to be contracted out was for a high temperature helium rig in which a complete seven element unit as used in the core of DRE could be tested. The rig, which was built by C.A. Parsons and its Industrial Research Laboratories, never operated satisfactorily owing to the difficulties of heating the flowing gas by conventional means. Perhaps the most valuable information to come out of this exercise was the realisation that it was much easier to create a source of heat at high temperature by nuclear fission than by any other means. After several years' work and the expenditure of about £125 000 the experiment was abandoned.

For the construction of DRE, the direct costs of which totalled £10.5M, 13% was paid to the Authority (mainly for Risley's services) and 63% to other British concerns. When, however, site work worth £3M and conventional plant, £700 000, are deducted, the UK share of reactor plant came to a little under 40% and it could be claimed that the most interesting items such as the pressure vessel, charge machine, gas blowers, helium purification plant, and heat exchangers, went overseas. Otherwise, the most novel contribution of a British company was probably the 37-way selector valve for tapping the fuel element purges, produced by Plessey Nucleonics Ltd.

Some of the other Signatories were prone to protest against the quantity of work going to the UK. Certainly the return was proportionately higher than the UK's contribution to the budgets, but the difference was no more marked than is to be found in other international collaborations. At CERN, for example, with contributions to the annual budget in 1975 of respectively 21.5%, 3.4% and 16.4% from France, Switzerland and the UK, the return in the form of contracts was 29%, 23% and 7.5%*.

One last gesture the Authority made to Dragon was to agree to delay the start of dismantling DRE until at least two years had elapsed from the time the Agreement was terminated. DRE was emptied of fuel, the reflector taken out, the system flushed and certain parts removed for examination. Otherwise the reactor was mothballed and at the end of 1979 still sat on its knoll at Winfrith, dominating the view from the entry gate, a silent monument to the 17 years' collaboration. Mechanical parts have been maintained in working order and if the need arose, the reactor could in principle be re-started. Nothing remains of the fuel fabrication lines (except the buildings) and the staff is now dispersed. No one has approached the Authority to suggest any renewed activity.

---

*Contract distribution figures are cumulative up to the beginning of that year. A few years before, when the British contribution was over 20%, the UK cumulative share of contracts was only 5%.

CHAPTER 25

DRAGON AND THE NON-SIGNATORIES

TWO NON-SIGNATORY countries had a particular impact on Dragon's affairs, the USA and Japan. The influence of the former was fundamental to the Project's evolution and demise, that of the latter seemed on several occasions to be on the verge of being important without it ever quite materialising. Whereas the USA had at onc time, the most significant HTR power programme in the world, and from the announcement of the Fort St. Vrain contract appeared to lead the commercialisation of the HTR system, the Japanese HTR programme was subject to almost continuous delay and remained confined to research and feasibility studies.

The USA and Japan occupied a special position in relation to the Agency also. The former was an observer to the OEEC, both became full members of OECD with the USA leading and successively, in reserve order, members of the Agency. Canada, observer to OEEC and full member of OECD, has concentrated all its nuclear effort on heavy water systems and thus had no reason to be concerned with Dragon's activities.

No contacts were made by the Project with eastern European countries, the only other rcgion where there has been an active independent development of nuclear energy. As the Soviet Union has concentrated on light water cooled reactors, with or without additional graphite moderation, and fast breeder reactors, there has been little overlap of interest apart from graphite research. Neither side made any effort to establish a relationship more close than was created by the normal scientific interchanges of open publication and international meetings.

JAPAN

The first tentative contacts between the Japanese Atomic Energy Commission (JAEC) and Dragon date back to June 1961 when the Chairman approached Huet and Penney (Chairman of the Board of Management) with a proposal for a collaboration that it was hoped might be along the lines of the Dragon – USAEC agreement, notwithstanding the very different level of nuclear development in Japan as compared with that in the USA. In the economic sector, Japan was emerging as a major world power and was already negotiating its entry into the OECD 'club' of strong free economy countries, that from September of that year would include Canada and the USA as full members in addition to the European constituents of OEEC. (Japan formally became a member of OECD on 28 April, 1964.) Its status in the nuclear energy sector was not, however, on a par with its overall industrial strength and in recognition of this, Japan was not seeking to join the OECD nuclear agency, preferring to channel its main international communication in nuclear matters through the International Atomic Energy Agency in Vienna. It could not, therefore, claim any sort of equality with the Dragon Signatories and had little to offer in exchange for any information received.

Japan was, at that time, seeking technical and economic data to help in choosing between gas-cooled and light water reactor systems, and to gain experience in the first had ordered in 1958 from a UK consortium, a 165MW(e) Magnox power reactor that was under construction at Tokai Mura. However, written into the Commission's long-term research and development programme was a project for a semi-homogeneous high temperature reactor which might be said to be the embryo of an HTR programme and which could become of interest to the Dragon Signatories in the future.

Rennie saw as the prime motivation in Japan's approach, the desire to engage expert help in the design of this reactor and in the development of the fuel. He saw little advantage to the Signatories in establishing an agreement and beyond making a trip to Japan to see the Japanese programme at first hand, he did not pursue the matter further. Some years were to elapse before discussions were renewed.

Soon after sending the letter of intent to GEC – Simon Carves for the Magnox station, a contract had been placed with GE of America for the construction of a 12.5MW(e) boiling water reactor, also at Tokai Mura. By the end of 1963, this reactor was in commercial operation and American industry was launching a new generation of nuclear power plants of capacity around 600MW(e) which promised economic generation. The Magnox reactor was still far from ready. Politically and geographically much closer to the USA than to Europe, there was nothing to discourage Japan from adopting the light water system as its main line. Work on the semi-homogeneous project was stopped and in its contacts with the Agency, the talk was of participation in Halden rather than Dragon.

SECOND PHASE

Towards the end of the decade, Japan's interest in gas-cooling was re-awakened. On the one hand, the Magnox reactor was performing well and giving an availability that was at times superior to any other generating plant; on the other, the possibility of using nuclear heat for steel production was coming to the fore. At the behest of the Ministry of International Trade & Industry and the steel industry, the Japanese Atomic Energy Research Institute (JAERI) had initiated a programme of research into the problems of nuclear process heat applications. The construction of a very high temperature heat source seemed to be indicated and this would inevitably be an HTR. Consequently, Hiroshi Murata, Vice-President of JAERI, as one of the promoters of HTR research, approached the Agency at the end of June 1969 to sound out the possibility of Dragon acting as a consultant to the Japanese project. It was by then practically certain that the Dragon agreement would be extended beyond March 1970 for another three years (only the Norwegian position remained unsettled) and this was, therefore, an opportune moment for discussions to begin.

There was no follow-up until the Spring of 1970 when Williams, during a visit to Japan, broached the subject again. This led to a meeting in London at the end of April, when a Japanese delegation headed by Murata sat down with Peirson, Marien (Chairman of the Board of Management), Shepherd as Chief Executive and Boxer from the Agency to consider in detail what form of co-operation might be relevant. JAERI's programme centred on the construction of a 50MW(th) HTR, the Multi-Objective, Temperature High, Experimental Reactor (MOTHER) that would be equipped with three loops, for studies respectively on chemical processing, steam cycle power generation and direct cycle power generation. The cost of this reactor was estimated at $55M and the programme envisaged a two-year design period followed by a four-year construction period. For a team without a great deal of experience in HTR technology, this was a most ambitious time-table and it could only be realised if a significant amount of assistance was given, particularly in the design stage.

Full membership of the Dragon Project was one possibility, but it was reasoned in Europe that if Japan was to benefit from all past and future work as covered by the 1959-1973 budgets (an investment of £38M), an entrance fee of some $10M should be charged, equivalent to the total contribution that would have been made up to that date calculated from the average GNP over the period. This sum was clearly not available in JAERI's budget, even though by working closely with Dragon and, for example, adopting a proven design of fuel element for the experimental reactor, economies estimated to be in the region of $15M could be made. Funding a full back-dated membership was out of the question, so it was suggested instead, that a collaboration agreement be negotiated. A down-payment would nevertheless be demanded in return for help on the reactor, in addition to the feed-back of information, because by the time an agreement could come into force, Dragon would already be half way through its latest three-year extension, so that very little information of value would be produced within the period covered. Peirson, in particular, was anxious that no assumptions should be made yet about the future of Dragon after March 1973, by which time the Japanese project would not have progressed very far beyond the design stage. Shepherd on the other hand was strongly in favour of providing design help as this would give additional stimulus and incentive to his staff. He proposed that a Japanese team be installed at Winfrith and that the Project devote ten professional man-years plus ancillary effort in support.

When the Board of Management came to discuss the proposal, it cannot be said that the majority of the Signatories welcomed the Japanese advance. This was partly because of a general reluctance to see others gain advantage from the European investment, partly from the disinclination of some delegates to find themselves in partnership with Japan and partly from nervousness over the consequences of Japanese industry entering the HTR field. Whatever form the collaboration might take, it was agreed that a significant cash payment should be demanded as an initiation fee, in addition to payment for any work done, irrespective of whether the Japanese project was taken to its logical conclusion or not.

Immediately following the Board meeting, Shepherd and Marien went to Japan in June 1970 to examine the Japanese programme and pursue discussions. Their conclusion was that as there would not be significant funding for the Japanese HTR programme until 1972, and as current funding was only sufficient for their existing internal activities, there was neither any basis for an equitable exchange of information, nor money available for the Project to undertake contract work for JAERI. Any move must await budget examinations that were in progress for the following year, as a result of which JAERI might be in a position to make some modest suggestions.

JAERI had been hoping that the Project would be prepared to review the reactor design in advance of its submissions to the Government, and was disappointed by Dragon's lack of response. All that transpired was the purchase from the Project of lkg of natural uranium in the form of coated particles; there was little further communication between the two parties until the end of the year. In response then to an enquiry from Shepherd, Murata explained that there was still no official authorisation by the JAEC for a national project to be established and proper funding had not yet been approved by the Government, pending a report in preparation by a special HTR committee comprising representatives of the iron and steel and petrochemical industries, and JAERI.

THIRD PHASE

By the beginning of March 1971, it seemed that the necessary internal agreements for the MOTHER project would soon be completed and Murata proposed a meeting for the end of April to consider a JAERI collaboration with Dragon based upon a six-point proposal. The main topics were:

1. Procurement of materials as used in DRE such as coated particle fuels and graphite;
2. Irradiation service for domestically made fuels and other materials;
3. A review by Dragon of Japan's multi-purpose HTR;
4. Exchange of information, notably DP reports for JAERI memos;
5. Attachment of JAERI staff to Dragon;
6. Meetings for the exchange of information.

It was recognised that to get rapid agreement on all these points would be difficult but it was hoped that an agreement could be drawn up covering some at least.

General if cautious approval was given by the Board of Management at a meeting in April with hesitations still being expressed about too much detail on manufacturing know-how being disclosed. In the working document prepared by the Project, it was suggested that Japan be asked to pay as down-payment (additional to the fees charged for specific work, calculated on a 'commercial' rather than 'cost' basis) the sum of £1 million. The Board agreed that this was a sensible figure and Shepherd and the Authority (acting for the Project) were invited to go ahead and obtain the best terms they could. It was moreover agreed between the Signatories that there would be no competition from individual countries, seeking to provide consultancy services to Japan.

It was apparent that there was by then a complete mis-match between the timing of the Japanese project and that of Dragon. Designs of MOTHER were at a very early stage and the real contribution from Dragon would be building up towards the latter part of 1972 when, according to the present Dragon Agreement, the Project would be running down preparatory to termination – unless, of course, another extension was agreed. Even so, an encouraging exchange of views with the Japanese took place early in May and the Project was asked to draw up an agreement. This was essentially a consultancy agreement on the design of MOTHER, to cater for the fact that the Japanese project would not be funded in a major way until 1972, whereas it was desirable that exchanges began in the Autumn of 1971 so that comments could be fed in at the conceptual design stage. The agreement then was drawn up in two parts: a preliminary consultancy service, to be followed by the full accord and the down-payment. From a business point of view, this was not entirely satisfactory but it was imposed by the phasing of the Japanese HTR appropriations.

Details of the contract worked out in June 1971, surprisingly proved unacceptable to the juridical services of the Communities, which considered that the Board had no authority to conclude a consultancy agreement with an organisation in a non-Signatory country. The juridical service of the Agency was not in agreement but the niceties of the legal arguments were cut short by a letter from Murata received early in August stating that the Japanese HTR programme would not after all be funded in 1972 to the extent expected, and the MOTHER project would be delayed by one or two years at least. The following month it was also made clear to the Authority's representative in Tokyo, that Japan would expect to be accepted as a full member of Dragon and would only be interested if a prolongation to the Project were agreed.

The year 1971 in Japan was a year of great uncertainty, with all nuclear projects coming under review. Opposition to nuclear power was growing in intensity and all sections of the nuclear programme were suffering. Japan had in effect emerged from the exploratory stage and its problems were by then those of the nuclear developed countries; siting was particularly difficult and safety and regulatory aspects were a source of great public controversy. It was no doubt consideration of these factors which helped convince the Government that the time had come to join the Agency, and from 20 April 1972, Japan became the first non-European member.

During the discussions on membership, Williams took the opportunity of a visit to the country in December 1971, to sound out current thinking in JAERI and he explored the question of whether there were real prospects of Japan wishing to become a full member of Dragon from April 1973 should an extension be agreed. A tentative entry fee of £2M was mentioned and a contribution to the forward programme of betwen 15 and 20%. Reactions were not entirely negative but it was pointed out that until the budgets in Japan for 1972-73 had been defined, it was difficult to take the matter further. In March 1972, the news was received that funds for the design of JAERI's own experimental HTR had been deleted from the budget. The possibility of Japan participating in an HTR project overseas in the near future was evidently remote. As Dragon's existing Signatories struggled to agree the terms of the new extension, relations between the Project and JAERI withered.

FOURTH PHASE

Contact was re-established by JAERI in the Autumn of 1974 and the possibility raised of a Japanese organisation – not necessarily JAERI –

becoming associated with Dragon. The reason for this was that the Japanese programme was moving forward under two headings: a research programme by the Engineering Research Association for Nuclear Steel Making (financed equally by industry and the Government) and JAERI's own programme wholly funded by the Government. The object of these parallel programmes was to pave the way for a decision around 1980 to go ahead with the construction of a dual purpose reactor for electricity generation and process heat.

It was contended that such a development pattern was closely allied to that envisaged for Dragon in the hoped-for five-year extension beyond 1976 so that an identity of interests in the years ahead could be anticipated. Collaboration should be seen as between equal partners and Japan should not be asked to pay an entrance fee. The £½M that had most recently been suggested was in any case out of the question as a mere £70 000/a had been set aside in the forward budgets to support any overseas agreements.

Co-operation on this basis held little attraction for the Dragon Signatories, searching for a major influx of new finance. Shepherd, on the other hand, was receptive to JAERI's argument, as the existence of the Japanese programme provided an explicit inducement for Dragon to be continued, and it was in the Project's own interest to support a liberal form of association. Towards the end, he tried to justify Japanese entry on the basis of its current research, but by then, Japanese involvement had become a peripheral matter, submerged in the attritional processes of closure. In Japan itself, while research has continued into materials and component performance at high temperature, by the end of 1978, no HTR reactor project had yet been approved. Had, therefore, the life of the Project been extended, it is still probable that the pattern of irregular unsatisfactory exchanges would have continued without a conclusion being reached.

THE USA

The launching of an HTR development programme in the USA came near to preventing the Dragon collaboration getting under way; the foundering of the HTR power programme in the USA precipitated the abandoning of HTR construction programmes in Europe, with consequentially the closure of the Dragon Project. In between, there was much useful co-operation between the two continents and the programmes on the two sides of the Atlantic provided mutual support, both technical and spiritual.

To be fair, the HTR proposal received by the United States Atomic Energy Commission (AEC) in response to its invitation to industry, was far from being the only perturbation to the discussions on the new joint undertaking that were held in the Autumn of 1958; it could still have tipped the balance. Germany, in particular, was attracted by the prospect of a power demonstration reactor being built instead of a reactor experiment, and was inclined to wait to see whether participation in the US project might not be preferable to the UK sponsored scheme. The UK at this time was already becoming impatient with the delays and it was touch and go whether it would be willing to wait until the New Year for a decision on Dragon to be made. In the event, the deadline that had been set down was not imposed and Dragon was able to come into existence.

At this time, the AEC was negotiating the details of agreements, finally concluded in the Summer of 1959, with Philadelphia Electric Company (PE) and the General Dynamics Corporation. These covered the extent of its support for a 40MW(e) HTR power reactor with helium cooling, graphite moderation and uranium/thorium fuel, probably wholly ceramic but possibly metal canned, to be built at Peach Bottom, York County, Pennsylvania. In essence, the AEC, in return for full information from PE on the technical, economic and financial characteristics of the reactor and its subsequent operational behaviour, was to pay up to $14½M for the necessary research and development to the plant designers, the General Atomic Division of General Dynamics, and would waive fuel charges for the first five years' operation (up to a ceiling of $2M). PE was to own and operate the plant, the cost of which was to be shared between the members of High Temperature Reactor Development Associates Inc., composed of 53 US electricity companies.

Dragon and GA's HTR work had a common root in the early Harwell studies and in the person of Peter Fortescue who had participated in Frederic de Hoffmann's original seminar at which the potential nuclear field was analysed. He had then been detached from Harwell to GA's centre at San Diego and was never to return. De Hoffmann himself was conversant with the work at Harwell, his intellectual and research style approach, coupled to his infectious dynamism giving him access to a wide circle of friends there. It was not surprising, therefore, that after exploring a number of gas-cooled reactor designs that were based on metallic clad fuels, the GA team, following the 1958 Geneva Conference, settled into the same general concept as Dragon. For its part, the AEC had considerable respect for the work done in Europe on gas-graphite systems and in one of its own centres, the Oak Ridge National Laboratory, operated by Union Carbide, construction was starting on a helium-cooled reactor, the EGCR, that otherwise had a specification similar to the Windscale AGR. At a more advanced level in gas-graphite technology, the Harwell work on the HTR was regarded as the most significant in the world.

DRAGON – AEC COLLABORATION AGREEMENT

In the early Autumn of 1959, the AEC approached the Authority with a request for an interchange with the Project under clause 9(c) of the Dragon

Agreement which provided for "agreements for collaboration in furtherance of the joint programme". There were no statutory impediments to the USA as Associate Member of OEEC participating as a full member but, as already noted, American policy at this period was to channel its formal collaboration with Europe through Euratom as far as possible and maintain a distance from the Agency. At the first exchange of views held at the end of September, both the AEC and the Authority expressed enthusiasm for a full exchange of information and personnel between the two HTR projects. Eklund, Chairman of the Dragon Board of Management was able a month later also to report that from conversations he had had with GA, "the Americans evidently are very eager to start a collaboration as soon as possible" and "very interested in the long-term assignments of staff".

Negotiations on the information exchange, despite being an area of considerable sensitivity, proceeded quite rapidly and in March 1960, the first 'exchange of letters' took place between the AEC and the Authority acting on behalf of the Project. Rather than conclude a formal agreement with the AEC and the Project the USA much preferred this arrangement and throughout the life of Dragon the practice was maintained.

Each party agreed to grant the other a royalty-free licence with the right to grant sub-licences on any discovery arising from the information exchange, for use in the production or utilisation of atomic energy. In addition, no party was to discriminate against any citizen of the other party in the granting of licences for atomic energy work. Moreover, all discoveries made by members of either party since 1 April, 1959, would be freely available for use in the two projects, Dragon and Peach Bottom. Information was defined in the agreement as all unclassified information arising from the Dragon Project and the AEC-controlled part of the Peach Bottom project. Implementation would be through the exchange of reports and visits and the holding of conferences. Apart from this, the AEC and the Authority had already agreed early in December 1959 that any patents held by one in the other's country and which ante-dated the Dragon Agreement should be made freely available to the other for Dragon and Peach Bottom.

Much more trouble was encountered in obtaining agreement on the exchange of staff, which the Americans were still keen to conclude, and a further year was to go by before an acceptable formula was found. Both Euratom and the AEC were tied by formal rules regarding the rights to inventions that came from detachments of staff to another laboratory. So much so, the Authority handed over the negotiations to Euratom and only came back into the discussion when matters had been more or less settled. Of particular difficulty was the distinction that had to be made between 'in-field' and 'out-of-field' application. To resolve this, it was eventually agreed to limit the detachment of US staff to Winfrith (i.e. exclude the contractors) and to receive an assurance from the AEC that the provision for free use by either party for the two projects would be honoured. For patents arising from work done in conjuction with Dragon, GA was prepared to waive its own out-of-field rights to get over the point that normally it received exclusive rights for out-of-field exploitation of a patent arising from AEC sponsored work.

A spin-off from the conclusion of these rather sweeping legal accords was to encourage the Signatories to be as liberal amongst themselves with regard to patents arising from the exchange. The Dragon Agreement gave exclusive rights to each Signatory in its own territory leaving no comparability rights to the other Signatories. Each Signatory, therefore, having already agreed not to discriminate against US subjects in the area covered by the Dragon/US exchange, agreed not to discriminate against the subjects of each other's country. Comparability was also demanded in information. It had been revealed that the USA was receiving internal documents denied to the Signatories – notably the reports labelled Dragon Project Staff Only. The explanation lay in the fact that these reports were interim documents, incomplete and soon to be superseded. Notification of their existence was restricted and they were only made available on a 'need to know' basis. Information of a more permanent nature was assembled in documents of different labelling that were destined for distribution. To a number of the Signatories' delegates, this still seemed an anomalous situation and so a small number of copies of internal Dragon reports were from then on lodged with the Signatories also.

For several years, the collaboration was channelled through the Oak Ridge Operations Office of the AEC in the person of David Cope. A keen promoter of gas-cooling and the links with Europe, he was amply supported by William Larkin, also a site representative at the Oak Ridge laboratory. Even when later the formal routing was moved to Germantown (AEC headquarters outside Washington) the two men continued to act as a centre of communication and their close personal contacts with the Project ensured a warm and fruitful relationship that extended beyond the purely technical. Oak Ridge was involved in basic work on fuel, graphite and helium as well as its own project, the EGCR. This ran into engineering problems and was finally abandoned at the end of 1965; it seemed pointless to continue with Dragon operating and Peach Bottom nearing completion. From then on, all the research and development work done at Oak Ridge in the gas-graphite area was concentrated on HTR applications.

Soon after the information exchange agreement was concluded, a party headed by Larkin spent a week at Dragon and a second party arrived the following month. Dragon was a little slower in responding, a small group going out in September to discuss physics problems and compare work on Zenith and the US equivalent, followed in October by a larger party led by Shepherd. It was on this visit that Huddle learned of the work of the

Battelle Institute on the coating in a fluidised bed of spheroidal kernels of uranium carbide by pyrolytic carbon to combat hydrolysis. Seizing on this he paid a visit to Columbus where Battelle proved to be singularly open and co-operative, so that on his return to Winfrith (as already described) he was able to initiate the work that led to the production of fission product retaining fuels. An auspicious and fruitful beginning. The General Electric Hanford Laboratories, which had great experience of graphite behaviour again proved to be extremely helpful and relations between this establishment and Dragon were always agreeable and rewarding.

With GA it was more patchy, and could become formal on occasions especially if it was a large group from Dragon that was visiting. In general, however, there was a spirit of friendly rivalry between GA and Dragon which continued even after the company had ostensibly progressed beyond the R & D phase to one of commercial discipline. Between individuals, personal friendships were formed and the strict letter of the agreement was then often forgotten. It must also be recognised that the agreement was more necessary to Dragon than to GA which had other routes. GA established in mid-1960 a centre in Zürich where, initially, Fortescue undertook advanced studies far from San Diego, using European talent. As a properly constituted Swiss company this meant that GA had access to Dragon information almost from the beginning, as well as a base for its European commercial operations. In any case, once GA had negotiated agreements with HRB, HOBEG and the French CEA, it had multiple access to Dragon information. This was still not the same as direct contact and it would seem that GA continued to appreciate the value of the exchange until matters got beyond them.

EXTENSIONS TO THE COLLABORATION

Following the first burst of enthusiasm when a great deal of information passed between the organisations, Rennie became somewhat dissatisfied with the return to the Project. Peach Bottom had been delayed and whilst initially an operating date in 1963 had been quoted, it was not until February 1962 that even the construction permit was issued. As as result, there was a natural tendency for the information flow to be one way. Much of the AEC information received was in the form of published reports available in the AEC libraries, whereas information from GA had practically dried up with the completion, apparently, of GA's work for the AEC. What was particularly frustrating was the refusal of both the AEC and GA to furnish a copy of the contract between them; there was no way of knowing to what extent GA was liable to provide information on work they were doing that was equivalent to that undertaken in Dragon. Exchanges had only been reasonably equitable because of the co-operation of Oak Ridge and Hanford which had been making available information that strictly lay outside the agreement.

Looking to the future, Dragon was, in 1962 concluding an extension which would prolong the life of the project until 1967, during which time the original programme would be restored to its full value and further work would be undertaken that would include assessment studies of power reactors. Such studies were not written into the AEC's own programme. Dragon intimated that it would welcome an extension of the collaboration until 1967, with the proviso that it would seek a modification to the existing terms to bring a better balance to the exchange.

The AEC was keen to continue and Dragon, having established the Project's policy at a restricted meeting of Signatories in London in November 1962, proposed that the terms should follow the original pattern but be much more precise. Assessment studies should be explicitly excluded, leaving open the possibility of their inclusion at some future date by mutual consent. The Signatories would have preferred a more formal and comprehensive document but at a joint meeting in Germantown in January 1963, the Americans made it clear that for domestic reasons the exchange of letters with the Authority was better continued. The two sides then sat down to settle those parts of their respective programmes that would be covered by the exchange. The AEC was prepared to include the Hanford and Oak Ridge work and it also revealed that it was on the point of placing a contract with GA for an assessment and design study for a 1000MW reactor. The AEC would, therefore, like to extend the exchange to cover the whole of the Dragon programme. It was accepted that it was necessary to define GA's liabilities in regard to providing information on Peach Bottom much more closely but the hope was that the exchange should be as complete as possible, even if special arrangements had to be made in regard to some detailed items.

Against this background and encouraged by an expected improvement in the flow from GA, Rennie could have presented a rather more positive picture to the Board, when it met in March, than he did. While willingly putting into the pool the Authority and the AGIP/Indatom studies that he deemed equivalent to the AEC's sponsored assessment by GA, he recommended excluding the Project's own work on assessment. At this distance, such insularity over what was to be no more than a study may seem exaggerated, but this was the spirit of the time and Dragon could be as possessive as any other organisation.

Proposals were submitted, therefore, to the AEC that there should be an exchange of conceptual ideas on power reactors but details of studies should be omitted. In reply the Americans accepted the principle and also proposed to exclude all but generally applicable work relating to the design and construction of an advanced prototype. By then, the AEC was contemplating sponsoring R & D which would be directly related to prototype construction; Dragon, it appeared, was not. Should it do so then an exchange on this subject could be contemplated. The tables then were turned, with the US proposing to exclude important aspects of its forward programme and Rennie protested that this went right across the spirit of complete exchange that had motivated them in January. The AEC's programme, it should be noted, was much more fluid than Dragon's, being worked out annually, while Dragon's was effectively defined several years in advance. A new complication appeared also in regard to patents. There was a general reluctance to waive rights on all future HTRs let alone out-of-field rights, and one more problem: according to revisions introduced in 1961, AEC contractors had the right to file patents outside the USA for discoveries made while undertaking sponsored work. By mutual consent it was agreed that the simple extension was now out of the question and a new agreement had to be formulated.

In the end, the new terms proved to be not so very different from the original. Two main areas were excluded from the exchange:

1. Detailed techniques of fuel manufacture (Dragon had set up its own production line and viewed the processes in a commercial light, as would an AEC contractor).
2. Detailed designs and assessment studies, associated with the development and construction of prototype, and large power-producing reactors.

It was further agreed that either party could break off the exchange if a major change of programme was introduced.

Surprisingly, the patent question did not cause too many difficulties; GA agreed to waive its overseas rights on AEC contracts and apart from a few drafting amendments, the patent document was identical to the first. Letters concluding the agreement were exchanged in June 1964 as Dragon was getting ready to load fuel and, more or less at the same time, arrangements were concluded to exchange the Authority and AGIP/Indatom studies for GA's Target report on their design study. Of them all, the AGIP/Indatom study was probably the most comprehensive. The agreement as it was then concluded was extended without change as the Dragon Project itself was extended, first to the end of 1967, then to the end of 1968 and back into rhythm to March 1970.

In the meantime an important change had been made in the balance of effort between the two countries. In the face of competition from other systems, GA (as already noted) was selected in 1965 by the Public Service Company of Colorado to build a power station generating 330MW of electricity at Fort St. Vrain, Plattville (North of Denver), Colorado. Peach Bottom had still not gone critical but those troubles that had been experienced were principally with conventional parts of the plant and there was a great deal of optimism. Moreover, there was the welcome evidence of the excellent performance of DRE as back up.

Changes had also taken place in the structure of the industry as a result of the acquisition in 1967 by Gulf Oil Corporation of the General Atomic Division of General Dynamics. The new company was named Gulf General Atomic with the contraction GGA, a contraction that remained current after the expansion of the organisation to become Gulf Energy and Environmental Systems. Later, at the turn of 1973-74, the Royal Dutch-Shell Group acquired a 50% share in the company and it was then renamed General Atomic Company and referred to again as GA or GAC. Overseas, operations were handled by General Atomic International.

Discussions on extending the Dragon/AEC agreements from April 1970 went smoothly and easily. With Fort St. Vrain under construction in the USA and THTR under construction in Germany and a UK prismatic HTR in the offing, there was a mutual interest in expanding the previous terms to take into account the forward programmes of the AEC and Dragon. In addition, it was decided to modify the patent rules governing any invention or discovery made since 1 April, 1959. The new clause provided that each party would grant royalty free licences to the other for use in the Government funded nuclear research establishments within the several countries, in DRE and Peach Bottom and, in addition, in Fort St. Vrain and either THTR or the first UK HTR according to European choice. The exchange of letters took place at the end of 1970 in an atmosphere of general optimism. It must be remembered that the patents referred to did not include GGA patents taken out as a result of their own work but then neither did the Dragon portfolio include details of the German and UK designs.

When it came to discussing the next extension GGA had a great deal on its shoulders and tentatively explored the possibility of using Dragon as an additional aid, initially in the form of a joint R & D programme and then as contractor undertaking specific jobs. A second US company also approached the Project with a view to obtaining consultancy services. This was Nuclear Utility Services Corporation of Rockville, Maryland that had been retained as HTR consultant by all the electricity companies installing or proposing to install HTR power stations. Both of these independent collaboration proposals drifted on and never developed into concrete action. As a result, the only significant change made for the last agreement was the substitution of "an HTR with prismatic fuel to be built in a Signatory country" for "the first HTR to be built in the UK". (The choice between this hypothetical reactor and THTR was never made.)

EXCHANGES

The exchange of documents followed an established routine and was supplemented by a personal interchange that would drift beyond the formal limits; there was little in the R & D line of lasting interest that did not cross the Atlantic in one direction or the other. For the most part, visits to each other's centres were of short duration and little use was made of the staff exchange agreement, particularly on Dragon's side. George Lockett spent three months in 1967 with GA, helping on their gas turbine studies and on their designs for large power reactors and Luigi Massimo also spent a few months with the company. The only one who spent a lengthy period in the USA was Jacob Flamm of Euratom (Germany) who spent two years from July 1964 at Oak Ridge, out of his stay of seven years with Dragon. It was not a very satisfactory detachment as his activities were much more restricted than were those of the Americans on detachment to Dragon. Short trips were in many ways more paying, as the information return was so much more rapid and it was also more agreeable for the staff. Americans and Europeans alike enjoyed their short trips across the Atlantic personally as well as professionally.

Altogether 10 people described as US secondments spent a significant time in Dragon, the record being held by Al Goldman who came from the AEC for 2½ years from July 1964 to work on fuel. One person only was detached from GA – Martin Kantor – who came for five months from June 1965, to observe the final commissioning and run up to full power of DRE. Odd man out was Leslie Graham, seconded to Dragon first from Hawker Siddeley of the UK and, after the company closed its nuclear activities, by the Agency. When the Agency withdrew its sponsorship, he was taken on by GA. He still remains a GA staff member, leading the metals programme at Flight Refuelling.

It is idle to try and draw up a blance sheet for the exchange; let it be noted simply that a visit to the USA stimulated the coated particle development in Europe; the idea of a porous buffer layer immediately around the particles to take the initial strain emanated from GA; the multi-layered coating (TRISO) was European and was adopted for Fort St. Vrain which was able thereby to get safety clearance; the study of the corrosion of primary circuit materials by helium contaminants started in the USA and was taken further in Europe; core physics theories grew together and the US provided Dragon with a number of computer programmes that were adapted for the Project's use; graphite research at Hanford and in Europe was mutually stimulating; Oak Ridge pioneered oxide fuel kernels (as against carbide) and the Sol-Gel process and its work on spent fuel processing was the most important done in this area. Because of the time lag between the two Projects, GA could benefit a great deal from Dragon in the design and commissioning of Peach

Bottom and much of its later industrial designs leaned upon Dragon ideas. Irradiations were made of Oak Ridge fuel and GA graphite in DRE before Peach Bottom became operational, and later, Dragon irradiated fuel went to Oak Ridge for reprocessing experiments. And so it went on; no catalogue would do justice to the complete exchange nor reflect accurately the mutual aid that comes from two projects working towards similar objectives.

Fort St. Vrain was the first US reactor to be contained within a concrete pressure vessel, and its core design resembled that naively put forward by de Havillands for an HTR powered nuclear ship in the Spring of 1959, namely a stack of integral block fuel elements, where the blocks as a whole are cooled. There may well be a connection as Frank Bell left de Havillands when that company closed down its nuclear design section and went to GA. At the same time, it can be said that practically all possible configurations were considered by HTR designers and the integral block was attractive for a dispersed highly enriched fuel. With a low enriched feed it is necessary to lump the fuel more and Dragon was also concerned by the thermal gradients that are produced in the structural graphite with an integral block. This was the main reason for their preference, initially supported by the UK consortia, for a pin-in-block design in which the fuel columns are individually cooled. Eventually all prismatic power reactor designers, with the exception of JAERI, adopted the GA type, when it was proved that the graphite could withstand the gradients, and there was less emphasis on minimising the enrichment level needed. In principle, it is more economic in the consumption of graphite, although a pin-in-block could compete if the blocks were re-used.

Dragon's switch from the highly enriched uranium/thorium to the low enriched uranium/plutonium system was not followed in the USA and so, in the middle 60s, there was a divergence of line that may not have been seen as fundamental but which led to quite different solutions of fuel element forms, as already noted, and different approaches towards optimisation, etc. As time went on also, the differences in motivations exerted an influence. Dragon was an R & D organisation searching for the best solutions. GA, at least on paper, became first and foremost a power station constructor which had an R & D support programme.

US INDUSTRIAL RECORD

In 1970, GGA was running in top gear and its salesmen were out in the USA and Europe alike. The first major break-through came the next year with the selection by Philadelphia Electric Company, the contented operators of Peach Bottom, of GGA to build two 1160MW(e) HTRs on a site at Fulton near Lancaster, Pa., followed soon after by Delmarva Power and Light Company which announced plans for two 770MW(e) for its Summit site at Middletown, Delaware. Then, as GGA teamed up with HOBEG and HRB in Germany and negotiated licence and collaboration agreements with the GHTR and CEA in France, Southern California Edison Company announced plans to build a twin 770MW(e) HTR at Vidal, with options for two additional very large units, while Louisiana Power and Light opened discussions on two 1240MW(e) HTRs for St. Rosalie near Alliance, La. In the space of the three years 1970-1972, the stranglehold of the light water reactors, it seemed, had been broken and GGA had eleven HTR reactors on its books with a firm hold on the emerging European market – a programme altogether of $10-15 000 million.

Peach Bottom first went critical on 3 March, 1966. The series of tests preliminary to its generating power on a regular basis went smoothly and a licence to operate at the full power of 50MW(e) was granted on 12 January, 1967. Electricity was fed to the system the same month and full power operation achieved for the first time on 16 May. On 1 June, 1967, Philadelphia Electric accepted it into the system as a commercial unit, 5½ years from the start of construction. Peach Bottom, having completed its useful life, was finally shut down in October 1974.

Construction work on Fort St. Vrain began in 1968 and the project seemed to progress satisfactorily, even well, during the first years. In July 1970, Marien having organised a visit to HTR installations in the USA by representatives of European utilities wrote in the report he prepared in collaboration with Shepherd and de Bacci (dated 22 January, 1971): "the most impressive fact about Fort St. Vrain progress was the ability of GGA . . . to keep construction schedules". The reactor pressure vessel had been completed in February 1970 and stressed the following month. While they were there, installation of core graphite, heat exchangers and circulators commenced. The visitors were also impressed by support installations they saw at La Jolla (near San Diego) including a new fuel manufacturing facility that was being equipped to feed four reactors of 1100MW(e) each.

When Theo House prepared his paper for the Board Meeting in December 1972 on discussions the Project had held the previous month in the USA, he noted that "construction of Fort St. Vrain was complete except for some minor items, clean-up work and modification of some systems". In the area of operation tests "the programme as a whole was 70% complete and should finish in March 1973". Reference was, however, also made to "numerous leaks in seals, joints, etc.", and to "considerable water ingress". Nevertheless, fuel loading was expected to start in March and be completed in about six months – indicating a slippage on the original schedule of 16-17 months.

Criticality was finally achieved in January 1974 and if the subsequent commissioning to power had followed the Peach Bottom rhythm, the two years of lateness would have been forgotten. As it was, the driving wheels on the blowers had to be replaced, the control rods were unsatisfactory, there were oil leaks, a water leak caused over 20 tons of water to enter the reactor and then following the fire in a BWR power station at Browns Ferry in March 1975, new regulations required 300 cables to be rerouted. (Already in 1972 GGA had been instructed to segregate cabling; it had then to be done again.) It was not until December 1976 that the plant was connected to the grid and it was 1978 before it was operated commercially and then at the reduced power of 220MW. Meanwhile, during 1975, all contracts for future stations had been withdrawn by either the customers or GAC itself, chiefly because of price escalations but also in part because of the Fort St. Vrain record. First the Vidal and Rosalie projects were abandoned early in 1975, and then in September, work on Fulton was officially stopped, followed by Summit.

Not that the Fort St. Vrain record was so exceptional in the USA. To placate the anti-nuclear groups and to make up for structural deficiencies in the safety and licensing bodies, from the end of the sixties onwards, continual changes were made in the safety criteria applied to reactors. These created delays and more important, great confusion, as one system would be added to another, original plans lost, the original reasoning forgotten, producing an increasingly chaotic outcome. Light water reactors were the main target and little thought was spared to see if new rules were really relevant to the HTR. Several light water reactors – supposedly proven systems that were being replicated – begun at the same time as Fort St. Vrain, took as long to start generating power (see page 266). Some of the designs of the AGR in the UK also took as long, exacerbated by there being so many prototypes at once which fragmented the effort put into getting them right.

At Fort St. Vrain the choice of steam driven circulators was a mistake and GGA persisted in the design when more experienced turbine subcontractors had turned the job down, whereas elsewhere a great deal of chopping and changing went on. There were also many small design errors. Observers attribute much of Fort St. Vrain's problems to the fact that GA failed to adapt its structure to its new role of engineer-constructor, retaining its traditional framework of an R & D organisation with an R & D approach of change and experiment instead of a firm system philosophy based on practical experience. Fort St. Vrain bore little resemblance to Peach Bottom; the big commercial designs bore little relation to Fort St. Vrain. Fort St. Vrain became regarded as a secondary occupation compared to the 1160MW projects and attention was allowed to stray. Design changes were made to one section of the plant without reference to the others and there was no strong central coordinating body.

For Peach Bottom, PE had appointed as engineer-constructor the Bechtel Corporation, while General Atomic was made responsible for the major components of the nuclear steam supply system. Bechtel was one of the most experienced organisations in the country with a well-established competence in power station construction of both conventional and nuclear types. GA had little experience of major constructional projects and its management structure was better adapted to research and development activities and to the supply of small reactors. GA's Triga reactor is considered by many to be the most successful nuclear research tool that has been developed, and large numbers were installed all over the world under the US Atoms for Peace programme.

GA without Bechtel had to learn like so many companies before it, just how difficult it is to get the engineering of a big nuclear power station right. This was not so unexpected and it was assumed that Gulf with its wide experience of large petro-chemical projects would bring in the necessary managerial experience and would transform the organisation into a cohesive industrial orientated group. Instead, Gulf's presence made little difference and in GGA the old R & D firm carried on. Royal Dutch-Shell's contribution was purely financial.

THE IMPACT

A few light water reactors going off programme made little impact on the reputations of the reactor systems they represented. What was so important in the case of the HTR was that GAC had captured the world markets and the entire credibility of the system was dependent on the basic designs developed by the company. The dreary record of Fort St. Vrain, added to the very large increases in cost that GAC tried to negotiate for the bigger stations, made a fatal combination in a country facing rapidly rising inflation and still reeling from the crisis of confidence that was consequent upon the Watergate affair and the resignation of President Nixon in August 1974. The American market disintegrated and the European market went with it. Europe had tied itself to the GAC wagon and when this broke down, the commercialisation of the prismatic HTR came to a halt everywhere.

The United States Energy Research and Development Administration (ERDA) which had replaced the AEC was distressed at this turn of events and made some effort to promote the continuation of Dragon. Its intervention was, however, too little and too late.

EXTRACT FROM OPERATING REACTORS - GENERAL INFORMATION IN THE 1977 EDITION OF "POWER REACTORS IN MEMBER STATES" PUBLISHED BY THE IAEA, VIENNA

Code	Station	Type	Contractor	Start of Construction	First Critical	Connection to Grid
US 267	Fort St. Vrain	HTGR	GA	September 68	January 74	December 76
US 272	Salem-1	PWR	West	September 68	December 76	1977
US 280	Surry-1	PWR	West	June 68	July 72	Late 72
US 285	Fort Calhoun-1	PWR	C-E	June 68	August 73	August 73
US 286	Indian Point-3	PWR	West	November 68	April 76	April 76
US 289	Three-Mile Island	PWR	B&W	May 68	June 74	June 74
US 293	Pilgrim-1	BWR	GE	August 68	June 72	July 72
US 296	Browns Ferry-3	BWR	GE	July 68	August 76	September 76
US 313	Arkansas-1	PWR	B&W	October 68	August 74	August 74
US 317	Calvert Cliffs-1	PWR	C-E	June 68	October 74	January 75

Abbreviations:

GA — General Atomic
West — Westinghouse Electric Corp.
C-E — Combustion Engineering Co.
B & W — Babcock & Wilcox Co.
GE — General Electric

CHAPTER 26

IN SEARCH OF A PROGRAMME

BY THE end of 1972 it seemed that the HTR had really arrived. In Germany, Brown Boveri and Nukem had concluded agreements with GGA and were hard at work preparing the design and fuel for a large prismatic reactor power station for VEW. Apart from this, work continued on the construction of the THTR prototype; the HTR helium turbine association (HHT) had been set up to pursue the development of the direct cycle system and Kraftwerk Union was preparing to enter the scene with a longer term development of the pebble-bed for process heat applications. In France, the Commissariat à l'Energie Atomique was well satisfied with the agreements made with GGA for co-operation and, while concluding further contracts to cover components and fuel, was supporting a joint study made by the industrial groups GHTR and Technicatome for an 1160MW power reactor based on GGA technology in which EDF and the Swiss were interested.

Even in the UK the HTR had not been abandoned and the Secretary of State for Trade and Industry when announcing the new consolidation plans for the industry in August, had referred to the HTR's prospects as being "in the medium term, and we shall explore the possibilities of an international collaboration development"! What that meant, apart from CEGB's membership of Euro-HKG was not clear, but it was apparent that the CEGB had retained a serious interest in the HTR and it seemed not impossible that a prototype would be constructed in the UK in the not too distant future. BNDC and TNPG had begun a review of their previously separate designs and were basing their core on an integral block system similar to that of GGA.

No other European country had specific reactor design or construction plans but all were aware of the mounting pressure on conventional energy sources and supply systems, a pressure that was to become a crisis at the end of 1973 when the price of crude oil was abruptly increased by a factor of four and serious questions were raised on the continuity of its supply. Key to many of the new fuels such as gasified coal or hydrogen was a nuclear source of high grade heat that only the HTR could provide. At more and more conferences on energy, on steel making, on the future of the chemical industry, systems would be discussed that had at their centre as a "black box", and HTR heat source. Of growing influence also were the so-called environmentalist groups. While much of their hostility was directed towards nuclear energy in any form, they saw the fast reactor as the principal menace, and even within the nuclear industry, the sodium-cooled fast breeder was recognised to be a difficult technology. The HTR, on the other hand, could claim to possess unique virtues from a safety point of view, that to many made it superior to all other systems.

As already noted, four electricity companies in the USA had, during 1971 and 1972, announced their intention to build large twin reactor HTR power stations while progress at Fort St. Vrain seemed good, even very good. GGA had also with AEC backing, launched a direct cycle development programme that was talked of as a medium term activity which should see an experiment or even a prototype started in just a few years' time.

If the fortunes of the HTR were rising towards their zenith, this did not mean that those of Dragon were also. Undisputed leader in the field in 1965, Dragon's star had waned as GGA's had risen. A measure (even if extreme) of Dragon's eclipse can be gained from the fleeting references to the collaboration in the French CEA's Annual Report for 1972 compared with the enthusiastic coverage given to the work flowing from the CEA's agreement with GGA. "Acceptance" of the Dragon extension shared a sentence with an expression of pleasure (*se félicité*) over the Agency's food irradiation programme **(38)**. At the working level, the CEA still had a substantial joint programme of development with the Project, justifying the signing in December 1973 of a formal collaboration agreement on the lines of Dragon's agreement with KFA, which served to formalise and co-ordinate the existing and planned joint undertakings and made also for a more equal balance with Germany. Official attention in France, however, was more and more directed towards the development programme being carried out with, and for, GGA.

Similarly in Germany, at the official level there was a cooling of interest towards Dragon, even if at the working level, the contribution that Dragon could make towards the realisation of the HTR power programmes was clearly understood.

DRE PERFORMANCE

DRE was still a unique facility with an impressive operational record that had been accompanied by a continuous expansion of its capacity for taking experiments. Of special importance had been the introduction of block elements at the beginning of 1968 which made it possible to irradiate fuel pins such as might be used in a power reactor, but development had not stopped there.

After the chemistry of the intermediate cooling water circuit had been corrected and the reactor restarted in March 1969, 130 days of operation at full power had been logged by October. In the process, some of the thorium bearing fuels that had been loaded early in the reactor's life completed the equivalent of 500 days at full power and various fuel pins accumulated up to 150 days. Both groups showed excellent qualities on examination.

By then the further 3-year extension to March 1973 had been agreed and it was decided to advance the planned shut-down of DRE in order to carry out at once the replacement of the heat exchangers and some components of the water circuit, as well as undertake a major refit on other parts of the plant. The reactor was restarted in March 1970 and after six months of almost continuous operation at full power was shut down so that a further programme of improvements could be completed. An ingenious new device was built into the roundabout where irradiated fuel elements were stored, allowing the removal and recharge of pin assemblies only, and so re-use of the graphite blocks. These could, as a result, be subjected to very long periods of irradiation. By this time, over half the experimental channels contained fuel forms in direct support of power reactor studies – a figure that was to rise progressively to 75%.

Five months of operation from January to May 1971 with effectively 100% availability, was completed by a series of gymnastics to check the reactor's response to special conditions and for a time DRE was run at a power of 21.5MW. Tests were also made on the effects of sudden changes in power demand and the reactor was seen to follow the load smoothly and comfortably. A particularly gratifying feature of the reactor's performance was the decrease in primary circuit activity after each major change in the core structure, demonstrating in a convincing manner the progress made in fuel development and the advanced state this had reached.

Availability in the next group of runs up to October 1972 continued to be very good and then just one week after the Project learned (on 20 October) that the terms of the extension to March 1976 had been agreed, difficulties were experienced in unloading the reactor. The decision was taken to change immediately the inner reflector rather than wait any longer. When DRE was built it had been calculated that this part of the reflector – the ring of graphite blocks around the core that contained no fuel but circular channels in which the control rods moved – might have to be replaced every two years owing to the distortions that would arise. Eight years had in practice gone by before the bowing of the blocks became excessive.

During 1972 a distinct change in emphasis could be distinguished in the Project's programme. Quite apart from the growth in the number of fuel samples prepared in the Signatory countries that were being irradiated in DRE, basic work on a coated particle for low enriched fuels had advanced sufficiently for the Project to feel able to extend the range and scope of fuel research to cover a broader range of requirements. This was reflected also in the extra-mural programme of irradiations in the reactors of the Signatory countries. At the same time a greater accent was placed on thorium loaded fuels in line with developments taking place in Germany and France although there was never time to mount a serious irradiation programme of thorium/highly enriched fuel pins in DRE.

In addition, the metals programme was gathering momentum. The experimental work in Norway was highly productive and DRE itself was used as a source of information on metallic components that had been exposed to HTR conditions. The Physics Meetings, too, had become more generally appreciated and there was much useful exchange of information among the relevant laboratories.

Even GGA was not unmindful of the value of the European work and during a visit made to the USA in November 1972 by a Dragon group that included Marien and Shepherd, the question of a closer co-operation once more was broached. It was proposed that a form of 'commercial' collaboration be explored in which the Project would undertake special work for the American company.

Dragon's relations with government agencies and institutes, and with industrial concerns in the Signatory countries, were of a highly diverse nature. In some instances, Dragon would commission work to be done and pay for it, in others agree a joint activity and pay for part, or supply some of the effort or equipment; again it might supply material for work initiated by the other party in exchange for information, or it might provide irradiation facilities. Yet another possibility was a mutual exchange of information on comparable activities. All such agreements had been concluded more or less on an ad hoc basis, but as Dragon's role became more and more that of a consultant and service organisation, it was felt necessary at the end of 1972 to formulate guiding principles. These laid down that if a Signatory organisation wished the Project to undertake confidential work, then payment would be required. If a non-Signatory organisation wished to commission confidential work it should be directed towards a commercial organisation in one of the Signatory countries. Work done by the Project for non-Signatories should be in the frame of a collaborative endeavour and the information derived should interest and be available to the Signatories. In effect, the formal policy enshrined the essential elements of the practices that had been built up over the years.

In the case of Dragon and GGA, both parties recognised that the American company already received a maximum of information via the AEC agreement and through its Swiss office, so that any specific work done for the company by the Project would have to be on a contract basis. For the Board of Management such an arrangement would represent a major departure from its recently defined principles and it was not without hesitation that the Project was authorised to go ahead and try and conclude terms. GGA then began to back away. One excuse offered was that such a direct link would by-pass the AEC and be resented. However, when questioned on the subject it seemed that the AEC was not even aware of the negotiations and would certainly not have been opposed to such a link. Most probably the original ideas had been generated by a few enthusiasts who knew Dragon well and when it came to the point of formalisation, the official side at San Diego had no time to bother. Discussions dragged on without a conclusion until Spring 1975 when they were abandoned by mutual consent.

Removal and exchange of the inner reflector provided a most instructive exercise in working inside an HTR pressure vessel, proving that the radioactivity was significantly lower than could be expected with any other reactor system. Special tools had been designed for the occasion and as problems arose, new ones were prepared in remarkably quick time. Lockett was in his element and his last major contribution to the Project before his retirement in September 1973, was to see the inner reflector change through to its conclusion.

DRE started up again in June 1973 with, for the first time in its core, a three high stack of integral block fuel elements, of the type used by GGA in Fort St. Vrain (but half the cross-section), made by the CEA in collaboration with BelgoNucléaire.

VHTR PROPOSAL

Whilst recognising that Dragon's main function would be to support the steam cycle power reactors that were to be built, Shepherd was not prepared to let the Project relapse into being an irradiation facility only. He was also much in sympathy with the hopes held out in some quarters that the HTR could be developed for direct cycle electricity generation and for process heat. International interest in the latter process was given voice at a two-day meeting of experts organised by the Communities in Brussels in December 1973 and consequent upon the meeting, Shepherd prepared a report on the participation of the Project in the future development of the HTR for presentation to the GPC in February 1974.

The experts had identified two separate areas — the simply high temperature area where outlet temperatures would not exceed 950°C and the very high temperature area with outlet temperatures over 1000°C. High temperatures required only an extrapolation of current techniques and would be adequate for direct cycle generation and for certain process heat applications such as the steam gasification of lignite and steam-methane reforming. Very high temperatures would require a much longer development effort and for this it was proposed that in the initial stages, Dragon act as host to a group of guest scientists to evaluate the requirements and the technical implications, and prepare the ground for a reactor experiment. This was to be the launching pad for a new 25-year development programme with Dragon and a VHTR experiment at its centre.

If the concept was grand, the investment asked was modest in the extreme. Dragon's share was to be the £60 000 already in the budget for advanced studies and would constitute about 10-15% of the total required for stage 1. Germany alone amongst the delegations was discouraging, intimating that all Germany's effort in this field was to be concentrated in the national projects, which in any case were open to other countries. Germany's HHT programme for 1972-74 was assessed at DM150M, say £30M, and the newly constituted GHT programme for process heat was additional. Anything that Dragon did would be negligible by comparison.

In the light of comments made at the GPC and in discussions afterwards, the paper presented to the Board of Management omitted all reference to a 25-year programme and no longer proposed the setting up of a special association similar to the Gas-Cooled Fast Breeder Reactor Association. Instead, the recommendation was made that the study should be covered by the advanced applications sector of the programme and should also not be restricted to the upper temperature range. A team of ten professionals was envisaged, of whom eight would be guest scientists and hence not a charge on the budget. The programme was approved with certain minor reservations but never gathered momentum, as when it came to the point of detaching scientists to come and work at Winfrith, the response from the Signatories was minimal.

Staffing had become a major problem for the Project. With the long delay in coming to agreement on the latest extension it had been impossible to maintain a healthy influx of seconded people. Even in July 1973 the professional staff of Dragon numbered only 69 compared with an authorised complement of 81 and despite the Project's adjurations to the Signatories there was little difference between the March 1973 and 1974 figures. Total staff went up from 99 to 113 but guest scientists dropped from 13 to 3. These numbers neglect the DRE operating staff provided by the Authority but even when they are taken into account, the total professional effort for the European project was less than was being expended in France and but a fraction of that going into the HTR in Germany and the USA.

In one respect it was as well that the personnel numbers sank below the agreed level, as inflation and the steady decline of the pound had already begun to erode the Project's purchasing power in comparison with the estimates prepared for the continued 1972 programme and the subsequent extension programme. Unwilling to touch the contingency sum in the first year of the new extension, the Board had to call for savings of nearly

£100 000 in the proposed budget for 1973/74 to compensate for rising costs. With no indexing of the Project's income, coping with the two factors of inflation and a depreciating pound was to remain a major preoccupation for the rest of the Project's life.

FUEL DISPOSAL AND DECOMMISSIONING

Two other financial problems were to worry the Project and the Signatories, notably the Authority, in the closing phase. Running the reactor implied a long-term obligation, as discharged fuel had to be disposed of in a way that ensured that it could never present a hazard in the future. Either it could be stored indefinitely without further treatment, or some processing carried out to simplify the long-term storage. With a closed fuel cycle, the useful fissile material has to be extracted and re-used, the fission products and, incidentally, radioactive components, i.e. the waste, being dealt with as a separate operation. Until President Carter introduced his non-reprocessing policy it had always been assumed that as a necessary resource conservation the fuel involved in a power programme would form part of a closed cycle. Certainly, it is difficult to justify a uranium/thorium system in economic terms unless the residual fuel and uranium bred from the thorium are extracted and re-used.

For DRE as a single experiment, the quantities had been insufficient to justify any major effort being put into even fuel reduction, and discharged fuel elements were, with the exception of small pieces needed for examination, stored at Winfrith in holes constructed inside the Dragon compound. Each new core added to the problem of indefinite storage that was a common charge on the Signatories, arising as it did from DRE operation. Winfrith had only a limited capacity and no licence for indefinite storage so that some bulk reduction was indicated and the Project thought it sensible to take advantage of the necessity by treating the matter as a research and development activity.

Already in 1971, the question had been raised as to what the Project's long-term policy should be, and the Authority's proposal at the time to store the fuel in the Windscale ponds at a cost of £54 000 per charge was not considered either technically satisfactory, or economically sound. In response to the Project's call for tenders, France made tentative suggestions that the fuel be treated at La Hague, but finally the only concrete proposal came from the Authority which offered to store the fuel in coffers that it was admitted would be costly to produce. Eurochemic could not be persuaded to take the problem on.

Again, the Signatories were approached, this time with a request for offers to take possession of the fuel, hoping that the US terms for buying back unburnt uranium would be an attraction. Some interest was aroused, but no satisfactory solution emerged and the Project entered the 1973-76 extension period with only an estimate from British Nuclear Fuels that if they were willing – and there was no commitment – a total charge of around £500 000 would be made for indefinite storage. Accordingly a Working Party was set up under Peter Koss of Austria to try and resolve the impasse.

Quite separate were the charges that would arise from shutting DRE down permanently. In the original Dragon Agreement (1959-1964) the Authority assumed ownership of the plant in recompense for the additional £3.6M it contributed over and above the shared investment of £10M. This was changed when the Agreement was revised to cover an eight-year period (1959-1967) with all charges shared. It was then agreed that should the Authority decide to continue operation of DRE after the international project closed down, the other Signatories would be compensated for the written-down value of the fixed assets on the site, whereas should DRE be shut down at the same time, the Authority as legal owner would assume responsibility for these assets but would make no payment. Come 1970 and the written-down value of DRE was zero, and agreements from 1968 onwards, merely required that the Authority, if it continued to run the reactor after closure, would give preference to its former partners desiring irradiation space. By 1974 the Authority was beginning to see Dragon's fixed installations no longer as an asset, but as a liability and of no more use once the collaboration ended.

Simply shutting down a reactor and making it safe in a quiescent state costs a substantial fraction of the cost of operation (fuel and power consumption excluded). Decommissioning involves going through a first stage of unloading fuel and coolant, and disconnecting the control system, followed by a second stage of reducing the plant to a minimum size and sealing off the primary circuit. With both these stages, continuing plant surveillance is necessary. Complete removal and restoration to a "green field site" is the final act of which there has to date been little experience.

REVISED BOARD OF MANAGEMENT

The Board of Management meeting of March 1973 was the first to be held following the entry of Britain, Denmark and Ireland into the European Communities and Norway's notice of withdrawal from the Project. It also marked the replacement of Donald Fry by Harry Cartwright as Director of Winfrith. It was the first following the restructuring of the Board, which produced an important change in the balance of the Signatories' representation. Whereas in the past Euratom had been represented by a spokesman from Brussels supported by delegates from France and Germany, the new five-man delegation consisted of one representative from the Commission, three national representatives and one group representative even though formally all were representing the Commission.

Newcomer to the assembly of nations was Ireland and a nice question was posed as to the country's rights to past information. Conversely, what should be the position of Norway in the future? As a member for 14 years it seemed unreasonable to place it in the general category of outsider yet it could not continue to receive the same advantages as full members. Ireland was the smaller problem as essentially its rights were covered by the package deal it had made with the Communities. Norway was less easy and the situation was complicated by the big extra-mural research contract with CIIR, which the Project was loth to see run down as so much experience as well as equipment was invested in that part of the metals research programme.

Williams argued that Dragon was not a simple commercial partnership, a withdrawal from which would clearly imply a cessation of rights, but represented a political attachment to international co-operation and as such, wider considerations should apply. A compromise proposed by Norway, which the Board finally accepted, was that during the extension period, for a flat payment of £1 000 the Institutt for Atomenergi would receive a catalogue of Reports and Technical Notes issued by the Project and requests for documents appearing in this catalogue would be examined individually by the GPC.

In June 1974, Dragon was approaching the half-way mark of the current extension and decisions about any further prolongation had to be taken in 12 months' time. The Board of Management therefore instructed the Chief Executive to undertake the initial exploratory work and prepare outline plans for presentation at its next meeting in November. Implicit in this instruction was an assumption that a prolongation was to be desired and the really fundamental question of whether there was a consensus for a new extension was never clearly debated. Immediately the discussion became centred on what the programme should be, assuming that the Project was to go on. Comments at the time for example, related to the need to include the advanced assessment study in the new programme in view of the poor response to the drive to recruit guest scientists during the current period. No reference was made to the clear statements made by the Authority that 1973-76 was to be a terminal period (perhaps because it had been said too often), nor to the reservations that had been noted by France and the abstention of Italy in the vote at the Council of Ministers in November 1972.

Shepherd, understandably, was most anxious that the Project should continue, even with renewed vigour, and in this, whatever the reactions from the Signatory countries, he had Marien's support, a support that did not end when Marien left the Communities at the end of June. Enlarging the Communities had led to an influx of new staff and left little opportunity for indigenous Belgians to advance further. Changes in the General Directorates and the introduction of new levels had effectively diminished his status and he resigned. Over the ensuing months, he acted as a consultant to Dragon and participated in the series of meetings in the Signatory countries during which the next extension programme was elaborated. His place as senior man for the Commission on the Dragon Board of Management was taken by Caccia-Dominioni, Director in the General Directorate for Industrial and Technological Affairs. On the GPC, he was succeeded by de Bacci, who had returned from Winfrith to Brussels again, where he took over Marien's position as Head of Advanced Reactors Division.

HTR PROJECTS FALTER

Three weeks after the Board of Management meeting, on 10 July, 1974 the UK Secretary of State for Energy, Eric Varley, announced to the House of Commons that "The Government had decided that the electricity boards should adopt the Steam Generating Heavy Water Reactor for their next nuclear power station orders" **(39)**. This came as no surprise to the Project as there had been clear indications of the steady deterioration in the position of the HTR since the rather buoyant period in mid-73, when in the second report of the Parliamentary Select Committee on Science and Technology that had appeared at the end of June, reference had been made to the advocacy of the HTR by the CEGB, the good prospects forecast by BNDC provided a lead station was built in the UK, the impressive scale of ordering in the USA and the convincing expression of faith in the system evidenced by the Royal Dutch-Shell Group's input of $200M into GGA **(40)**. In spite of the Authority's contention that the HTR was further away from commercial development in the UK than the SGHWR, the firm recommendation had been made that "the major UK research and development effort be channelled into the HTR".

How different was the report from the same Committee early in 1974 when, following the oil crisis, it analysed the CEGB's proposal to launch a programme of nine twin-reactor stations, each of 2400-2600MW(e) capacity and implicitly based on PWRs **(41)**. By then, both the CEGB and NNC (the new single combined consortium) were agreed that the PWR should form the mainstay of Britain's reactor programme in the medium term "though the CEGB do see some role for the . . . HTR". "In our view, common sense would indicate that until the HTR and Fast Breeder Reactor were available on a commercial basis, the way forward should be to use one of the British nuclear technologies which is already proven. . . . we note the enthusiasm of the South of Scotland Electricity Board for the SGHWR."

The Nuclear Power Advisory Board which had been asked by the Government to advise on the choice of thermal reactor systems went further, narrowing the choice to either a PWR or SGHWR and expressing doubt that a parallel major effort on HTRs would be possible, because it would be necessary, as an insurance policy, to maintain an effort on the other water system, whichever of the two was chosen as the main stream.

There was clearly a latent wish to remain in the HTR field and when NNC presented to Euro-HKG at Schmehausen on 2 May, 1974, its design for a low enriched 1320MW(e) HTR power station, the preamble included the statement that whilst further plans in the UK were not yet finalised, it was hoped that when the Government decision was taken, a large HTR would be included in the programme. "On present plans the design . . . will be offered to the CEGB this year with site work commencing in 1976." Even when the Government did take its decision in favour of the SGHWR, the HTR was not entirely forgotten and the announcement to Parliament included the statement that the Government "is asking nuclear organisations to pursue the prospects of international co-operation in developing the High Temperature Reactor, which has considerable potential".

At first, the last statement was not considered to be an empty phrase and Stewart, technical advisor of NNC (and former member of the Dragon Board) suggested to Shepherd that he put together proposals for a Super Dragon project that could be a European collaboration for the construction of a prototype that did not depend on a GAC licence. Dutifully this was done but there was no machinery that could respond to such a proposal and there was little support for a UK inspired design that the British Government was not prepared to adopt. Nevertheless in the Summer of 1974, even if the UK decision on the SGHWR represented a set-back, there was no reason to think that the country would be wanting to cease its international work on the HTR.

There were other than chauvinistic reasons for wanting a design that was not based on the GAC system (apart from the continuing difference in approach to the fuel cycle). The news from America was not good. Hosegood and others from Dragon had been dismayed by what they saw at Fort St. Vrain in October 1973, and in their report subsequent to their visit, they made clear their concern at the way the project was being managed. Royal Dutch-Shell's injection of capital seemed to make no difference and although criticality was achieved in January 1974, and from a physics standpoint the reactor behaved well, engineering problems were serious. As the months went by, and no attempt could be made to go to power, confidence in several important aspects of GAC's designs deteriorated steadily. At the same time, doubts were being expressed on the soundness of the 1160MW designs, and GAC's image was becoming tarnished.

PRELIMINARY ENQUIRIES ON AN EXTENSION PROGRAMME

Shepherd's primary concern though was with Dragon and its future and he went to Brussels to explore the form a new programme might take. At this first preliminary meeting the problems of staff recruitment and the damaging effects of long drawn out negotiations on extension agreements were evoked once more and from the discussion emerged the plan to try this time for a five-year extension rather than three. This would add stability and confidence on the personnel side and should at the same time allow programmes to be planned without a permanent uncertainty regarding the time and budgets available.

Armed with this proposal he began his tour of the Signatory countries in September 1974, beginning with Scandinavia, where he found little response to his theme that the hopes for HTR development in Europe were now pinned on a collaboration between the Signatories to build a steam cycle prototype centred on Dragon. Denmark was discouraged by the UK decision on the SGHWR but, contributing by then only through the Communities, was prepared to go along with a majority view. AB Atomenergi expressed a marked preference for the accent to be placed on advanced applications, as it had little interest in any immediate new power reactor plans, being almost wholly occupied with the problems posed in Sweden by the opponents to nuclear power of any sort. Sweden had no irradiations to make in DRE, being more interested in Dragon's irradiations at Studsvik, and was only concerned in more long-term evaluations.

Before visiting Brussels again, Shepherd was informed by Allen that the Authority would not, in spite of warnings issued in 1972, dispute the percentage contributions, but would agree to the same partition, modified only to take into account the increased contribution of the UK to the Communities' budget. This declaration, that was tacitly confirmed a month later was accepted as the definitive position, and subsequent delays and statements indicating the contrary were at first interpreted as bargaining gambits only.

Discussions with Euratom at this time centred largely on technical questions related to the problem of allowing for inflation, speculation on the probable attitude of Signatory countries and the possible participation of other countries, notably Japan. It was believed that the members of the Community would eventually accept a broad compromise programme despite their sharp differences in attitude.

Clarification of the French positition was given at a meeting in October when it was stated that the national programme was now firmly geared to a succession of light water reactors followed by fast reactors. The HTR might have a long term future, hence the agreements with GAC, but in view of the deficiencies perceived in the GAC designs, it was believed that HTR development should go on at a reduced pace. EDF was not interested in the direct cycle but was not opposed to a European syndicate building a steam cycle. French reactors, it was explicitly said, had proved superior to DRE for their test purposes and collaboration with Dragon on irradiations in Pégase would be ceasing with the shut-down of this reactor. In other words, enthusiasm for an extension was practically zero.

The Italian attitude was less easy to analyse. On the one hand a delegation from Brussels and from the Project was told by CNEN that Dragon was well regarded and should continue until HTRs came into their own, supporting steam cycles as a first necessity, but for choice concentrating on advanced systems and in particular process heat applications. On the other, they heard that it was seen as a political problem in which Italy's attitude would be determined by the UK attitude towards the Communities, and the share out of scientific funds amongst the Community states. The opinion was held very strongly that any project must be whole-heartedly backed by its host country to remain valid.

In Germany, doubts about GAC's 1160MW design were strengthening the awareness of the Government and KFA of the value of Dragon research and they were keen to see explicit reference to future support for the German programme, more than just the provision of fuel irradiation facilities. Particular emphasis was laid, for example, on the need for a thorough examination of the experimental duct liner that had been installed in DRE and for an intensification of experiments relating to the behaviour of materials and components. There was, however, strong opposition to Dragon getting involved in any advanced studies that would cross the lines of direct cycle or process heat research in Germany. Dragon's role was to provide data for the current power projects.

At a meeting between the Communities, the Authority and the Project held at the end of October 1974, with Marien in attendance, Allen approved the principle of a 5-year extension making the reservation that it might be difficult to obtain the agreement of the British Treasury. The Authority view of the programme was that the research and development aspects were more important and should take precedence, even if continued operation of DRE was still necessary. For the first time the issue of terminal costs was also raised, as the Authority felt that the Project was entering a new phase and these costs should be shared. Work on shutting down the reactor permanently should be included in the programme because of the useful information to be gained. Finally, coming to the question of percentage contributions, Caccia-Dominioni remarked that any dispute about these would put a new extension in jeopardy; it seems that this was not contested.

Belgium and the Netherlands, without making commitments or conditions, indicated support for an extension; Austria, being more interested in HTR technology than HTRs as such, expressed a preference for general studies, with emphasis on process heat; Switzerland, while wary of the prospects for steam-cycle HTRs, was strongly positive and had already prepared proposals for a heat exchanger experiment to determine the behaviour of a particular alloy in the system. It was firmly in favour of a balanced operational and R & D programme.

Against this conflicting range of advice, Shepherd drew up a first document comprising, as principal elements: a five-year programme with a concentration on 'generic' studies, and streamlining the DRE irradiation programme to support these; maximum use of the DRE primary circuit as a test bed for HTR materials; a basic materials development programme for steam cycle, direct cycle and process heat HTRs; a generic HTR safety programme, and an assessment programme to identify design features and development requirements for direct cycle and process heat applications. This was not a consensus programme so much as the best mixture that could be devised. A great deal of effort had been put into its preparation and the Board of Management at its November meeting, recognised the care with which the diverse interests had been accommodated. What still remained was to sketch in an annual budget figure that resembled the existing expenditure rate in real terms, and from there on it was a question of detailing the effort to be spent on each aspect of the programme and dealing with technical problems such as provision for inflation.

WORKING PARTY'S REPORT

According to tradition, a Working Party was set up that consisted on this occasion of Koss, Chairman, de Bacci from Euratom and Thorn from the Authority. They were advised by the Project and by representatives from France, Germany, Italy, the Netherlands and the Agency.

When the Working Party met for the first time, the sharpness of the differences was made apparent. France and Germany expressed interest in DRE only, and were not prepared to entertain a British proposal that the overall budget pressure be eased by introducing sponsorship of experiments in the reactor. At the same time, Germany in particular was resistant to the budgetary implications of that part of the programme that would satisfy the countries primarily interested in advanced aspects, despite the smallness of the sums involved.

Understandably, the result of the Working Party's deliberations was not a consensus either. Conditioned by the deteriorating power reactor situation, it was an uneasy compromise that followed closely Shepherd's original document and put the accent on generic studies. The programme that was set out while still continuing the current irradiation pattern envisaged a steadily increasing emphasis on the development of materials and technology for more advanced applications, a change that would affect all areas of the Project's work. An essential aspect was to be the allocation of a small team to a phase by phase programme of advanced HTR work, to identify and quantify crucial materials requirements and other problems requiring further intensive development. It was considered essential also that advanced fuel cycles which conserved uranium resources be studied.

It was not the task of the Working Party to consider whether the programme should indeed be carried out, rather to determine a programme should an extension be agreed. Ready acceptance was given to the five year principle and the overall cost was assessed at £18.7M, at the prices obtaining in January 1975. Inflation it was proposed should be calculated at 15% per annum for staff and 10% per annum for other costs which would inflate the overall budget to £27.4M. After three years, the rate experienced in practice would be reviewed and, if necessary, the Signatories would be asked to make good any forecast deficiency in the final total budget. From the inflation provisions it was suggested that 3% and 2% be set aside for contingency. The single controversial note that was sounded was a warning that the Authority would be raising the issue of cost policy on decommissioning DRE.

At the meeting of the Board of Management in March 1975, when the Working Party's report was examined, it was generally agreed that in a new terminal extension budget, it was reasonable to increase by £450 000 the sum that had been allocated for making the reactor safe, in order to cover decommissioning. To be discussed was whether this sum should be found from within the budget proposed or whether it should be added to the existing total. Italy contended that the programme did not provide a sufficient return to Italian industry for the investment the country would have to make, and proposed cutting extra-mural irradiation and post-irradiation examination work to the benefit of metals and fuel work – eliciting a predictable reaction from Sweden. France reverted to a previous stand taken during the Working Party sessions that the budget should be cut by about 10%. Otherwise the minutes of the meeting state that "a large majority of Board members (France excepted) was in principle unreservedly in favour of an extension . . . and of five years".

CHAPTER 27

THE LONG ROAD TO CLOSURE

AT THIS same meeting of the Board of Management in March 1975 a budget should have been approved for the next 12 months. However, when the notional terminal budget drawn up by the Project came to be discussed, Frank Chadwick, representing the Authority (the legal guardian of Dragon), felt obliged to point out that a large discrepancy existed between the sum of £150 000 set aside for fuel disposal and the only valid offer the Project had received (from the Authority) which came to £585 000 as of January 1975. In the Authority's view it was essential that the budget made provision for all the Project's liabilities and took proper account of the diminished resources available to the Project resulting from inflation. Anticipating the serious risk of insolvency that Dragon was facing, the Authority had obtained the sanction of the UK Government for the current Agreement to be extended immediately for one year on the existing financial basis. Agreement by the other Signatories on these lines would provide an interim solution and allow negotiations on the longer extension to proceed against a stable financial background.

Coming so quickly after the general consensus that Dragon was to be prolonged for another five years, Chadwick's proposal was dismissed out of hand as being liable to put at risk the whole programme for the sake of academic accounting niceties. To get such a proposal through the communities' Council of Ministers would require as much effort as the 5-year programme and so was not to be entertained. Indeed, so little attention was paid to the idea, it was not even mentioned in the minutes of the meeting.

No alternative plan for balancing the books for 1975-76 could be advanced by Koss, speaking on behalf of the Fuel Disposal Working Party. Nearly £1M had been earmarked in the 1976-81 programme for fuel disposal and new technical solutions were in sight as CNEN might take some of the fuel if research undertaken by the CEA proved successful. If, however, there was to be no extension to the Dragon Agreement, it would be necessary to accept the Authority's offer for bulk reduction followed by permanent storage at Dounreay. To satisfy the accountants, the Board concluded that the cost difference to which Chadwick had drawn attention should be offset against the value of realisable assets, in particular, unused uranium. No figures, however, were available, and the decision on the budget had to be deferred to a special meeting in April. As a result, Dragon entered its final year with neither budget nor programme approved.

Initially a book value of £145 000 was ascribed to the unused uranium-235 which, when added to the value of known receipts showed credit to the Project of £339 000. Cuts made in the programme amounted to £21 000 so that with the £150 000 already set aside, a total of £510 000 could be allocated to fuel disposal. Allen was urged by Caccia-Dominioni to accept this as an accounting figure, near enough to the latest real cost estimate which, when updated to the end of DRE operation had been re-assessed by the Authority as amounting to £625 000.

Allen in his reply again made reference to the UK's willingness to extend the programme "on the same financial basis as the current programme", reiterating that as "all public expenditure is currently under review by the Government in preparation for the forthcoming [UK] budget [it] meant that approval could not be given for a commitment beyond one further year, i.e. to March 1977".

He went on to propose that de Bacci (Chairman of the GPC) and Shepherd should re-examine with the Authority the valuation of future Project income plus realisable assets. During the resulting exchange, it was concluded that the value of the Project's unused fuel (that had always been prudently underestimated) could be raised to £285 000 and the GPC could then pass on to the Board a programme that was correctly funded, and which the Board could approve.

At the special meeting in April it was hoped to be able to finalise details of the programme for the new extension so that formal decisions could be taken in June in accordance with the planned procedure. It was not an easy meeting and arguments over the minutes were only resolved by the new Chairman, Pictet from Switzerland, going to London to help put together the comments. One of the main difficulties was the handling of decommissioning costs. The apparently open-ended commitment that the Authority had put forward could not be approved as such even though the principle of including them in the programme was accepted (and confirmed by the Commission of the Communities in May). It was agreed that a sum of £600 000, to be adjusted for inflation as the Project proceeded, be set aside, of which half would be found from cuts in the programme and half by an addition to the draft budget bringing the total sum for the five years to £19M.

The Board had thus an agreed programme and an agreed budget, the one dissenting voice being that of Italy which gave notice that it reserved its support for a further extension based on the propositions formulated. Reacting to the statements made by the Italian delegate at the Board meeting in March that the extension programme as currently conceived would not produce an adequate return by way of Dragon contracts, de Bacci, Shepherd and Graham had journeyed to Rome to discuss, in particular, an enlargement of the contracts with Fiat on the materials testing side and with AGIP on the development of a liquid process for producing fuel kernels. However, while the companies had been keen to see the extension and their contracts go forward, Albonetti had reaffirmed his opposition to the programme as it was presently being drafted. It was evident that the Italian position had not changed.

DELAYS IN UK DECISION

Not too much dismay was provoked by the warning from Chadwick when the Board approved the 1975-76 termination budget, that cash flow problems might become serious, as there was likely to be at least three months delay before the UK could make its decision on a further extension, and it looked as if the June deadline would not be met. Once again the UK was in the throes of an economic review and a nominal question mark hung over its future relations with the Continent pending the holding in June of a referendum on Britain's membership of the Communities. The Government had stated that it would feel no obligation to act on the result but presumably if there was a massive vote to withdraw, policies towards European projects could be affected. In any case the Signatories had no reason to expect any fundamental change in the Authority's position regarding Dragon in view of its previous statements and the vigour with which the decommissioning question had been argued. Moreover, as the Project's administrative angel, the Authority could itself sort out any temporary shortage of funds.

Shepherd had more reason to be worried but preferred to be optimistic. He had been alerted at the beginning of the year to the hardening of the Authority's attitude towards a continuation of work on the HTR. When giving evidence to the Committee of Public Accounts on 22 January (the report of which was published on 12 May), Hill had been categoric that the Authority was obliged to concentrate all its research and development effort on the systems to which it was committed **(42)**. On the other hand, private communications from industry and the CEGB indicated that the HTR might yet be seen to have advantages, notably on safety, and it was still necessary to press on with the direct cycle and process heat.

When the Board met on 27 June, the UK referendum had been held (5 June) giving a two thirds majority to the supporters of the status quo, and Benn had replaced Varley as Secretary of State at the UK Department of Energy. The Authority, however, was not in a position to make any statement and neither was Euratom. In Brussels, when the subject of

Dragon had been raised in the Group for Atomic Questions* following the submission of the Commission's proposals to Council on 27 May and their subsequent transmission to the Group, the subject had effectively been frozen by the UK's request for a delay of several months.

Even in the Board, support for the approved programme was far from whole-hearted. The French delegate informed the members that his Government had been advised by the CEA that the Project was not considered to be essential or the most appropriate means for developing the HTR. Reservations had also been expressed on the decommissioning proposals as they were currently worded. Nevertheless, France would not stand out against the other Signatories if they wished to go on. Italy, too, was cautious, being unwilling to make any commitment pending the British decision, although it was generally assumed that if the majority was in favour of a prolongation, Italy would not in the end oppose. In the circumstances the only useful progress the Board could make was in the wording of the draft agreement, modifications to which were to occupy Otto von Busekist of the Agency for the rest of the year.

Shepherd voiced his disquiet at the effect the delay would have on the remainder of the current programme and particularly on the staff position. Already inflation had taken its toll on the reserves and the longer the programme was maintained at the present level, the more drastic would the later adjustments have to be. If the Project was to be stopped, then an orderly closure would require immediate action – action that would seriously prejudice continuation if that was to be the decision. Economies already put in hand could cover the situation until the end of July, following which, if no additional funding could be guaranteed, desecondment notices would have to be sent to the staff. Should the programme be maintained until the end of September, a shortfall of £365 000 would arise.

Recognising the position, the Authority offered to pay its share of the extra cost to maintain the Project's contracts and staff strength up to 31 March, 1976, but Euratom in particular saw great difficulties in making a similar promise as again, the full decision-taking machinery would have to be invoked. The only way to avoid issuing notices on 1 August was to find

*The Group for Atomic Questions is a second level political committee which prepares the dossiers on atomic energy affairs for the Committee of Permanent Representatives (COREPER). COREPER is made up of delegates of ambassadorial status from each member country of the Communities. All significant decisions in the Communities are taken by the Council of Ministers in which each delegate has the power of veto. The Commission prepares proposals and first passes them to the Council which then normally passes them on to the appropriate advisory group. If COREPER can then unanimously recommend action, subsequent acceptance by the Council is virtually automatic. If not, then the dossier can still be sent to Council for a decision but is liable to come back, particularly if the subject is not of great moment.

additional resources from within the Project itself and to this end Shepherd was asked to obtain market guarantees for the unirradiated fuel stocks.

On the eve of the meeting, a presentation had been made by Pictet, on behalf of the Board and GPC, to Huddle in recognition of his tremendous contribution to the Project. Huddle had seen the sands running out early the previous year, and had decided to take early retirement. Shepherd had finally let him go at the end of December.

Allen was again not present at the meeting. Having been a regular participant since taking over from Peirson in 1971, his first appearance in 1975 was in October. From this it could be inferred that from March onwards, the policy of the Authority was no longer its own and the UK position was being determined by the Department of Energy which was primarily concerned with the Government's wider negotiations with its partners in the Communities in the face of a tightening of the money supply at home.

Benn inherited in the Department a situation in regard to the country's forward nuclear policy that was as confused as it had ever been. Doubts were being expressed over the implementation of the SGHWR decision and Benn was in need of advice on what this might mean for the HTR. This advice was formalised at a top level meeting held in July 1975 at which were present the heads of the Authority, industry and the generating boards. It would seem that at this level, Dragon had no supporters willing to defend its case. Even those who believed that the HTR might have a long-term future accepted the view that Dragon might no longer be the appropriate vehicle. The Authority was adamant that it could not pursue both the SGHWR and the HTR and it was fully committed to the former. In view of the subsequent about-turn on the SGHWR, there was perhaps already a sense of desperation in the single-mindedness with which it was promoted. The CEGB, despite the views of senior technical staff that replaceable cores with coated particle fuel and helium cooling should not be forgotten, could not contemplate an unproven system and was not at all interested in very high temperature research or the thorium cycle. Industry when not fully occupied with its present commitments was only anxious to enter the light water reactor field. Only the steel industry was prepared to express enthusiasm for higher temperature research as indicated by the evidence that had been furnished in March to the Sub-Committee on Energy Resources of the UK Science and Technology Committee, and to the Department later by Robert Barnes, Director for R & D of the British Steel Corporation and active member of the Steering Committee of the European Nuclear Steel Makers Club. However, the market for HTRs in the steel industry could not justify a national programme and there would be little HTR activity before at least one power station had been built. Otherwise the Authority had concluded after a 5-year study on process heat applications that there were no other markets within sight.

The unanimous conclusion was that the HTR should be dropped and Britain's part in any future Dragon extension limited to its contribution through the Communities. Essentially it was not the cost as such that counted although attempts were made later to justify the conclusion on these grounds; it was a fundamental policy recommendation to abandon the HTR.

ALTERNATIVE EXTENSION PROPOSALS

At virtually the same moment, the Dragon Board of Management was meeting in special session in Paris to go over again the terms of the draft agreement and consider also an alternative proposal for a three-year extension that had been put forward by Euratom's Scientific and Technical Committe on 2 July. Asked to comment on the Commission's recommendations of a five-year extension, this Committee, because of the uncertainties surrounding the development of HTRs in the next few years (and in spite of a persuasive presentation of the potentialities of the system set out in a paper prepared by Shepherd) had advised that the Project be extended for a three-year term. Even de Bacci, as the Commission's representative, wished this proposal to be regarded as a fall-back solution, only to be entertained in the event of budgetary pressures. Nevertheless, Shepherd was asked to re-examine the hasty analysis he had prepared in order to have in reserve a set of proposals which was clearly beyond criticism. Austria and Switzerland had already obtained approval for a five-year extension and could, if required, give their agreement to a three-year extension provided an adequate research and development programme was maintained.

This was now the end of July and Albonetti, perhaps more aware of the complexity of procedures in the Communties than other country representatives, forecast that it would be impossible to get any decision before the end of the year. He gave warning that Italy did not favour an extension with the programme foreseen for either the five-year or three-year period and expressed astonishment that the Project could be in financial difficulties so long before the end. But it had to be recognised that with August a 'dead' month, a decision by the middle of September was out of the question and he therefore proposed that a survival budget be prepared to take the Project through to the end of 1976 should there be no other solution. The Board while considering his views to be unduly pessimistic instructed Shepherd to draft such a plan to be kept in reserve as a last resort.

Maintaining the viability of the Project was already bordering on the impossible. Shepherd had been modifying the programme as much as he could, preparing to stop all external contracts (with the exception of those at Fiat and CIIR which he proposed continuing until 30 September), in order to delay issuing notices to the staff and he had negotiated with KFA the sale of enriched uranium for a sum equivalent to over £500 000. This was not the first time that he had had to cope with two incompatible situations and he was encouraged in his fight for survival by a message from the Director of Advanced Nuclear Energy Systems of ERDA urging him to keep the Project going as long as he could. Everything depended on the UK, and the Authority, in a bid to keep the programme alive, renewed its offer of joining the other Signatories in providing extra funds proportional to the 1973/76 contributions. This was impracticable on the time-scale now facing the Project and the Board could not do otherwise than authorise Shepherd to proceed with the run-down plan, entailing issue of the first three months desecondment notices on 12 September.

The situation in the United States was steadily deteriorating. No sooner had one problem been cleared up at Fort St. Vrain than another appeared. Cost estimates for the big stations were escalating and electricity companies were refusing to renegotiate the terms of their HTR options with GAC. Although ERDA was trying to find ways of maintaining an HTR presence, the imposing US HTR programme of the previous year had almost completely disintegrated. In Europe, GAC's associates had backed away. VEW had still formally to renounce its plans for the Schmehausen prismatic HTR and was still intimating that it might go ahead within an international consortium, but Euro-HKG was not constituted to initiate a rapid rescue operation of a system that was being rejected in the countries of its origin.

UK REQUIRES REDUCTION IN CONTRIBUTION

The plight of the Project was fully understood by Pictet who assumed a quite personal commitment to find some way through the fog that had descended on it as a result of the UK silence. Already in June he had been in touch with Christopher Herzig, Under-Secretary, Atomic Energy Division at the Department of Energy and UK delegate to the OECD Steering Committee for Nuclear Energy, to impress on him the gravity of the situation that was developing. Accompanied by Shepherd and Williams on 18 August he again went to Herzig to argue the case for an extension of the Project. Underlining the urgency, were the letters to be sent out eight days later to all the Signatories advising them that the salaries of the staff could not be paid beyond 6 December with the exception of the minimum number necessary to close the doors.

Herzig explained the UK position in terms of the difficult economic situation in which the country presently found itself and for the first time the statement was unequivocally made that the UK would be seeking a reduction in its financial contribution. Herzig said that the UK nevertheless recognised the special position of the host country and was able fully to agree that Dragon must not die out of indecision. Ironically, just a few hours before, the Communities (the UK included) had agreed to collaborate on a hydrogen study programme to supplement existing Ispra research – a programme that included a high temperature materials programme and would no doubt require, if only implicitly, a high temperature heat source.

On 12 September, Pictet visited Brussels and during discussions with Euratom came to the conclusion that there was no longer any possibility of having the 5-year programme agreed in the weeks remaining. It was necessary to provide for an interim period to give time in which a different financial structure could be elaborated. Twelve months was considered to be necessary and it was proposed that the UK be asked to continue its financial contribution at the existing rate of 48.3% for that length of time.

Benn, who had remained silent on energy affairs since his European tour, made soon after taking over the Ministry, came to Brussels on 19 September, aware that at least the German Government and KFA were in favour of the Project continuing for the sake of THTR, even if German industry was apathetic. In discussions with Commissioner Spinelli, the history of the recent Dragon negotiations was reviewed and it was fully recognised that if no decision was taken before the end of October, the Project would be finished. Benn on his side made it clear that in view of Britain's diminished interest in HTRs, the UK contribution was too high. Spinelli countered by urging the Minister to agree to the 3- or 5-year technical programme, and authorise the existing financial arrangements for one year more, to give time to examine the problem and determine what contributions might be forthcoming from the Japanese and also the Americans. Another possibility was a special contribution from Germany; help was at hand if only there were time. Benn remained irresolute. The matter was set down for the 30 September meeting of the Group for Atomic Questions and much discussion went on in the corridors of Brussels and also Vienna.

NO INTERVENTION FROM NEA

The Nuclear Energy Agency throughout the negotiations maintained a low profile, partly because Pictet believed that no help could be obtained from the Steering Committee – whatever action was taken would be in Brussels – and partly because Williams was not convinced that the essential problem was one of procedure. Huet in the sixties saw the bricks and mortar of joint undertakings as the manifestation of international collaboration and the proof of the Agency's vitality and relevance. Williams in the seventies saw the bricks and mortar as incidental to the main task of the Agency, which was its inter-governmental role of providing the means for collaboration between the Governments of Europe and those of countries outside Europe with a developed nuclear capacity. The HTR was seen to be (if anything) an industrial activity. Dragon had no industrial backing and was becoming wholly concerned with R & D that had no obvious purpose.

Halden was different, as its research programme was determined by the participants on the basis of their own requirements which were in turn generated by real power programmes. Halden provided a service. Dragon had never been content with a passive service role, wanting instead to lead and promote, and Williams believed that the programme that had been worked out for the new extension was largely Project inspired. It did not satisfy a genuine need on the part of the majority of the Signatories, and was therefore unlikely to be an instrument for promoting collaboration between countries; it could even be a source of dissention.

Had there been a majority wish for a continuation, the Agency would have been happy to go on as Dragon's sponsor in the same way that it was indefatigable in serving the Signatories as they swung from one emergency solution to another. In the circumstances however, Williams saw no value in trying to establish another route for negotiation – the machinery was there in the Board of Management. A parallel could not properly be drawn with the 1967 impasse that was resolved through the Steering Committee calling a top level meeting. Then, problems arose because of a structural situation, with France not attending Euratom meetings, putting the budgets of a major partner, incidentally, into limbo; this time the other of the major partners was wishing to reduce its participation and a number of the others were showing signs that closure was to be preferred.

NEGOTIATIONS IN BRUSSELS

At a meeting of the Group for Atomic Questions on 30 September, the British position was at last made clear. Responding to the proposal made by the Commission that a one-year extension out of five should be agreed, the UK delegate referred to a re-examination of the place of the HTR, that had led to the conclusion "that against the background of low priority for the HTR, the benefit to the UK's nuclear programme of the proposed extension would not justify us in continuing to bear the substantial costs which we would incur as the largest single contributor. Accordingly, the UK is not in favour of an extension of the Project. However, we are conscious of our special position as hosts to Dragon and certainly would not wish to see the Project brought to an end if the other Signatories wished to continue work at Winfrith under a new financial regime. We have all along been aware of the risk that exhaustion of funds during the current extension might lead to premature termination of the Project before a decision on a further extension could be reached. It was for this reason that in April the UKAEA made it clear that they were ready to bear their share of the costs to prevent closure before 31 March, 1976."

The UK statement went on "We understand from the Commission that the removal of experiments from the reactor prior to shutdown will have to begin on 6 October and, as is well known, desecondment notices already issued to about 75% of the staff will take effect on 6 December. This timescale may well be too short to enable new arrangements to be concluded to continue the Project if the other Signatories desire this. Accordingly, we are prepared to provide support to enable our partners to have more time to formulate alternative arrangements. The UK is willing to continue to bear its present share of the costs of the Project to the end of June 1976 if the other Signatories will do likewise" (i.e. a three-month's extension, not one year). "This offer to our partners is conditional upon its acceptance by 30 November at the latest, given that desecondment would take effect on 6 December. If this offer is accepted, a decision on the future of the Project will need to be made by the end of March at the latest . . . This would give the necessary three-months' period for desecondment if termination were then decided upon. If continuation is decided upon under a new financial regime, the UK would expect this regime to come into effect from 1 April 1976."

For the first time since 1962, the UK was dictating terms. Over the previous 10 years, it had been the other Signatories either individually or through Euratom that had introduced new conditions and the Authority had always finished by agreeing to pay more, working up from 40% to 48.3%. A tradition had been established and the reversal of the roles was strongly resented. A month of informal discussions in Brussels culminated in the Commission's 12-months' and the UK's 3-months' extension proposals being set before COREPER on 23 October. No progress was made and the conclusion was drawn that stoppage must now be envisaged. Alternative extension periods of six and nine months had also been aired but when the Board of Management met on 29 October, the Authority made one more thrust, intimating that it would expect decommissioning costs to be included in any period that exceeded the 3-months' offer. A source of controversy already in the projected 5-year extension, such a demand could only make the possibility of unanimity even more remote.

Shepherd would not give up hope, even if he did refrain from the intensive lobbying of the press and politicians undertaken by his senior staff and notably Hosegood. Reacting to the news that all GAC orders had been rescinded, he tried to impress on the Signatories the uniqueness of Dragon and the enhanced importance of the European collaboration. Karl Krebs, the German member of the Dragon Board of Management appeared to share this view but there was no other reaction from inside Germany, even from KFA, although it was to Germany that the Project looked for the saving gesture. Nothing could happen overnight in any case and the UK was urged to be more reasonable and recognise the longer timescale needed for the negotiation of a new financial structure which might include participation by the United States and Japan.

American interest was indeed real, and just prior to the Board meeting, representatives of ERDA had been at Winfrith to discuss how the USA might participate at least in kind in a Dragon continuation. ERDA, it seemed, was extremely distressed at the prospect of the collapse of the US programme being followed by the shutdown of Dragon. So much so that Thomas Nemzek, Director of the US Division of Research and Development spelt out in a telex on 10 November the intention of ERDA to assist, even though it could not hope to produce constructive ideas before March 1976.

Shepherd gave this telex a wide distribution, and for a time it seemed that Germany would be influenced sufficiently to act as Dragon's sponsor, morally if not financially. Herzig had been in Germany and had been arguing the case for a larger German share, the first result of which was a tentative German motion that appeared on the agenda of COREPER on 5 November but was not debated pending a meeting in London on 18 November between Herzig, Pictet and representatives of the Commission and the German Government.

The plan that emerged was that Germany would propose to the Communities a nine-months' extension, which would allow either an orderly run down from 1 April if no sign of a new financial structure was in view, or would prepare the way for a longer term extension. Herzig would recommend to his Government acceptance of the terms, and it was hoped to conclude the arrangement in Brussels on 27 November. It seemed to the Project that at the eleventh hour rescue was in sight, although in view of the categoric statements of Italy, convinced that any relaxation of the UK position was designed to off-load the decommissioning costs on to the other Signatories, and the unwavering disinclination of France, the optimism that swept through the Project was scarcely justified. Shepherd and his staff though, were searching the horizon for any reprieve and it is a tribute to the hold of Dragon on its crew that although the ship had been sinking by then for over four months, there was no rush for the life-boats and the dominant thought was how to patch the holes.

When the German proposal for a nine-months' extension was put to COREPER on 27 November, the UK delegation made a statement reiterating that it was not convinced that an extension to the end of 1976 would disclose new proposals for financing Dragon that could not be made by March. If, however, there was a unanimous wish to go beyond June the UK would agree to a nine-months' extension provided that there was a general agreement by the end of March that there was a clear prospect of a new financial regime being established and the principles settled. If that proved to be impossible there was no point in going on beyond June and the three months' interval should be used to run the Project down. If it seemed that a new regime could be established, this would be back-dated to 1 April and had to be worked out by 30 September, 1976 at the latest. The Commission felt bound to comment that it could not guarantee the participation of the other parties and was doubtful that a new scale of payments could be arrived at by March.

No progress was made and when COREPER reconsidered the matter a week later, the only new information before the delegates was that several delegations excluded any possibility of the Communities' contribution being increased. The majority of delegates, therefore, concluded that the proposal did not provide a basis for continuation. One further move remained, namely to go back again to the Council of Ministers and it was generally understood that Germany would be raising the Dragon question under 'Any Other Business' at the end of the 15 December meeting. That meeting was particularly unproductive, tempers were short, progress on all subjects extremely slow and the agenda was abandoned half way through. Germany, the reluctant promoter of the nine-months' extension had concluded that there was much less substance in the offer from ERDA than had been thought, only the smaller countries were keen to see an extension and against so much opposition voiced at so many meetings on the lower levels, further effort especially on such a day, was pointless.

Too late came outside reactions to the cries for help. On 17 December Dragon was debated by the European Parliament in Strasbourg which resulted in an invitation being sent to the Commission to present to the Council urgently, a proposal for a six-months' extension. The following day, Dragon was debated in the British House of Commons at the instigation of Arthur Palmer. Misinformed about the exact evolution of the situation in Brussels and believing that Germany had been persuaded to withdraw the proposal, he was forced to accept the uninformative response of the Under-Secretary of State, Alex Eadie, that the matter was not on the agenda. Eadie recapitulated the UK attitude and laid the blame for the absence of an agreement on the other COREPER delegates **(43)**.

Desecondment notices, it will be recalled, expired on 6 December and although not all the seconded staff had returned to their bases – Euratom could not find places immediately for some of its staff and the UK people were still all at Winfrith – the Dragon programme was finished. DRE had been reloaded in September but had never been restarted and early in November it had been decided to unload the core again in preparation for decommissioning. No irreversible actions were taken at that time but come December there was no alternative but to start dismantling the fuel. The new heat exchanger that had been installed and the various experiments in the reactor core were never run. When the Board had met on 5 December, it had agreed that there was now no chance of securing extra funds under the present Agreement to avoid the abandonment of the programme, accepting that resources were no longer available to round off experiments and report on the considerable amount of technical knowledge and results that had been amassed. The rearguard team to be retained by the Project, consisting of 14 administrators and nine scientific staff, would be fully occupied in disposing of the non-fixed assets and arranging the hand-over of buildings, etc.

SOME RETRIEVAL BUT NO REPRIEVE

KFA by then appreciated the full import of the waste that five months of indecision had precipitated and the impact that cancellation of further work would have on the pebble-bed programme, returning to the fore with the abandonment of the German prismatic projects. Although AVR was able to provide some information on fuel ball behaviour, this was at power densities inferior to those obtaining in a power reactor. In the fields of materials, general HTR technology, physics and safety, a great deal had still to be learned and KFA had been relying on a continuation of the Dragon programme and a full evaluation of past experience. This could now only come from a thorough examination of components irradiated in Dragon, and analysis of the data including that already accumulated but not yet written up in an assimilable form.

Krämer with a team from KFA visited Winfrith on 19 December with a view to seeing how much could be retrieved and whether certain parts of the Project's programme could be continued up to the end of March and beyond. The cost of the necessary effort came to about £½M and Shepherd was urged to try and retain the staff necessary to carry out the work while Krämer sought confirmation of a deal. An immediate transfer of £50 000 was arranged and Pictet as Chairman of the Board authorised the Project to go ahead pending more formal arrangements. For the Project, the reprieve could only be temporary and Pictet, just before Christmas, travelled specially to Winfrith to explain the position to the remaining staff members. In a moving address, he conveyed the Board's profound regrets at the failure to reach agreement on an extension and, on the Board's behalf, expressed sympathy with the staff over the distressing period they had passed through, finally thanking them for their devoted services to the Project over the years.

When the GPC met on 8 January for what was to be its last meeting, the main item on the agenda was the supplementary programme for the continuation of the metals programme, in particular, continuation of the experiments at CIIR and Fiat and the selective programme for data extraction. Since the Winfrith meeting with KFA, ERDA had agreed to contribute £22 000 and Euratom through the Joint Centre at Petten, £10 000 to the metals programme, making a total of £82 000 that was available for the first three months of 1976. In addition, KFA had set aside £80 000 for a further continuation up to the end of the year. For the data extraction work, KFA allocated the remarkable sum of £356 000 for work that was to be mainly completed by 31 March. The seconding bodies were prepared to co-operate and Cartwright as Director of Winfrith undertook to be as accommodating as possible in keeping such facilities as hot cells available. Again the loyalty of the staff was demonstrated and when the Project did formally terminate on 31 March, 66 staff members and 12 guest scientists were still on the roll.

As to the proposal to continue beyond March, the immediate question was in what legal form. The only practical solution, it was generally agreed, was to retain the existing structure and seek an extension of the Dragon Agreement up to the end of the year with no commitment of funds by Signatories, only a contribution from interested parties. The Agency went away to sort out the legal aspects while the Commission prepared a proposal for the Council. An intimation of the response of the political sectors was received almost immediately in the form of a telex from Albonetti reiterating the position of the Italian Government that it was against any extension of any sort.

The Commission, faced with the invitation of 17 December from the European Parliament, was pleased to be able to proffer a new approach that implied that the maximum technical advantage would be drawn from the 17 years of Dragon collaboration. Parliament's demands could not have been met directly because, as the Commission was obliged to report to COREPER, it was no longer possible to contemplate operating DRE. Fuel dismantling and the removal for examination of certain parts of the primary circuit quite excluded a new start up before 1977 even if the UK were willing to keep the reactor in mothballs, rather than reduce it further.

The proposal as outlined to the Board of Management on 3 February was for two parallel programmes. One on data retrieval costing £1.3M, financed entirely by the Communities, but open to contributions from countries outside the Communities, and a metals programme lasting through to the end of the year costing £332 000 contributed by: KFA, £155 000; ERDA, £67 000 and the Petten centre, £110 000. Of this total £82 000, as already seen, had been committed to the first quarter. In view of the fact that all results would be available to all Signatories, the Board expressed its gratitude to KFA for its timely action and endorsed the proposals, subject only to the dissent of Italy. Austria, Sweden and Switzerland indicated that they would recommend contributing and the Authority undertook to co-operate. Yet another draft agreement of von Busekist was studied and with minor modifications approved. Warnings were, however, sounded that no commitments could be made in the Board for the reaction of Governments within the Communities.

Moral support came from the Committee on Energy, Research and Technology of the European Parliament which met to consider the situation in the light of the fact that even though the Project as such could not be kept going, alternatives had been formulated. This led to the adoption of a resolution which was unanimously approved by Parliament on 27 February, that regretted that its original proposal could not be implemented and urged the Council of the Communities to agree in good time the nine-months' extension for the metals and information retrieval programmes. Council was further invited to take no more decisions to close down research projects before the European Parliament had given its opinion, and a protest was also lodged that the decision on Dragon had essentially been taken by the Permanent Representatives.

This no doubt stimulated Benn on 3 March to reply to an invitation that had been sent on 23 January by the Technical Commission of the European Parliament to come and explain the UK position **(44)**. He justified the UK action, as did Eadie in his answer to the House of Commons, by the argument that it was the other members of COREPER that turned down the earlier nine-months' extension proposals. Two anomalies are apparent in his letter. He stated that the UK was "very disappointed to have these proposals rejected" whereas the statement to COREPER on 27 November referred to a "lack of conviction" that the extension to the end of 1976 would disclose anything new. He also stated that the UK had offered to continue to pay its existing very substantial share of the costs up to the end of 1976. Yet the COREPER statement explicitly required that for an extension beyond June to be acceptable, it would only be under new financing terms that would have to be back-dated to April. Even for the data retrieval and metals programme, the UK had declined to offer a contribution in excess of its contribution through the Communities.

That apart, the adjurations of the European Parliament were ignored and when the Committee for Atomic Questions came to consider the proposals presented by the Commission, no agreement could be arrived at, either on the form of the extension structure or on providing the 90.36% of the data retrieval budget that was required. When the Board of Management met on 23 March for its last meeting, COREPER had still not discussed the proposals and even the meeting fixed for the following day had been postponed. To complete the sadness of the occasion the Board received the news that David Peirson had died while on a climbing holiday in the Lakes on 21 March. Dragon survived him by just 10 days.

It is appropriate that COREPER should have the last word. On 30 March, 1976 the Commission had to report that to its profound regret, the Council had now no possibility of adopting any position concerning Dragon before the Agreement expired, in which case, further discussion on the subject was futile.

IN CONCLUSION

FORMER members of the Dragon Project are inclined to make the criticism that insufficient attention was paid to publicising either Dragon itself or the HTR. Such was the hold of the system on the staff, they find it unthinkable that given full information, the Signatories, either together or severally, would not have adopted it for their forward nuclear programmes. The HTR is not unique as a nuclear power system in generating loyalties of this nature. People become committed to the system to which they have devoted their professional lives, and strongly biased towards the technical merits that it possesses.

Dragon was by no means reticent in making known in the technical world the results of its work, nor the hopes that it held for the HTR. During the construction and commissioning of DRE, nine symposia were organised by the Project at which the different scientific and technical aspects of the development were discussed before international audiences. From then on, the rate of holding Dragon symposia dropped, but the Project staff participated actively in meetings all over the world, presenting, for example, six papers at the American Nuclear Society's Topical Meeting on Gas-cooled Reactors at Gatlinburg in May 1974. The staff contributed widely also to specialist meetings organised by learned societies and made regular contributions to the scientific and technical literature. Taking at random the Annual Report for 1973-74, 21 papers are listed as having been published by the staff during the year.

Dragon's presentations were not limited to dry scientific data; the potentialities of the HTR for steam cycle electricity generation, for direct cycle generation, for steel making, for process heat and for secondary fuel production were energetically put forward. Two films were made by the Project. The first was a factual account of the construction of DRE which Rennie had had the foresight to organise early on in the Project so that the various stages of construction could be filmed in sequence. The second, shown for the first time at the 1971 Geneva Conference, was one of the hardest hitting commercial films that has been seen at international gatherings. Each sequence was opened by a type of 'one armed bandit' or 'fruit machine' ringing up the commercial pluses of the HTR that were then elaborated in detail. Dragon was even accused of being undignified in producing such a direct advertisement for the economic advantages of the HTR.

Nor was the larger public forgotten. Occasions such as the tenth anniversary were used to interest as wide an audience as possible and Bruce Adkins, in charge of public relations in the Agency, amply supported by the public relations offices of the Authority in Winfrith and London, was assiduous in drawing the attention of the press to Dragon's successes. A number of booklets were produced and the press was encouraged to visit Winfrith.

The Project kept open house and some hundreds of visitors, mainly technical, passed through its doors each year. Visits of members of governments were treated seriously and the opportunity was taken to promote the Project and the HTR as much as possible.

Rennie was perhaps more of a 'salesman' than Shepherd, but Shepherd never missed an occasion to present in his own persuasive style, the technical arguments in favour of continued development of the HTR and the continuing role of Dragon in this development. Hosegood could be quite fanatical on the subject and other senior staff were no less dedicated to spreading the word.

Dragon probably did more promotion of the prismatic HTR than some of the Signatories would have wished. None had an obligation to do anything more than participate in Dragon's research programme and the Project's efforts to encourage the commercialisation of the HTR could be interpreted as interfering in the nuclear power policies of the individual countries. Dragon was not constituted to lead a crusade for the system so much as build a reactor experiment and provide the Signatories with data.

Of this there was no lack, and the jibe has been made that Dragon was the most prolific producer of paper of any of the Agency's projects. For this it has to be commended. Rennie adopted the practice from the beginning of supplying the Signatories with all the technical material on performance and design that could be useful, and this policy was followed throughout the Project's existence.

INFORMATION TRANSFER

The official channel of communciation for technical information was through the Dragon Project Reports of which the last, summarising and evaluating the technical achievements of the Project, carried the number 1000, not just because it was tidy but because over 980 had preceded it. In addition, technical material in less polished form was transmitted to the centres directly interested, through the medium of Dragon Project Engineering Documents and Research and Development Documents, combined later into the series Dragon Project Technical Notes. A monthly report was produced on DRE operation and the Authority's monthly reports, safety documents and the voluminous records covering the detail of DRE operation could be consulted. Over and above these were the reports from the various working parties on physics, fuel, materials and safety; even the internal documents of the core programme working party were made accessible. Dragon's Annual Report published by the Agency was packed with technical data, minimum space being devoted to administrative matters.

For each irradiation experiment made in DRE a special dossier was created that was opened by the sponsor, who provided details of the experiment and the performance to be expected. A record of the specific irradiation conditions would be added and subsequently the sponsor would furnish the results of the post-irradiation examinations. The complete file was then transmitted to all the Signatories so that the full technical history could be followed.

Distribution within the Signatories depended upon local practices. For those with no HTR programmes, Dragon reports tended to sit on the library shelves – a practice which in the case of IFA caused adverse comment as they were then open to anyone working at Kjeller, whether they were from a Signatory country or not. In consequence they were locked away from public view. In general, reports were available on a 'need to know' basis. Euratom had its own list of companies and organisations within its member countries, built up from requests for information that carried the approval of the relevant Government. Once the lists were established, the Community was sending out between 400 and 500 copies of each official technical report that appeared.

In addition to all these channels, there was the personal exchange. Dragon staff travelled widely, in particular to those laboratories carrying out extra-mural research for the Project and to companies manufacturing equipment for the Project, Dragon staff assisting in development and following manufacture at the companies' works. In reverse, technical staff from the Signatory countries participated in the work of the technical working parties and the PSC and, as already noted, there was a continuous stream of visitors to Winfrith. Above all was the experience gained by the staff on secondment. Every organisation within the Signatories with a recognised interest in HTR technology was able, eventually, to gain access to all the information available in the Project that was relevant to its activities and which the Project could legally divulge.

Working documents for the Board of Management and the GPC were grouped under Dragon Project Documents and were nominally confidential to the Signatories. Over 1300 were produced, and in many instances, they were far from being of an administrative character. They included technical review information relating to current work and quarterly progress reports which were instituted to ensure that topical information was available as early as possible. The Signatories, it is clear, were made fully conversant with all aspects of Dragon's activities. How they used this information was their own affair.

OPEN PUBLICATION

For open publication of technical material, the authorisation of the Board of Management was required although discretionary powers were delegated to the Chief Executives who, in turn, for the most part interpreted their mandate in a liberal manner. It was not, however, until Shepherd's time that a major declassification was made, after which regular reviews were carried out to determine which documents should be released. Even so, almost 200 remained classified at the end since when all have been released. Certain of the Signatories were more resistant than others to the principle of open publication on the grounds that documents might contain information that was confidential to a particular organisation.

The Authority is a case in point, yet when Peirson outlined his proposals for a joint undertaking to develop the HAR he put forward the principle of open publication as adopted at CERN. This was never seriously debated. One argument that is put forward against open publication is that certain organisations would have been less free in their communications with the Project. As it was, Dragon staff had access to a great deal of information that was confidential to a particular laboratory. They were not, however, free to pass this on and although the collaboration between the Signatories and Dragon itself was close, few organisations would regard transmission of private material to another European country as any less objectionable than to a country not included in the Project. Nevertheless, it has to be acknowledged that, despite the awareness of the immense gulf that exists between knowing how to do something and having the capabilities and the resources to carry it out, traditional attitudes towards commercial secrecy still persist and in the late fifties in nuclear energy could be considered obsessive.

One may still regret that the principle of open publication was not more thoroughly examined, or rather failing that, that an explicit effort was required to classify a document rather than have it declassified. At the very least, it would have saved time and the shift in approach might have led to a wider public debate on the promises of the prismatic HTR. This in turn could have catalysed the emergence of a European approach to commercialisation and stiffened the resolve of electricity companies and governments when it came to prototype construction.

Similarly in regard to patents. Publication is every bit as effective in avoiding the need to pay royalties for one's own inventions as patenting, and far less expensive. Dragon's list of patents is impressive when considered as a measure of its innovative capacity. The financial return has been zero. It would be interesting to cost out fully the expenditure implicit in the effort made in working out international agreements, the time spent by the Standing Committee on Patents (the records of which weigh tens of kilogrammes), the work involved in drawing up specifications and finally the cost of the patents themselves. For all this there has been no return and none is foreseen. Atomic energy commissions in many countries have had the same experience – Dragon was just another example of the application of ritual thinking to the completely new circumstances surrounding the industrialisation of nuclear power, with its long lead times between R & D and commercial operation.

Would countries have been less willing to join the collaboration had information been open? It seems improbable, as it is a generally accepted maxim that participants receive back from a collaboration in proportion to what they put in. If it was money only, this at least gave the right to bid for contracts, so offering to industry and research organisations the possibility of working in advanced technology and gaining experience not to be had otherwise. To go further required additional investment. Experiments could be made in DRE only by developing the technical competence necessary and preparing the relevant rigs. Knowledge in depth came not from reading documents but from working for and with the Project and seconding people.

JUSTE RETOUR

Dragon's statutory requirements to distribute evenly its favours were only loosely expressed, and the Project was reasonably adroit in reconciling its needs with the aspirations of the Signatories. Although pressures were exerted from time to time, the *juste retour* was not made an issue, for which the Signatories can take credit. Even if significant changes took place in the distribution of contracts following the 1970-73 and final extensions, there was no pressure to cancel the big materials research programme at CIIR when Norway ceased to be a Signatory. On the whole, the balance that was maintained, while generally being to the advantage of the smaller countries was healthy. Little was put out that did not carry full technical justification and which did not bring a response greater than purely commercial accounting would have indicated to be justified. At no time were suitable staff turned away. Rather the problem was one of recruitment and persuading the Signatory countries to exploit the possibility of attaching Guest Scientists who were particularly welcome as they imposed no burden on the budget.

STAFF

The practice of seconding staff, while being a fundamental element in the process of information transfer, brought other advantages to Dragon. It meant that specialists could be engaged for limited periods in numbers that reflected current needs, making for a smooth transition as priorities changed, for example when construction of DRE was replaced by the experimental phase. Moreover, the ability to adjust staffing levels to accord with budget limitations gave a flexibility to the management that was of crucial value when inflation was making in-roads into the Project's purchasing power. Secondment allowed the Project to get under way rapidly, and it simplified closure. On the other hand, it diminished the sensitivity of the Signatories to the effects of delays in reaching decisions on extensions and termination, as they had no collective responsibility for careers that were being jeopardised, and no worries over pension rights and other personal problems. For the staff, there was undoubtedly more security of employment, even though the uncertainties that accompanied every extension negotiation made life uncomfortable. If staff had been directly contracted to the Project and the behaviour of the Signatories not been modified, it is difficult to believe that they would have been willing to continue to work for so long under the threat of imminent stoppage.

It must, however, be questioned whether it would not have been wise to appoint some permanent or semi-permanent staff in senior positions. In the first place, this might have avoided the progressive domination of these positions by British subjects and given a better international balance to the organisation; in the second, it would have removed the ambiguity inherent in the position of Chief Executive. By remaining Authority staff members, both Rennie and Shepherd became too involved in the formulation of British policy regarding the HTR. Their total loyalty to Dragon was not in doubt but the propriety of their dual role was dubious. In practice, the relationship between the Chief Executives and one of the Signatories was distorted. Within the Authority, the Chief Executives became staff members, below Director level, with no executive responsibilities and so to be treated as relatively junior people trying to promote a personal cause. Yet in Dragon, they were required to harmonise the demands of all the Signatories of which the Authority was just one. It is not surprising if at times in this position they made efforts to impose Project view-points on the Authority that they had failed to get accepted in another capacity. At the same time, Authority representatives would resent the presentation of plans which were not in line with British policies, and which the Chief Executives had ostensibly helped to formulate. To a lesser extent, the position was similar for Dragon's Division Heads and the dichotomy was probably a major factor in the hostilities that developed.

A further question mark hangs over the advisability of continuing the principle of having a British Chief Executive following Rennie's departure. It is generally accepted that the principle of starting with a British Chief Executive, as was imposed by the Authority, was sound, because it ensured the speedy establishment of the international organisation on a British site. The whole of the administration had to be made compatible with that of the Authority and this could best be achieved by the appointment of a man already conversant with Authority practices. Once that administration had been set up, the same requirement was no longer evident. Without in any way calling in question Shepherd's personal suitability, it is perhaps unfortunate that the principle was not re-examined in 1968, and consideration given to the appointment of a Chief Executive from another Signatory. In this way, the international character of the Project would have been better preserved and attitudes in France, Germany and Italy might have been modified accordingly.

FIXED-TERM EXTENSIONS

The rhythm of three-year (or less) extensions was manifestly unsatisfactory. Even the initial Agreement for five years was inadequate to cover the building of a reactor experiment and the extraction of any useful information from its operation. Eight years might have been long enough to prove the idea of an HTR unfeasible but it was already apparent by the time that DRE operated, that the HTR with coated particles was a promising system and the DRE as the only versatile HTR test reactor in the world would have a continuing role to play over many years. Devising, working up and proving a fuel element is itself a lengthy process and this is only part of the research required to demonstrate the basic characteristics of a new reactor.

Signatories were, however, afraid of an open commitment and did not seem able to contemplate any alternative to the process of negotiating extensions of a fixed term whose length bore little relevance to the inherent time scales of the work to be done. Until Shepherd was able to obtain some relaxation in 1969, even the books had to be closed at the end of each extension period. Only if negotiations had gone completely smoothly would the Chief Executives have avoided the need to operate a see-sawing programme. As it was, they were required to run the last period of an extension to cater for either closure or continuation and in the first period after an extension was agreed set about restoring the former level and trying to bring the personnel up to strength. It says much for their resilience that they were able to cope and continue to keep the Project productive.

Inflation made the problem more acute. Costing of the programme to be followed in the last year of a three-year extension period had been made at least four years before so that when inflation rose more rapidly than had been estimated – as it always did, and latterly by a large margin – the Project's purchasing power was cut. Shepherd (Rennie did not have to face this problem) made allowances in his programme planning but he could not work miracles and by 1975, the difference between the real cost of the established programme and the budget, required him to start running down nine months before termination, and no funds were left to cover the last four. With another three-year extension, about 15% of the budget would have been mortgaged in advance. What was lacking was any quasi-automatic method of compensating the drop in value of the pound sterling, to give the Project a stable income in European terms.

More insidious, however, than any of the points raised so far was that through fixed term extensions, Dragon was at regular intervals thrown back into the political arena of the Communities to risk being slain in a totally separate cause. Out of phase with Euratom's programmes (which at times were yearly) Dragon could escape unscathed but when in phase, its future prospects were compromised. In the Board of Management, programmes, budgets and percentage contributions could be agreed for the general good, but back in Brussels, political considerations were uppermost and the Project was left unheeded as Governments with the power of veto manoeuvred over differences that in actual money terms were minute.

Unfortunately no principles had been established to control percentage contributions, apart from basing those of the five independent countries on their net national revenues – with no provision for adjustments to be made to cover changes in national wealth. Huet's concept of equality between Euratom and the Authority was arbitrary and was only made to apply initially by ignoring a quarter of the budget. Even then it was difficult to see the logic of a distribution which required Sweden with no HTR programme to pay 4.5%, and Germany with a net national revenue some five times as big and a very direct interest in the results of the Dragon programme to pay 13%. Fixed-term extensions led to what principle existed concerning the relation between the Authority's and Euratom's contributions being forgotten, to give place to an undignified procedure of bargaining that ignored the Project's needs and incidentally reflected little credit on the negotiating countries.

It would clearly have been in the interests of Dragon and therefore the Signatories if a rolling programme of budgets had been fixed for a period of three or four years ahead, with largely immutable principles governing percentage contributions and quasi-automatic rules for taking care of inflation. Provision for adjustment could still have been made by defining a period of minimum notice. This would have brought stability to the Project's structure, continuity to its programmes, and allowed changes to be made progressively rather than in a series of steps. The fear is expressed that with such a system, a project rolls on for ever – a sorry comment on the decision takers, if spasms have to be built in so that the patient will be sure to die of one at some time.

COMMERCIAL APPLICATION

It must be remembered that Dragon was not borne out of any collective desire to develop the HTR as a power reactor. It was conceived as a scientific collaboration supported by a political will for co-operative effort. DRE as it was first designed could not have formed the basis for a central power station programme, while ship propulsion had only a small and temporary following. It was even suspected in some quarters that the Authority would not have offered its HTR research project as a joint undertaking if it had believed that the HTR had genuine commercial potential. This was almost certainly unjustified cynicism but the fact remains that a system which depended on continuous purging to limit the primary circuit activity, and which employed a coolant of doubtful availability, was never considered to be commercially viable by any but the most dedicated.

Coated particle fuels, concrete pressure vessels, an easing of helium availability and (for Britain and France) the ability to operate on the uranium/plutonium cycle with low enriched fuel, totally transformed the commercial prospects. Regrettably no new initiatives were then taken to pursue commercialisation under an international umbrella. At the end of the fifties, of course, it had been firmly established that the separate Signatories would be able to exploit the system if they wished, on their own account, implicitly in competition. Yet by the time that DRE had operated at full power, it should have been evident that only through a concerted effort could a new thermal reactor be introduced. The prospects for nuclear power had changed greatly over the previous ten years. Oil was cheap and plentiful and the growth in unit size of stations which necessitated huge investments being locked up in a single installation, was forcing electricity companies to limit their choice to proven systems backed by huge industrial strength. Successively in Europe national nuclear power programmes became channelled into light water reactors backed by American technology. Only the UK in Europe maintained its independence but the UK at regular intervals came to the brink of abandoning gas-cooling and joining the common stream. Germany continued with its parallel development of the pebble-bed but as a solitary national effort and at a slow pace.

Although Dragon's technical programmes were orientated towards industrialisation, they were never given a single European purpose that could have spawned a European approach. Dragon was not an organisation that could have been adapted to constructing a prototype power station but it could have become the centre for the definition of the basic core design and materials for a European prismatic HTR. The very versatility of the HTR with its wide choice of fuels, power densities and operating temperatures made concentration on a single concept that had won wide European support essential. Instead, the Project felt obliged to do everything – steam and direct cycles, thorium and plutonium, a range of fuel forms, engineering design and so on; all this with a staff of 100 people.

Unfortunately in the second half of the sixties when the time was ripe for structural changes to be made, Euratom was in a turmoil and some of its member countries were intent on loosening rather than strengthening cooperation in nuclear energy. Dragon existed, was widely respected in both political and technical circles and for the Signatories the problem was to safeguard what had been achieved rather than search for radical changes. Moreover, at this time, the UK seemed intent on going ahead with prototype construction alone so might be expected to reject any closer collaboration. The net result was that the UK choice of fuel for Oldbury had the support neither of the Project, nor of the other Signatories and the construction of Oldbury was indefinitely postponed.

OECD limited its intervention to keeping the Project in being; the technical aspects of Dragon's future were left to the Board of Management. This, however was composed of representatives with no direct responsibilities for power generation and Dragon's long term role was never discussed in the context of the wider issue of the transition from R & D to joint commercialisation. It has already been noted that the OECD's Electricity Committee was an organisation separate from the Agency and this contributed to the failure to get to grips with the fundamental problem.

Although Euro-HKG made some effort to coordinate European design effort, its activities were remote and confined to the exchange of information. No direct contact was established with Dragon and no attempt was made to influence Dragon's research programme to accord with a European objective. Something bolder than a 'lunch club' was needed to get the HTR off the ground. Only at the level of the Dragon working parties was real progress made towards creating a European model.

It is a sad reflexion on Europe's confidence in its own capabilities and its power of collaboration that the nearest the countries came to building a prototype prismatic HTR was via the isolated approach of the UK and the collaborations negotiated separately by France and Germany with GGA. For too many years, Europe has suffered from an inferiority complex in regard to American technology that has been nourished by the mutual mistrust that still exists between the different countries of the continent. Yet Europe has a far greater experience in gas-graphite technology for power generation than the USA and all the resources necessary to commercialise the system. GGA had, of course, a far more aggressive sales policy than any of its European counterparts and a confidence in its own capabilities which far exceeded its competence. This would have been recognised had there not been an instinctive desire to look for partners outside Europe rather than work towards a European co-operation that could capitalise on indigenous know-how. It is too easy to put the blame on the narrow nationalism of the politicians; atomic energy agencies, the electricity companies and the reactor industry – Inter-Nuclear notwithstanding – all chose to shelter behind the difficulties rather than make the effort to surmount them.

In Peirson's words written on the occasion of the 21st anniversary of the creation of the Authority, "There is surely no sadder example of a good European idea being brought to commercial fruition on the other side of the Atlantic." **(45)** No satisfaction is to be drawn from the fact that the commercial fruition was an illusion, the lesson is still there to be learnt.

CLOSURE

All the Signatories deplored the untidiness of Dragon's end. It was, of course, inadmissible for the UK to remain mute for so long and then seek to renegotiate its percentage contribution after the date for deciding on an extension had expired, but it was not the first time a Signatory had done this and no country made any real attempt to find out if a token reduction would have sufficed. Dragon went by default because the major countries preferred to let procedural delays in Brussels take their course, instead of coming to a clear decision to close. A courageous Council of Ministers would have grasped the nettle and subsequent run-down could have been carried out in a more ordered manner to the benefit of both the Signatories and the staff.

It is clear that at ministerial level, Britain, France and Italy were happy for the Project to end and Germany was at best apathetic. Britain had given up the HTR, France had become discouraged by the collapse of the GAC contracts and Italy had long since abandoned interest in gas-cooling. Germany was thrown into confusion by GAC's deficiencies and was giving up the prismatic HTR and gathering its resources around the pebble-bed once more. With the exception of Switzerland and to a lesser extent Sweden, no country in Europe wished to pursue the prismatic HTR as a system.

At the working level, the picture was different. DRE was still a unique facility and its forward irradiation programme was well supported. New rigs

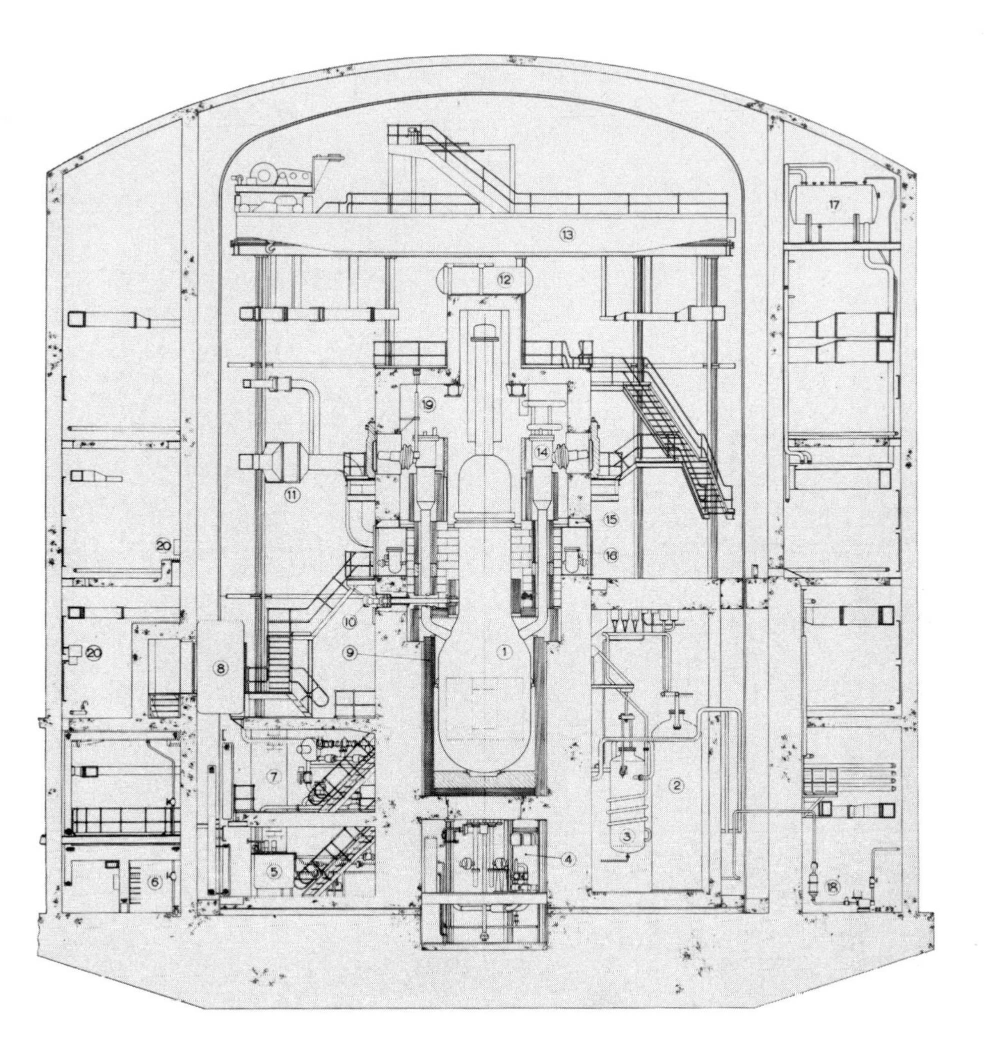

REF No.	DESCRIPTION
1	PRESSURE VESSEL
2	DUMP TANK (HELIUM)
3	HEAT SINK
4	ARRANGEMENT OF EQUIPMENT ON PIT WALL
5	SUMP TANK
6	PUMP, VALVES ETC.
7	COOLING FANS ETC.
8	EMERGENCY AIR LOCK
9	THERMAL SHIELD
10	VIEWING TUBE

REF No.	DESCRIPTION
11	NITROGEN RECIRCULATION SYSTEM
12	STEAM DRUM
13	25 TON CIRCULAR CRANE
14	PRIMARY HEAT EXCHANGER
15	STEEL SUPPORT RING FOR UPPER BIO. SHIELD
16	P.V. ABSORBER ROD DRIVES
17	TREATED WATER HEAD TANKS
18	EMERGENCY RECIRCULATION EQUIPMENT
19	HEAT-EXCHANGER IONISATION CHAMBERS
20	ELECTRICAL SWITCHGEAR ETC

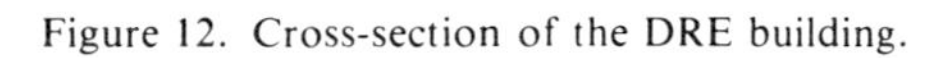

Figure 12. Cross-section of the DRE building.

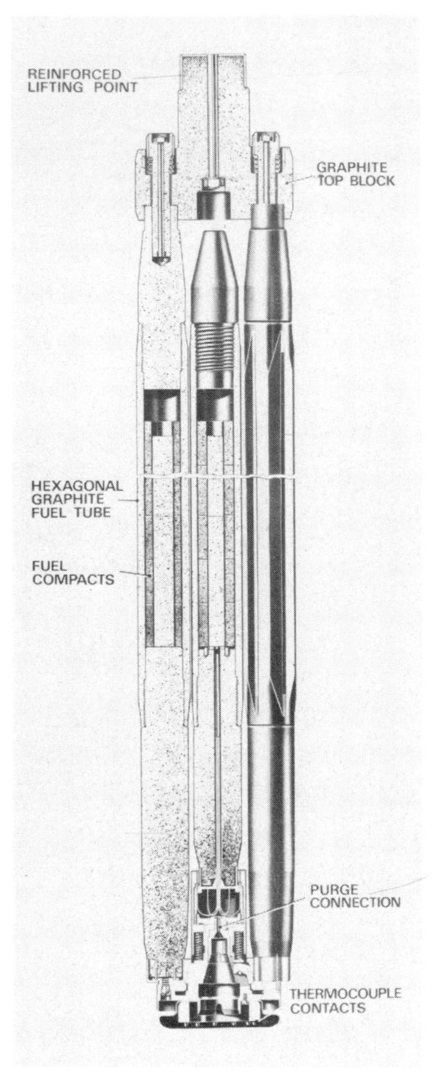

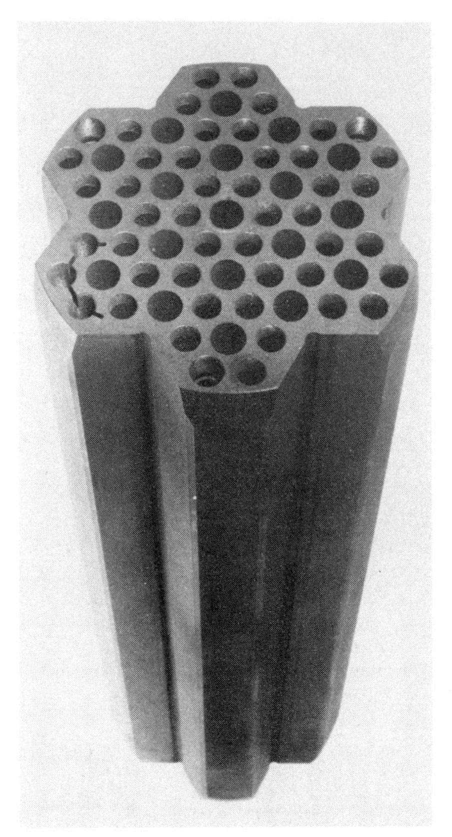

Figure 13. Cross-section of a seven rod fuel element.

Figure 14. Typical block element before the loading of fuel.

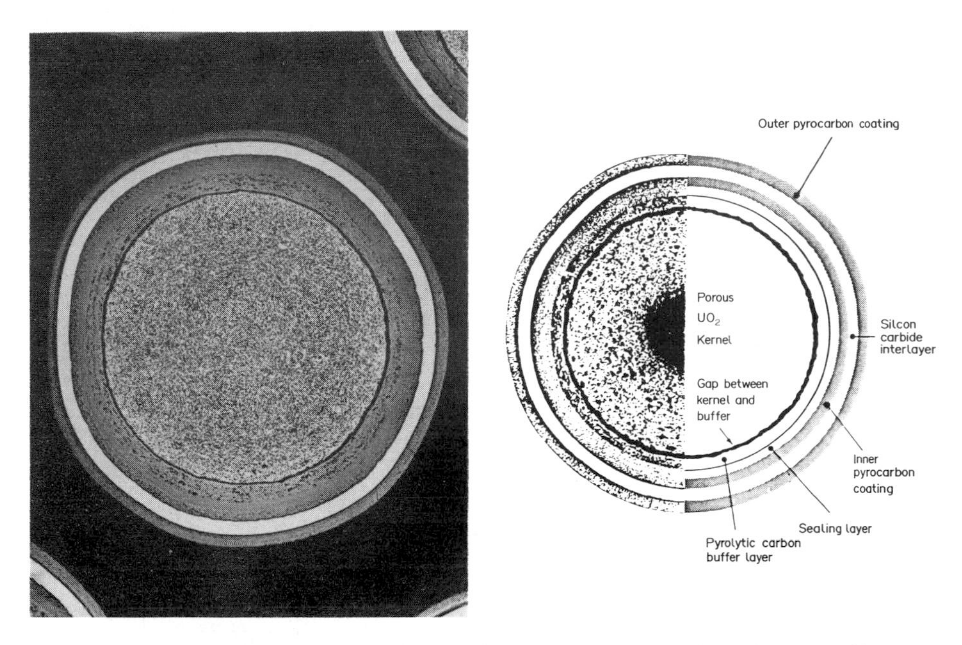

Figure 15. Photomicrograph of a typical uranium dioxide coated particles with, alongside, an explanatory sketch indicating the nature of the different coatings. Overall diameter is approximately one millimetre.

Figure 16. Different types of fuel body showing a variety of annular rings and pins with, above, fuel balls as used in the pebble-bed design.

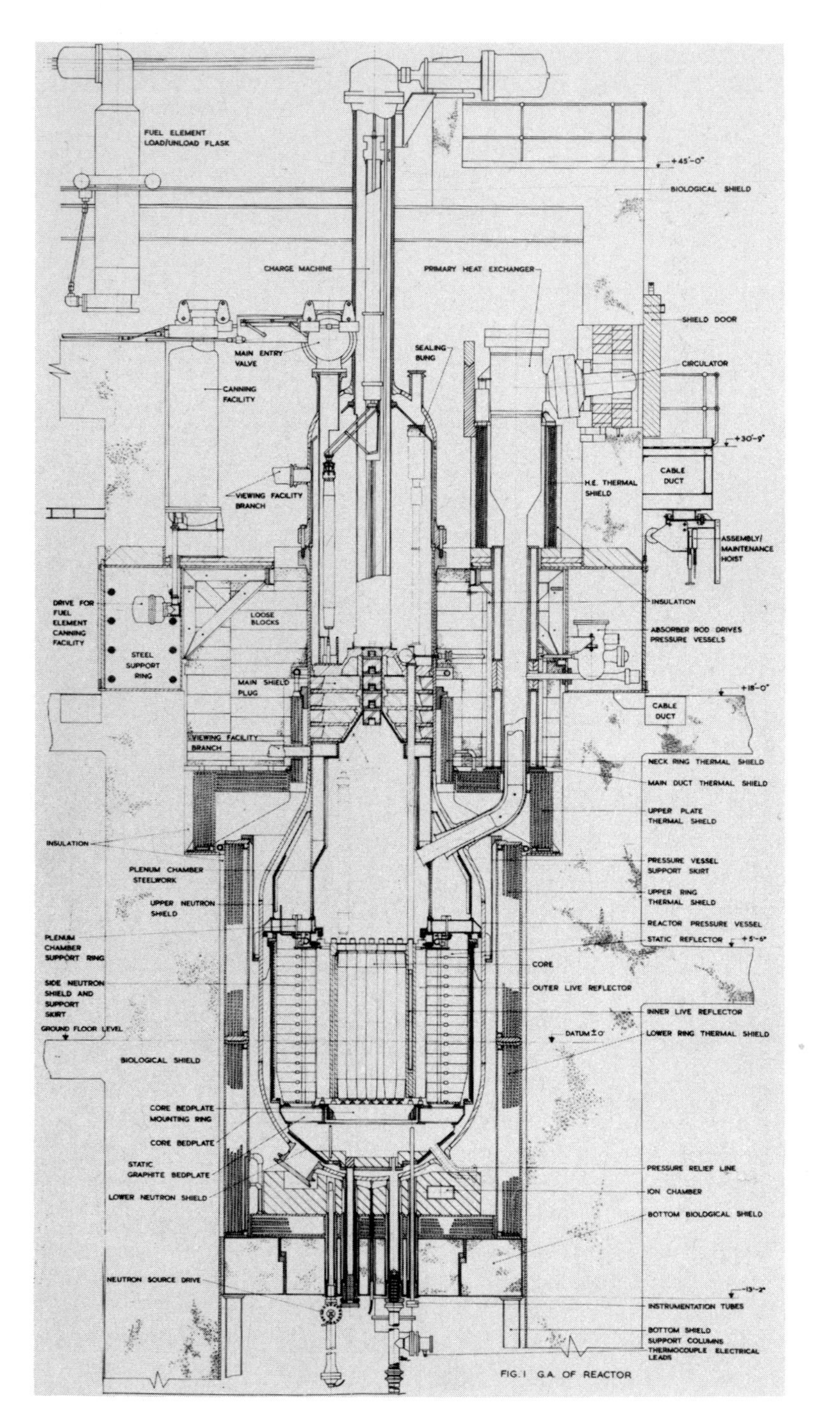

Figure 17. Cross-section of the Dragon Reactor.

and experiments were ready for installation prepared both by the Project itself and sponsors in the Signatory countries. The materials programme had extensive backing, as did the studies on physics, safety and fuel cycles. Representatives of the Signatory countries in the Board of Management had agreed a programme and budgets for a further five years. Much interesting research remained to be done – but to what end? Technical people will argue that Britain's new policy in regard to the HTR was wrongly conceived (as was proved by the subsequent recantation on the SGHWR) and that GAC's shortcomings were misleading. This is irrelevant, however. Dragon's role was to serve the Signatories and three of the biggest Signatories at the highest level, had decided that Dragon no longer had a purpose. No other country was prepared to assume added responsibility to keep it in being so that termination was logical.

The untidiness arose principally from a general unwillingness to see a successful collaboration come to an end, and no country wished to be identified as the one to wield the axe. So matters were allowed to drift to their dismal conclusion. Had Dragon been a failure, either technically or politically, its demise might have been cleaner. Project staff and other mourners should take consolation from the fact that it was Dragon's past record and its own will to live on that prolonged the agony.

DRAGON'S SUCCESSES

Despite the lacunae and the absence of any European project to build a prismatic prototype, Dragon over its 17 years must nevertheless be judged to have been a successful technical and human endeavour. The Project's reputation in scientific and technical circles bears witness to the quality of the research and development; DPR 1000 and all the other Dragon Project Reports provide a lasting proof. DRE was a tangible testimony to the high standard of the engineering and the competence of the design and operating staff. And this outstanding technical achievement was realised by a regularly changing international team. Scientists, engineers and administrators from many nations worked in harmony together in a common cause; national loyalties gave way to Project loyalties and nationalism had no place. In the 1980s, this may cause little surprise, as it is now well understood that an international team with clear objectives and properly led will work together at least as well as a national team. This was far from being an established fact at the end of the fifties, only 15 years after the nations of Europe had been at war with each other. Dragon was one of the first practical demonstrations that people from the different countries of Western Europe can work together in peace if allowed to do so.

The harmony in Dragon was evident in more subtle ways than merely accomplishing the tasks that were set. There was no language problem for example. There were problems of communciation at times of course but no requirement to make gestures to other languages as a sop to national pride, irrespective of the value to the Project itself. With Britain the initiator and Winfrith the site, it was clear that English would be the working language. This was, however, a matter of fact rather than of principle and the British staff responded by doing their best to enlarge their own linguistic abilities in order to help communication. Units also were never an issue. Dragon began life in feet and inches and DRE was built in these terms without the f.p.s. and m.k.s. systems assuming any political importance. Overseas contractors raised no objections (preferring in the main to receive specifications in the original units rather than in a converted form).

At a social level also, Dragon was a success – again not regarded as automatic in the sixties even if it elicits little surprise today. Dragon's modest size helped – everyone was in touch with everyone else – and the countryside location of Winfrith meant that although the staff had homes scattered over a wide area, they were not lost in a great impersonal conurbation. Staff wined and dined each other, recipes were exchanged, old prejudices fell away, horizons were widened and Dragon became identified with a warm camaraderie, absorbing the Authority's own unhierarchical style, and enriching it with the cultural attributes of the overseas seconded staff.

COST EFFECTIVENESS

Calculations on the cost effectiveness of Dragon form but a part of the estimation of the value of the Project. Even so, with a total revenue from contributions of £47 335 000 (compared with Germany's expenditure of say £300M on the pebble-bed and the Authority's annual expenditure of about £40M on the fast reactor) the very least one can say is that Dragon's total cost was not large. Moreover, all the Signatories would agree that, within the programmes laid down, the money was properly spent. Both Chief Executives were scrupulous in their respect of the budgets approved by the Board of Management, and the Authority proved to be an impeccable housekeeper, handling the Project's affairs with dispassionate responsibility.

Research and development is of course cheap compared with building a power station, when as much as £250/kW(e) may be spent in R & D for that station alone – outside the more generic development costs and the specific component costs. Investment in Dragon altogether was thus equivalent to the incidental R & D on a power station half the size of Fort St. Vrain. For this the Signatories had demonstrated for them all the essential core information for a wide range of HTRs, much of the basic data on fuel fabrication, helium technology and primary circuit materials. It is difficult to believe that the output from a purely national project of the same size would have been greater. Even considered in the narrow context of providing data for the only ongoing HTR power station project in Europe, the THTR, Dragon can be regarded as a sound investment, and in the development of coated particle fuels, Dragon may yet prove to have made one of the most important contributions to nuclear technology of the past 30 years.

FUTURE OF THE PRISMATIC HTR

Is there a long term prospect for the prismatic HTR? That question cannot yet be answered. Fort St. Vrain is operating in the USA at about two thirds of its design power and the record would suggest that the behaviour of the fuel and coolant is as satisfactory as in Peach Bottom and DRE. Examination of the fuels and materials irradiated in DRE and analyses of the accumulated results of the Dragon research which have continued in the laboratories of the former Signatories since March 1976 have reinforced the conviction that the essential science of the prismatic HTR has been proved. The operation of Britain's AGRs indicates that the basic engineering problems of large gas-cooled reactors with high thermal efficiency can be resolved while the residual mass transfer troubles experienced with the carbon dioxide coolant are not relevant to helium technology.

There remains, therefore, a prima facie case for the system. However, only when Europe has built its own demonstration power station with an output of several hundred megawatts – not necessarily 1200 – can reasoned decisions be taken on whether the system should play a role in central power station generation in the European continent. In the longer term, there remains the potential need for a source of high temperature process heat for the production of synthetic gas and liquid fuels. If that source is to be nuclear, the HTR would seem to have no serious competition.

APPENDIX 1

CONTRIBUTIONS OF SIGNATORIES AND EURATOM COUNTRIES

		(1959-64)[1]	1959-67	1967	1968-70	1970-73	1973-76	Total
DRAGON SIGNATORIES	Republic of Austria	1.36 %	462.5 k£	29 k£	82 k£	129 k£	174 k£	1.85 %
	Danish Atomic Energy Commission	1.47	500	31	90	140	-	
	Institutt for Atomenergi, Norway	1.21	412.5	26	73	50	-	1.19
	Aktiebolaget Atomenergi, Sweden	3.24	1 100	68	196	309	419	4.42
	Swiss Federal Government	2.42	825	51	147	231	313	3.31
	UK Atomic Energy Authority	58.4	10 200	634	2 081	3 276	4 066.5	
	Euratom	31.9	11 500	714	1 779	2 800	4 427.5	
EURATOM COUNTRIES	Belgium	3.16	1 138.5[2]	70.7	176.1	277.2	209.4[3]	3.96
	Denmark	-	-	-	-	-	60.2	1.73
	France	9.57	3 450	214.2	533.7	840	1 190.1	13.16
	Germany	9.57	3 450	214.2	533.7	840	1 465.5	13.73
	Ireland	-	-	-	-	-	15.0	0.03
	Italy	7.34	2 645	164.2	409.2	644	752.3	9.75
	Luxembourg	0.064	23	1.4	3.6	5.6	7.1	0.09
	The Netherlands	2.20	793.5	49.3	122.7	193.2	256.8	2.99
	United Kingdom	-	-	-	-	-	471.1	43.79
	Total		25 000	1 553	4 447	6 935	9 400	47 335 k£

1. Calculated for the full programme of 13.6 M£ wherein the Authority retained ownership of the assets.
2. Until 1973 the proportional contributions to the Euratom research budget were:
Belgium: 9.9%; France and Germany: each 30%; Italy: 23%; Luxembourg: 0.2%; the Netherlands: 6.9%.
3. Approximate figures covering the transitional period in the enlargement of the Communities.

APPENDIX 2

Chronological Summary

1948	*April*	Creation of the Organisation for European Economic Co-operation
	May	Signature of Treaty setting up the Council of Europe
1951	*April*	Signature of Treaty setting up the European Coal and Steel Community
1953	*December*	Eisenhower proposes to the United Nations the creation of an international atomic energy agency
1954	*January*	OEEC Council resolves to study energy problems in Europe
	September	Formal establishment of the European Organisation for Nuclear Research (CERN)
	November	UN resolves to set up International Atomic Energy Agency
1955	*February*	UK announces first civil nuclear power programme
	March	First meeting of European Atomic Energy Society
	May	Germany regains independence in civil atomic energy
	June	ECSC countries in Messina decide to study economic and political union; special emphasis given to atomic energy
		OEEC Council calls for study on atomic energy
	August	First Geneva Conference on Peaceful Uses of Atomic Energy
	December	First meeting of HTR study group at Harwell
1956	*February*	ECSC countries agree basis for Common Market and Euratom
		OEEC Council agrees to set up Special Committee for Nuclear Energy
	July	OEEC Council agrees to set up Steering Committe for Nuclear Energy
		Egypt nationalises Suez Canal
	October	First meetings of REP and REX
		Outbreak of the Middle East war
		UN adopts statutes of IAEA
1957	*January*	Construction of Zenith authorised
	March	Steering Committee accepts REX report
		- UK offers site for joint OEEC undertaking
		- HAR co-operation studies begin
		Signature of Treaty of Rome setting up Common Market and Euratom
	April	BBC/Krupp awarded contract to design pebble-bed reactor
	May	Report of the Three Wise Men
	July	Site work begins on the Atomic Energy Establishment, Winfrith
	October	Fire at the Windscale production reactor
	November	EAES meeting in Rome where a report is given on Harwell HTR research
1958	*January*	Euratom and Common Market come into being
	February	European Nuclear Energy Agency comes into being with Huet as Director
	March	Meeting of top level experts at OEEC in Paris
		- British HTR offered as joint undertaking
	April	Proposed joint undertaking on HTR dubbed Dragon
	May	OEEC Council endorses proposal for joint research on HTR
		De Gaulle comes to power in France

	September	Second Geneva Conference on the Peaceful Uses of Atomic Energy
	November	GA/PE proposal for Peach Bottom HTR received by USAEC
		Signature of Euratom-USAEC co-operation agreement
1959	*February*	Formation of AVR to construct pebble-bed power station at Jülich
		Terms of Dragon Agreement approved
	March	Signature of Dragon Agreement
		C.A. Rennie appointed acting Chief Executive
	June	First formal meeting of Dragon Board of Management with Eklund in chair; Rennie confirmed as Chief Executive of Project
	August	AVR places contract with BBK for pebble-bed power station
	December	Zenith commissioned
1960	*February*	Eklund symbolically turns the first sod on DRE site
	March	Exchange of letters between Authority, acting for Dragon, and USAEC establishing collaboration on HTR development
1960	*April*	Formal dedication ceremony to mark start of construction of DRE
	October	Dragon visit to USA that launched Dragon's coated particle fuel development programme
	December	Signature of Convention on the transformation of OEEC into the Organisation for Economic Co-operation and Development (OECD)
1961	*Early*	Decision taken to set up DRE fuel production facilities at Winfrith
	April	GPC approves programme of accelerated work on coated particle fuels
	August	UK applies for membership of the European Communities
		Construcion of AVR plant begun
	September	OECD officially comes into being
	End	Formation of Operations Group for DRE started
1962	*February*	Construction of Peach Bottom begun
	March	DRE pressure vessel installed
	July	GPC invited to approve coated particle fuel for DRE first charge
	December	New Dragon Agreement concluded covering an 8-year period from 1 April, 1959 to 31 March, 1967
1963	*January*	De Gaulle rejects British entry into the Communities
	May	Use of coated particles for DRE first charge authorised by GPC
	December	Huet leaves ENEA; Saeland appointed new Director General
1964	*Janaury*	THTR Association formed
	March	Signature of Dragon-Nukem agreement for collaboration
	April	Japan becomes 21st full member of OECD
		First driver fuel elements for DRE handed over to Operations Group
		Dragon symposium with presentation of first assessment studies
	August	Fuel loading of DRE begun
		DRE critical
	Aug./Sept.	Third Geneva Conference on the Peaceful Uses of Atomic Energy
	October	Inauguration of DRE
1965	*March*	Agreement between USAEC, PSC and GA for construction of Fort St. Vrain power station
	April	Fusion of the Councils of the European Communities
	May	ENEA/Dragon symposium on gas turbine HTRs
		CEGB selects AGR for next round of British nuclear power stations
	June	DRE begins approach to operation at significant power
	July	Signature of Dragon-THTR collaboration agreement
	Nov./Dec.	DRE Operating at 10 MW power level

1966	*February*	Board adopts programme for continuation of Dragon until December 1967
	March	Peach Bottom critical
	April	DRE at full power of 20 MW
		GHH receives order for HTR gas turbine plant at Geesthacht
	May	Signature of Agreement to extend Dragon to end of 1967
		BNES symposium on HTRs and the Dragon Project – Introduction of concept of low-enriched uranium/plutonium
	Summer	Brown Boveri, Baden establishes HTR design office
	August	AVR reactor critical
	November	UK Government announces intention of making a new application to join European Communities
1967	*January*	Corrosion of heat exchangers in DRE detected
	March	Norway confirms participation in Dragon
	May	Application of Denmark, Ireland and UK to join Communities
		Peach Bottom at full power
		Dragon/THTR/Euratom symposium on HTR power reactor studies
	June	Signature of collaboration agreement with Brown Boveri, Baden
	July	Merger of the Commissions of the European Communities
	October	Gulf Oil takes over General Atomic forming GGA
	November	Conditions of arrangement extending Dragon to December 1968 agreed
		De Gaulle vetoes enlargement of the Communities
	December	Euratom's participation in THTR Association ends; KFA takes over collaboration agreement with Dragon
1968	*February*	AVR reactor at full power
	March	Rennie announces resignation to Board of Management
	April	Signature of Dragon-KSH/GHH collaboration agreement
		First heat exchangers of DRE replaced
	July	L.R. Shepherd succeeds Rennie as Chief Executive of Project
		Extension of Dragon Agreement to March 1970 with terms back-dated to January 1968 approved
		Introduction of block-type elements into DRE; first irradiations of low enriched fuels begin
	August	Inter-Nuclear formally constituted
	September	Start of construction of Fort St. Vrain
	November	Signature of extension of Dragon Agreement to March 1970
		UK adopts HTR for its third generation of thermal reactors; prototype HTR project at Oldbury planned
1969	*April*	De Gaulle resigns as President of France and is succeeded in June by Georges Pompidou
	June	Dragon celebrates its 10th anniversary
		Extension of Dragon Agreement to March 1973 approved
	October	Second group of heat exchangers of DRE replaced
		First meeting of Programme Sub-Committee
		Willy Brandt becomes Chancellor of Germany
	November	Announcement of Government nuclear power programme for the French VIth national plan to run from 1971-1975
	December	HKG receives firm offer for THTR power station at Uentrop
		Community countries agree in principle on enlargement of membership
1970	*July*	THTR consortium receives letter of intent for THTR station

1971	*May*	DRE run at power of 21.5MW
		Agreement reached on major issues regarding enlargement of Communities
	August	GGA concludes (with Philadelphia Electric) first of a series of contracts for large HTR power stations
	September	Fourth Geneva Conference on the Peaceful Uses of Atomic Energy
	October	Formation of HRB
		THTR consortium receives order for THTR power station from HKG
	End	Geesthacht project abandoned
1972	*January*	Signature by Denmark, Ireland, Norway and UK of Treaty of Accession to European Communities
		Formation of Euro-HKG
	April	Japan becomes member of Agency whose name changes to Nuclear Energy Agency
	July	Signature of HOBEG-GGA licence agreement
	August	Official announcement of UK nuclear plans including the indefinite postponement of the Oldbury HTR prototype
		First of a series of agreements between GGA and the French CEA and GHTR
	September	Norwegian referendum excludes Norway's membership of Communities
	Late	Formation of HHT project in Germany
	December	Signature of the extension of the Dragon Agreement to March 1976
1973	*January*	Denmark, Ireland and UK join European Communities; Ireland becomes one of Dragon's Signatory countries
		GGA acquires 45% share-holding in HRB
	March	First meeting of enlarged Board of Management
	Spring	Replacement of inner reflector of DRE
		GHT project established in Germany
	April	Norway ceases to be a Dragon Signatory
	June	First irradiation of integral block fuel element in DRE begins
		Royal Dutch-Shell announces intention to put money into GGA
	October	OPEC decide to cut oil supplies and dramatically increase oil prices
	December	Collaboration agreement between Dragon and CEA formalised
		Agreement between Royal Duch-Shell and GGA concluded leading to formation of GAC
1974	*January*	Fort St. Vrain reactor critical
	July	UK announces choice of SGHWR for next round of nuclear stations
	September	Energie Ouest Suisse calls for tenders for HTR power station
	October	Planning hearings concluded on VEW's 1160MW(e) power station at Schmehausen
		Peach Bottom plant closed down
1975	*Early*	GAC tries to renegotiate its contracts for large power stations; Vidal and St. Rosalie projects abandoned
	April	Dragon Board agrees programme and budget for a 5-year extension
	June	UK referendum shows big majority in favour of remaining in the Communities
		Wedgwood-Benn becomes British Secretary of State for Energy
		Dead-line for agreement on the next extension expires
	August	UK makes known its demand for a reduced contribution to Dragon Project

	September	De-secondment notices issued to Dragon staff
		Philadelphia Electric announces that work on Fulton project suspended
		DRE shut down for the last time
	December	De-secondment notices expire
		Information retrieval and continued metals programmes funded
		Pictet addresses remaining staff on behalf of Board
1976	*March*	Dragon collaboration comes to an end

APPENDIX 3

Abbreviations and Glossary

Agency Contraction of either the European Nuclear Energy Agency or, as it became on 20 April, 1972, the Nuclear Energy Agency

AGR Advanced Gas-cooled Reactor; carbon-dioxide cooled, graphite-moderated reactor with enriched uranium oxide fuel canned in stainless steel developed in the UK, after the Magnox reactors

APC Atomic Power Constructions Ltd.; one of the UK industrial consortia during the 1960s

Authority Contraction of the United Kingdom Atomic Energy Authority used throughout this volume

AVR Arbeitsgemeinschaft Versuchsreaktor GmbH; Düsseldorf group that commissioned the 15MW(e) pebble-bed power station at Jülich

BBK Brown Boveri/Krupp Reaktorbau; constructor of the AVR reactor

BNDC British Nuclear Design and Construction Ltd.; UK consortium formed in 1969 out of the English Electric – Babcock and Wilcox – Taylor Woodrow group

BNES British Nuclear Energy Society

BWR Boiling Water Reactor; light water cooled and moderated reactor with no intermediary heat exchanger before the turbine

CEA Commissariat à l'Energie Atomique; French atomic energy commission

CEGB Central Electricity Generating Board of England and Wales

CEN Centre d'Etude de l'Energie Nucléaire; Belgian nuclear energy research centre

CERN European Organisation for Nuclear Research in Geneva

Charge The loading of DRE covering a group of related runs

CIIR Central Institute for Industrial Research, on the outskirts of Oslo

CNEN Comitato Nazionale per l'Energia Nucleare; Italian national commission for nuclear energy

CNRN Comitato Nazionale per Richerche Nucleare; organisation that preceded CNEN

Core Heart of a nuclear reactor containing the fuel. Core followed by a number was used to distinguish the particular loading of DRE within a given charge

COREPER	Committee of Permanent Representatives in Brussels answering to the Council of Ministers
Direct	-power cycle where reactor cooling gas drives power turbine with no intermediate heat exchangers
DRE	Dragon Reactor Experiment, the 20MW(th) reactor that formed the basis for the Dragon collaboration
Driver	Highly enriched fuel used in DRE, primarily to maintain the reactor power
EAES	European Atomic Energy Society
EDF	Electricité de France
EGCR	Experimental Gas-Cooled Reactor at Oak Ridge with a power of 85M(th), 30MW(e) that was never operated
EIR	Eidgenössisches Institut für Reaktorforschung; Swiss federal institute for reactor research
ENEA	European Nuclear Energy Agency
ENEL	Ente Nazionale per l'Energia Elettrica; Italian national electricity board
ENI	Ente Nazionale Idrocarburi; Italian national fuel and power company
Enrich	Increase the proportional concentration of fissile material in a nuclear fuel
EOS	Energie de l'Ouest Suisse; Swiss electricity company based on Lausanne
E.P.U. U/A	OEEC unit of account
ERDA	Energy Research and Development Administration; USA national body taking over part of the functions of the former US Atomic Energy Commission and subsequently absorbed into the US Department of Energy (DOE)
EUA	European Unit of Account; currency of calculation in the European Communities, formerly equivalent to about 1 US $
Eugene	Name given to HAR project considered in 1957 by OEEC countries for joint development
Euro-HKG	Association of electricity companies which acted as a European information centre for HTR power station experience
Fast	Distinguishes neutrons that are not deliberately slowed down in a moderator; hence Fast Breeder Reactor (q.v.)
FBR	Fast Breeder Reactor which is fuelled by high enriched material, has no moderator and breeds new fuel

Fertile Nuclear material which converts to a fissile material when irradiated with neutrons

Fissile Nuclear material composed of atoms, the nuclei of which can split on absorbing a neutron, giving rise to two massive unstable nuclei endowed with high speed and carrying most of the energy liberated in fission, plus additional neutrons

Fission Action of splitting; of a nucleus, normally caused by the impact of a neutron

Five Used to describe the smaller countries not in Euratom who were Signatories to the Dragon Agreements: Austria, Denmark, Norway, Sweden and Switzerland

FOM Foundation for Fundamental Research on Matter, at Utrecht

GAAA Groupement Atomique Alsacienne-Atlantique, French nuclear design and construction company

GA General Atomic Division of General Dynamics Corporation

GAC General Atomic Company formed when the Royal Dutch-Shell Group acquired a 50% interest in GGA

GAQ Group for Atomic Questions; political committee in Brussels serving COREPER

GCGR Gas-Cooled Graphite-moderated Reactor

GGA Gulf General Atomic formed when Gulf Oil Corporation took over GA from General Dynamics

GHH Gutehoffnungshütte Sterkrade; German company involved in the development of direct cycle systems

GHT Gesellschaft für Hochtemperaturreaktor Technik; KWU's company to exploit the pebble-bed for process heat applications

GHTR Groupement Industriel Français pour les Réacteurs à Haute Température; French industrial consortium formed in 1970 that in 1974 became the SHTR

GKSS Gesellschaft für Kernenergieverwertung in Schiffbau und Schiffahrt; German company concerned with nuclear powered ship developments

GPC General Purposes Committee of Dragon that reported to the Board of Management

HAR Homogeneous Aqueous Reactor where the fuel is either dissolved in water or is in the form of a fine suspension; only in the core vessel can a critical mass be formed

HHT Hochtemperaturreaktor mit Helium Turbine Grosser Leistung; German national project for the development of direct cycle systems

HKG Hochtemperatur-Kernkraftwerk GmbH; group of electricity companies formed to commission the THTR

HRB Hochtemperatur-Reaktorbau GmbH; German nuclear consortium led by Brown Boveri; GGA acquired a 45% shareholding in 1973

HTGCR High Temperature Gas-Cooled Reactor normally abbreviated in Europe to HTR

HTR High Temperature Reactor characterised by ceramic fuel, in a ceramic container (graphite), graphite moderation and helium cooling

IAEA International Atomic Energy Agency of the United Nations centred in Vienna

IFA Institutt for Atomenergi; Norwegian national atomic energy commission, Signatory to the Dragon Agreements

IKO Instituut voor Kernphysisch, Onderzoek; Dutch institute for nuclear physics research

JAEC Japan Atomic Energy Commission

JAERI Japan Atomic Energy Research Institute

KEMA Research organisation of Dutch electricity companies

KFA Kernforschungsanlage; German national and state nuclear research centre at Jülich

KSH Kernenergiegesellschaft Schleswig Holstein; German state power company formed to commission the 25MW(e) direct cycle HTR of GHH

KWU Kraftwerk Union; German light water reactor consortium

LMFR Liquid Metal Fuelled Reactor, a projected reactor in which the fuel, present as a molten liquid, is in continuous circulation. A system posing huge technological problems that has not been pursued

Magnox Fuel canning material of the first British natural uranium power reactors; name used to characterise the system

Moderator Material contained in the core of a reactor to slow down the neutrons produced in fission so as to increase their chance of causing further fissions; water, heavy water and graphite are the most commonly used moderators

MOTHER Multi-Objective, Temperature High, Experimental reactor project of JAERI

MTR Materials Testing Reactor

MW Megawatt = 1000 kilowatt; MW(e) indicates electrical output, MW(th) indicates thermal output

NEA Nuclear Energy Agency of OECD

Neutron Uncharged component of nucleus; two or three neutrons are normally released when a nucleus undergoes fission and these under suitable conditions can cause further fissions in fissile material or make new fuel in fertile material

NKA Nordic contact organisation for atomic energy

NNC National Nuclear Corporation; UK holding company replacing former separate nuclear consortia

NPC (before 1969) Nuclear Power Constructions; design and construction organisation of the English Electric – Babcock and Wilcox – Taylor Woodrow consortium in the UK

NPC (after 1973) Nuclear Power Company; design and construction group subsidiary to NNC

NWK Nordwestdeutsche Kraftwerk; German utility organisation that considered a Peach Bottom design for Wiesmoor

OECD Organisation for Economic Co-operation and Development; formally came into existence on 30 September, 1961 replacing OEEC (q.v.)

OEEC Organisation for European Economic Co-operation formed on 16 April, 1948

ORNL Oak Ridge National Laboratory, operated for the US AEC by Union Carbide

OSGAE Oesterreichische Studiengesellschaft für Atomenergie; Austrian atomic energy commission

PE Philadelphia Electric Company, owners of the Peach Bottom plant

Pebble-bed HTR with fuel elements in the form of balls (pebbles), developed in Germany

Prismatic -HTR has fuel elements consisting of rods or tubes inserted in graphite blocks (prisms) that are stacked together to form the reactor core

PSC Programme Sub-Committee of the GPC

PWR Pressurised Water Reactor; light water cooled and moderated reactor where steam to drive the turbines is raised in a secondary circuit

RAE Royal Aircraft Establishment at Farnborough in the UK

Reactor Assembly of nuclear fuel and other components in a vessel where a chain reaction can be produced

REP Reacteurs de Puissance; Working Group of the OEEC Special Committee to report on power reactors

REX Reacteurs Expérimentaux; Working Group of the OEEC Special Committee to report on experimental reactors

RWE Rheinisch-Westfahlisches Elektrizitatswerk; German utility

SCK = CEN

SGHWR Steam Generating Heavy Water Reactor; UK reactor system where the moderator is heavy water and the coolant light water that flows over fuel elements suspended in vertical pressure tubes

SHTR Société pour les Réacteurs Nucléaires HTR; French consortium which in June 1974 replaced GHTR

Six Refers to the original six members of Euratom: Belgium, France, Germany, Italy, Luxembourg, the Netherlands

Sol-Gel process Liquid route for the production of spherical fuel kernels devised by Oak Ridge National Laboratory (ORNL)

Thermal -power: heat output from a reactor; -reactor: a reactor that includes a moderator, and fission is induced primarily by slow neutrons

THTR Thorium-Hochtemperaturreaktor; 300MW(e) pebble-bed nuclear power station under construction at Hamm-Uentrop in Germany

TNPG The Nuclear Power Group: UK nuclear consortium formed in 1960 from a merger of the AEI-John Thompson group with the Nuclear Power Plant Company

TRISO US term for a multi-layered coated particle with at least an inner layer of porous pyro carbon to take up fission products, a dense pyro carbon shell for strength, a fission product barrier of silicon carbide and an outer shell of pyro carbon forming a further pressure vessel

UKAEA United Kingdom Atomic Energy Authority

USAEC United States Atomic Energy Commission

UPC United Power Company; UK nuclear consortium formed by the merger in 1960 of the interests of G.E.C.-Simon Carves and APC; reverted to APC in 1965

VEW Vereinigte Electrizitätswerke Westfalen; German utility, the leading company in HKG and Euro-HKG

VHTR Very High Temperature Reactor; a short-lived study by Dragon of a reactor experiment to explore direct cycle and process heat conditions in an HTR

WAGR Windscale AGR; UKAEA's 100MW(th), 27MW(e) experimental reactor to test the AGR concept

Zenith UKAEA's zero energy reactor for physics research on HTR core configurations at high temperature

APPENDIX 4

References

CHAPTER

1. **1.** For a detailed history of co-operation in Europe from 1946 onwards, see e.g. Broad R. and Jarrett R.J. *Community Europe Today* (Oswald Wolff, London) 1972.

2. **2.** Trends in Europe's energy demands prior to 1959 are reviewed in *Towards a New Energy Pattern in Europe,* Report prepared by the OEEC Energy Advisory Commission under A. Robinson (OEEC, Paris), January 1960.

3. *Some Aspects of the European Energy Problem: Suggestions for Collective Action,* Report of Louis Armand (OEEC, Paris), June 1955.

4. *A Programme of Nuclear Power,* Cmnd 9389 (HMSO, London) 1955.

5. For a detailed history of the development of atomic energy in the UK and the country's relations with the USA up to 1952, see Gowing M., *Independence and Deterrence, Britain and Atomic Energy,* 1945-1952 (Macmillan, London) 1974.

6. A summary of the history of CERN and its structure is included in Goldsmith M. and Shaw E.N., *Europe's Giant Accelerator* (Taylor & Francis, London) 1977.

7. Weinberg A.M. "Survey of Fuel Cycles and Reactor Types" *Proceedings of the First International Conference on the Peaceful Uses of Atomic Energy,* Vol. III (UN) 1956, p. 23.

3. **8.** For a history of developments in atomic energy up to 1962 with particular relevance to French interests, see Goldschmidt B., *L'Aventure Atomique* (Fayard, Paris) 1962.

4. **9.** *Possibilities of Action in the Field of Nuclear Energy* (OEEC, Paris) January 1956.

10. *Nucleonics,* 14 (1956) 3, pages 34 and 42.

11. *Joint Action by OEEC Countries in the Field of Nuclear Energy* (OEEC, Paris) September 1956.

5. **12.** *Experimental Reactors, Proposals for their Co-operative Development in Europe,* Steering Committee for Nuclear Energy (OEEC, Paris) March 1957.

13. Armand L., Etzel F. and Giordani F., *Prospects for Nuclear Power in the Euratom Community* (Euratom, Brussels) 1957.

14. *Nucleonics,* 15 (1957) 10, page 64.

15. Oak Ridge National Laboratory, "The Homogeneous Reactor Experiment No. 2", *Proceedings of the Second International Conference on the Peaceful Uses of Atomic Energy,* Vol. 9 (UN) 1958, p. 509.

16. *First Report of the Activities of the Agency* (OEEC, Paris) September 1958.

6. **17.** *Accident at Windscale No. 1 Pile on 10th October 1957,* Cmnd 302, 1957 and Cmnd 471, 1978 (HMSO, London).

7. **18.** The basic technical document around which the Dragon Agreement was formulated was a Harwell Report, H.T.G.C./P.101A, "The High Temperature Gas-Cooled Reactor" by L. R. Shepherd and G. E. Lockett, April 1958. This was distributed to OEEC delegates via the ENEA under the code SEN(58)27. The first public presentation is to be found in: Shepherd L.R., Huddle R.A.U., Husain L.A., Lockett G.E., Sterry F. and Wordsworth D.V. "The Possibilities of Achieving High Temperatures in a Gas-Cooled Reactor", *Proceedings of the Second International Conference on the Peaceful Uses of Atomic Energy,* Vol. 9 (UN) 1958, page 289.

19. Frisch O., *What Little I Remember* (University Press, Cambridge) 1979, p. 159.

8. **20.** Engelhard J., *Forschungsbericht K72-73, Kernforschung,* (Bundesministerium Für Bildung und Wissenschaft, December 1972) and *Atomwirtschaft,* May 1968, an issue devoted to the AVR project.

21. *Nuclear Engineering,* 3 (1958) 12, p. 538.

9. **22.** *European Nuclear Energy Agency: Statutes, Convention on Security Control, European Nuclear Energy Tribunal, Acts Relating to Joint Undertakings,* (OECD, Paris), 1957-1963.

10. **23.** See reference 2 and for detailed information on electricity consumption, the regular publication of OEEC "The Electricity Supply Industry in Europe", which was followed by OECD's "The Electricity Supply Industry in OECD Countries". For a wider coverage, see the annual publications of the UN Economic Commission for Europe (UN, Geneva): *Annual Bulletin of Electric Energy Statistics for Europe* and *The Electric Power Situation in the ECE Region and its Future Prospects,* Series ST/ECE/EP. For the UK only, the Electricity Council publishes annually a *Handbook of Electricity Supply Statistics* and the CEGB a *Statistical Yearbook.*

12. **24.** Thorn J.D., *The Civil HTR Reference Design Study,* DPR/135. AGIP Nucleare/Indatom Joint Group, *1250MW(th) HTGC Preliminary Design Study,* DPR/255.

25. Comment on Euratom's difficulties are to be found in e.g. Guéron J. "The Lack of Scientific Planning" and "Atomic Energy in Continental Western Europe", *Bulletin of the Atomic Scientists,* October 1969, p. 10 and June 1970, p. 62.

26. *Journal of the British Nuclear Energy Society,* 5 (1966) 3, pages 223 and 234-450. Helium reference p. 258.

27. Hosegood S.B., *Reference Design Assessment Study for a 528MW(e) Thorium Cycle HTGCR,* DPR/467; Lockett G.E., *Preliminary Assessment Study for a 528MW(e) Low-Enriched HTGCR,* DPR/468. See also DPR/475.

13. **28.** *Report from the Select Committee on Science and Technology, UK Nuclear Reactor Programme,* Session 1966-67 (HMSO, London), p. 10.

15. **29.** See reference 23 and *Power Reactors in Member States,* an annual publication of the IAEA Vienna.

30. Summarised in the *Rapport Annuel* for 1969 of the Commissariat à l'Energie Atomique, p. 3.

31. See reference 23 and *Nuclear Engineering International,* 17 (January 1972) 188, p. 5.

32. Commissariat à l'Energie Atomique, *Rapport Annuel,* 1970, p. 48.

16. **33.** *Hansard,* 8 August 1972, col. 1492. Comment in *Nuclear Engineering International,* 17 (September 1972) 196, p. 655.

19. **34.** Data concerning the nuclear programmes in Western Europe have been collected from numerous sources and then checked in the relevant countries. These sources have included the reviews of national programmes presented to the Geneva and other international conferences and the annual reports of the national atomic energy commissions. For this chapter, as elsewhere, the volumes of *Nuclear Engineering (International)* since the first publication in April 1956 have been consulted constantly. Note the special issues dealing with particular countries, e.g. France, September 1972; Germany, February 1973; Italy, September 1967 and October 1973.

An American's view of developments in Europe is to be found in: Nau H.R., *National Policies and International Technology – Nuclear Reactor Development in Western Europe* (Johns Hopkins University Press) 1974.

23. **35.** Mattick W., "Die Hochtemperaturreaktorlinie", *Atomwirtschaft,* September 1973.

24. **36.** Break-down of expenditure on gas-cooled systems, Marks II and III, was begun in the *Annual Report of the UK Atomic Energy Authority* for 1968-69 (p. 12). Figures are backdated to 1967-68.

37. A detailed account of the industrial development of atomic energy in the UK, recording the changes in policy and organisation, is given in: Pocock R.F., *Nuclear Power – Its Development in the UK* (Unwin Bros, and the Institution of Nuclear Engineers) 1977.

26. **38.** Commissariat à l'Energie Atomique, *Rapport Annuel,* 1972, Tome 1, p. 68.

39. *Hansard,* 10 July 1974, col. 1357.

40. *Second Report from the Select Committee on Science and Technology, Nuclear Power Policy,* Session 1972-73 (HMSO, London) p. xiii.

41. *First Report from the Select Committee on Science and Technology, The Choice of a Reactor System,* Session 1973-74 (HMSO, London) pp. iv-vii.

27. **42.** *Third Report from the Committee of Public Accounts,* Session 1974-75 (HMSO, London) p.3.

43. *Hansard,* 18 December 1975, col. 1716.

44. *Hansard,* 10 March 1976, col. 218.

28. **45.** Peirson D.E.H., "Twenty-one Years On", *Atom,* 225 (July 1975), p. 103.

NAME INDEX

SUBJECT INDEX

COMPANY INDEX